Mammals of the Pacific Northwest
From the Coast to the High Cascades

Mammals of the Pacific Northwest
From the Coast to the High Cascades

Chris Maser

 Oregon State University Press

Cover design by David Drummond
Cover photo: Oregon red tree vole, *Arborimus longicaudus,*
   photograph by Ronn Altig and Chris Maser

The paper in this book meets the guidelines for permanence
and durability of the Committee on Production Guidelines for
Book Longevity of the Council on Library Resources and the
minimum requirements of the American National Standard for
Permanence of Paper for Printed Library Materials Z39.48-1984.

Library of Congress Cataloging-in-Publication Data
Maser, Chris.
Mammals of the Pacific Northwest : from the coast to the
high Cascades / Chris Maser.
Includes bibliographical references and index.
ISBN 0-87071-438-4 (alk. paper)
1. Mammals—Oregon. 2. Mammals—Northwest, Pacific.
I. Title.
QL201.M37   1998
599'.09795—dc21   98-7570
                                                        CIP

Oregon State University Press
101 Waldo Hall
Corvallis OR 97331-6407
541-737-3166 • fax 541-737-3170
osu.orst.edu/dept/press

Contents

Introduction

WHEN I WAS ASKED to write this book for the Oregon State University Press, I was thrilled. My first conscious memory, at age two, is of a sand crab along the ocean's edge. My next searing memory is of a garter snake that would not come out of the bush into which it had crawled to escape the clutch of my four-year-old hands. Ever since, I have wanted to study animals as a scientist. I have now spent over thirty years in a consummate love affair with science—mostly studying mammals in the wild to understand their habitats and habits and how they intersect.

I shall not present you with a scientific treatise in the strict sense; instead I shall do my best to acquaint you with the mammals through the love, respect, and fascination with which I have over many years learned to know them. I do this in the hope of giving you a greater appreciation of the mammals with whom we share the Earth.

Because I am presenting a general natural history covering a wide range of information, I have dispensed with referencing the literature in the text, including my own studies, to keep this book as simple and readable as possible, but I have included a section of selected references. Having said this, however, I want it clearly understood that I have liberally used my own previous work and also that of many others. Therefore, *this book should in no way be construed as solely my work*. Where I have used published works of which I was the major contributor, I have occasionally written in the first person because the experiences about which I am writing are my own. Where I have had a major hand in a comparative study that has been published in a journal, I have, when it seems in the reader's interest, used it freely.

For anyone interested in a current scientific review of the mammals of Oregon, as well as their known geographical distributions within the state, I recommend *The Land Mammals of Oregon* by Verts and Carraway. This book also includes the contemporary scientific thinking about the taxonomic relationships of mammals in Oregon.

Natural history inquires into the secrets of an animal's life, not only how it lives as an individual but also how it relates to other individuals of its own kind, to other kinds of animals, and to its environment as a whole.

One of the early natural historians was Vernon Bailey, who in 1936 published *The Mammals and Life Zones of Oregon*, the first comprehensive work on mammals in the state. From 1962, when I entered graduate school at Oregon State University, until I left active research in 1987, I spent many years following Bailey's footsteps around the state as I too studied the mammals of Oregon. Although I had more sophisticated tools at my disposal than did Bailey and therefore learned things beyond his knowledge, I did not in any way improve on the quality of his work. To this day, I hold in awe the dedication, accuracy, integrity, and insight of Vernon Bailey's field work. Sadly, I never had the opportunity to meet him; I should very much have liked to.

Natural history, in my experience, seems to have been carried out primarily by two kinds of people—those who were gentlemen in the true sense of the word (such as Kenneth L. Gordon, Murray L. Johnson, Tracy I. Storer, and Walter P. Taylor) and those who were lovable characters (such as Bill Hamilton Jr. or "Wild Bill," as he was affectionately known, and Robert M. Storm—the only one of my role models still living).

It is with a real sense of loss that I watch the era of natural history drawing swiftly to a close, an era that allowed a softer personal touch into our relationship with Nature. I say this because the natural history that I knew was truly a science of forest, meadow, and fen, of mountain, desert, and sea, where I and others lived for weeks at a time out in the elements with the creatures we studied. It was a science of mutual relationships in a slower, gentler, quieter period in human history when there was ample time to reflect on a sunrise, a drifting cloud, a passing thought. It is thus in the spirit of natural history as I knew it that I pen this book as a tribute not only to the era of science that I loved so much but also to the men who helped to shape that era and who, as gentleman and character alike, shared it with me.

I have, to the best of my ability, written this book for a general audience. For this reason, I have included information acquired from places beyond the geographical scope of this book to help you, the reader, round out your understanding of the species as a whole. If I were to limit my discussion to data derived solely within the boundaries of western Oregon, where I have done most of my work, I would have far less to share with you, because one person can only do so much in a lifetime, and the mammals of Oregon have seldom been studied in the field over long periods of time by resident mammalogists who lived with them and got to know them intimately in their own habitats.

In addition, the coverage of individual species in this book varies. Some species have been more widely studied and/or have been studied over longer periods of time. Consequently, the data available on any given species are uneven and there are always some data of which I am unaware.

To make the text as readable as possible, I provided an appendix of common and scientific names of plants and animals mentioned in the book, as well as a glossary of terms.

I have endeavored to avoid as much jargon as possible and to make the format as simple as possible by moving from the general to the specific as follows:

Order

Family

Genus

Species

In so doing, it is necessary at times to be consciously repetitious in that similar information may appear in more than one place in the text so that each species account will stand on its own. From those of you with some familiarity with the mammals, I ask your forgiveness and your patience. For the sake of those readers who are in fact unfamiliar with mammals, I have assumed that all readers possess little or no knowledge of our global traveling companions.

IN ORDER TO BE ABLE to talk to one another about mammals—or any other organisms—we must agree on what to call them. In everyday conversation, we use a common name, like *mouse* or *bat*. Scientists need names that are more precise, and they use a system that was first proposed in a book published in 1758 by the Swedish naturalist Carl von Linné, better known by his Latinized name Carl Linnaeus. His system divided living beings into five kingdoms, and then into smaller and smaller groups as follows: phyla, classes, orders, families, genera, and finally species, of which about 1.5 million have been named, though scientists estimate that there are perhaps as many as 30 million on Earth.

The mammals are in the Kingdom Animalia, Phylum Cordata (sub-phylum Vertebrata), and comprise the Class Mammalia, which Linnaeus created from the Latin word *mamma*, meaning the breast, for those animals that produce milk and nurse their young. The Class Mammalia is divided into 21 orders, only seven of which are represented by terrestrial mammals living in western Oregon.

The system of nomenclature that Linnaeus invented combines two names—that of the genus and of the species—to describe a given organism. As an example, the small tree-dwelling red tree vole of western Oregon, is called *Arborimus longicaudus*. The generic name *Arborimus* is derived from the Latin words *arboris* (tree) and *mus* (mouse); the specific name *longicaudus* is derived from the Latin words *longus* (long) and *cauda* (tail). This accurately describes the "long-tailed tree mouse," which also happens to have reddish fur and is thus commonly know as the red tree mouse or vole.

The white-footed vole also lives in western Oregon. Although it is related to the red tree vole, it is primarily a ground-dweller, though it can and does climb. The relationship between the two is indicated by including both animals in the same genus *Arborimus*, but the latter is identified as a separate species by the specific name *albipes*, derived from the Latin words *albus* (white) and *pes* (foot). While the specific name identifies the individual kind of mammal, the generic name identifies the close relationship among two or more kinds of individuals. So, the scientific binomial, say *Arborimus albipes*, describes both the individual and collective properties of the animal.

Just because a mammal has been given a scientific name does not mean that the evolutionary relationship implied by the name is accepted by everyone, though many are. In part, this is because each person who studies mammals weighs the reported evidence independently against his or her own measure of experience in the field as a whole and/or with that particular mammal. Coupled with this is the fact that each person sees the data and its interpretation through his or her own particular lens.

I am no exception. I, too, am subjective without recourse because I am human with all the frailties that encompasses. But, while I may respect the quality of another persons' work and the integrity of her or his perceptions and interpretations, I reserve the right to disagree. It is possible, after all, that the ideas of one who disagrees with conventional wisdom may prevail in the end. Having said this, however, I hasten to add that I do not know what is right, of course, but neither does anyone else. Science is predominantly a process of hypothesizing, testing, and testing again. We are never completely certain of the accuracy of a hypothesis; we just know it has not been disproved—yet.

THE WORD "SPECIES" has one significance to a student of taxonomy and another to the student of evolution. To the student of taxonomy, the concept of a species is a practical device designed to reduce the almost endless variety of living things to a comprehensible system of classification. To the student of evolution, a species is a passing stage in the stream of ever-changing habitats to which the species must adapt or become extinct.

Before 1935, scientists based most definitions of species on the degree to which they were distinct in form and structure. They paid little attention to evolutionary relationships. In 1937, scientists revisited the definition and began to emphasize the dynamic aspects of species—their potential for change. Today, species are thought of as groups of natural populations that can and do interbreed with one another but are reproductively isolated from other such groups, which means that, even if members of the two species were put together, they could not produce fertile offspring of both sexes.

For example, Oregon and California red tree voles live primarily in Douglas fir trees in western Oregon and northwestern California. The two species meet occasionally in the vicinity of the Smith River just south of the Oregon-California border and may interbreed now and then. But hybrid males, those produced by such a union, are sterile. Hybrid females, on the other hand, though fertile, can breed only with a male of one species or the other. Because the habitat is not well suited for either species, there are not enough voles to populate the area and the hybrid females are therefore unlikely to find a mate; thus is the integrity of the two species maintained.

DARWIN HIMSELF WAS of the opinion that evolution was both continual and gradual. His theory says that evolution proceeds by mutation and natural selection. Mutations, which are simply random "mistakes" in the repetition of the genetic code passed from parent to offspring, are produced by all species at a more or less constant rate. Most individuals with mutations are eliminated through time, because they are "faulty" in some respect and unable to reproduce or adapt as well as "normal" individuals.

But occasionally a mutation arises that renders the individual more, rather than less, fit to survive and reproduce. When this happens, the individual is given a chance to pass its mutant genes on to its own offspring, which in turn passes them to its own offspring, and so on until the mutant trait becomes both dominant and "normal" in the population.

So through the combination of random mutations and natural selection, species continue to evolve, weeding out the less fit in favor of the more fit, until they occupy all available habitat niches in the biosphere, which keeps changing, so that the species must continue adapting.

Darwin probably adopted these two basic but unnecessary assumptions—that evolution is both gradual and continual—more out of innate conservatism than weighty scientific evidence. He thought that Nature made no "great leaps," which for him resembled the uncomfortable, sudden changes, such as revolutions, that transform human society. The dominant personalities of Darwin's time abhorred the revolutionary process of wholesale transformations, clinging instead to the idea of tiny, continual changes. Then, a hundred and twenty years after Darwin's O*n the Origin of Species* was first published, Stephen Jay Gould and Niles Eldredge, two American paleobiologists, wrote a seminal paper introducing the theory of evolutionary leaps. In their theory, these leaps, although dramatic, occur relatively infrequently.

Evolutionary stability, it appears, is the normal course of events in the persistence of species over long periods of time. Paleontologists have long dismissed the lack of evidence of evolutionary change, assuming that it simply reflected gaps in the fossil record. But the fossil record, although perhaps imperfect, does not prove that evolution is a continual process.

In fact, evolution, as it now appears, proceeds through leaps of "speciation" (the sudden dominance of new species) rather than through a slow, gradual, continual perfecting of existing species to fit changing conditions. In this new theory of periodic leaps of speciation, evolutionary change, rather than affecting individuals as survivors and reproducers, affects the entire system, which is composed of living organisms as they interact within their environments. Evolution occurs when a dominant species fails to adapt successfully to changes in its habitat, so that it can be challenged by a new species, which may have emerged "haphazardly" at the edge of the cycle of dominance. Thus the dynamic equilibrium is broken as the old species is suddenly replaced by the emerging new species in a leap of evolution. So, according to this most recent theory, new species are selected in sudden bursts of evolution during periods of critical instability for dominant species.

Some ancient species, such as opossums, are unlikely to become extinct because they live in environments that vary so much from day to day, month to month, and year to year, that they are unlikely to meet anything in the future they have not already survived in the past. Some other organisms are in much more severe danger of extinction. Perhaps they are the only surviving species of a taxonomic group that was once considerably richer. Perhaps they have not changed in millions of years. Most likely they are adapted only to specific habitats threatened with drastic modification.

Extinction carries two meanings, one local and one global. A local extinction refers to a particular population, such as the red squirrel on Mt. Graham, a "mountain-top island" in the desert of Arizona. A global extinction, on the other hand, refers to an entire species (all the red squirrels in the world). Local populations may—and often do—disappear, either temporarily or permanently, without implying the extinction or even the near-extinction of the species. A species, on the other hand, is composed of the sum of its populations, so the loss of populations will affect the species as a whole and may imply danger to its survival. Global extinction happens at the exact moment when either one of the last breeding pair dies. The minimum size of a population is thus critical to its continued survival in the face of change; the smaller a population is, the more susceptible it is to extinction from a variety of causes.

WHY DO WE NEED such a variety of species anyway? Is the loss or one or two species very important?

One marvelous effect of a wide variety of species is that they increase the stability of ecosystems by means of feedback loops—the means by which processes reinforce themselves. Strong, self-reinforcing feedback loops characterize and strengthen many interactions in Nature and have long been thought to account for the stability of complex systems. Ecosystems with such strong interactions among components can be complex, productive, stable, and resilient under the conditions to which they are adapted. But when these critical loops are disrupted, such as by the extinction of species, these same systems become fragile and easily affected by slight changes.

Although an ecosystem may be stable and able to respond effectively to disturbances to which it is adapted, it may be exceedingly vulnerable to the introduction of foreign disturbances to which it is not adapted. Nevertheless, ecosystems have a certain amount of redundancy built into them, which means that more than one species is able to perform similar functions. This is a type of ecological insurance policy that strengthens the ability of the system to retain the integrity of its basic relationships. Redundancy means that the loss of a species or two is not likely to result in such severe functional disruptions of the ecosystem as to cause its collapse because other species can make up for the functional loss. But there comes a point, a threshold, when the loss of one or two more species may tip the balance and cause the system to begin a noticeable and irreversible change.

Each species of living organism performs a specific function that in one way or another benefits the whole ecosystem. Diversity of plants and animals therefore plays a seminal role in buffering an ecosystem against disturbances from which it cannot recover.

As forests become poorer in species of plants through the conversion of forests into biologically simplified tree farms, the number of species of birds, mammals, and other creatures will also decline. The complex, interconnected, interdependent feedback loops among plants and animals will gradually simplify. Some of the species of which the feedback loops are composed will be lost forever—and the feedback loops with them. This is how the evolutionary process works. Ecologically, it is neither good nor bad, right nor wrong, but these changes may make the forest less attractive, less usable by species, including humans, that used to rely on it for their livelihoods and for products.

Functional Relationships Among Species in Western Oregon

Most people are probably aware of predator-prey relationships and especially those of the charismatic grazing megafauna of the African plains, thanks in large part to television. But few of us have any concept of the intricate relationships between the small, secretive mammals of the forest and the health of those forests, which stand as a symbol of western Oregon and the Pacific Northwest. With this in mind, let's take a peek into the incredible web of life in our western Oregon coniferous forests.

Photo 1. Northern flying squirrel. (USDA Forest Service photograph by Jim W. Grace.)

Photo 2. Mycorrhizae, the symbiotic relationship between certain fungi and the roots of plants. In this case, the fungus forms a mantle, a covering over the root tip of lodgepole pine. (USDA Forest Service photograph by C.P.P. Reid.)

All things in Nature's forest are neutral; Nature assigns no extrinsic values to this mammal or that, to this plant or that. Each piece of the forest, whether a bacterium, mushroom, mouse, or 800-year-old Douglas fir tree, carries out its prescribed function, interacting with other components of its habitat through their prescribed interrelated processes. None is more valuable than another; each is only different from the other. The northern flying squirrel is a primary example of the dynamic interactions of small mammals, fungi, water, nutrients in the soil, and trees in the coniferous forests of western Oregon.

THE NORTHERN FLYING SQUIRREL (photo 1) is common in conifer and mixed conifer-hardwood forests from the Arctic tree line throughout the northern conifer forests of Alaska and Canada, south through the Cascade Mountains of Washington and Oregon and the Sierra Nevada mountains of California almost to Mexico, the Rocky Mountains to Utah, and the Appalachian Mountains to Tennessee.

THE SQUIRREL

It is seldom seen because it is nocturnal. It is primarily a mycophagist, a fungus eater. In northern Oregon, it eats mostly truffles, which are the reproductive bodies of belowground-fruiting fungi, and epiphytic lichens (that grow on trees). In northeastern Oregon, truffles are the principal food from July to December; from December through June, the lichen *Bryoria fremonti* is the squirrel's predominant food, as well as its sole nest material. In southwestern Oregon, truffles are the major food throughout the year, and lichens are not important in the overall diet.

THE TERM *mycorrhiza* (photo 2) which literally means "fungus-root," denotes the symbiotic relationship between certain fungi and the plant roots on which they live. Fungi that produce truffles (photo 3) are probably all mycorrhizal. Woody plants in the families Pinaceae (pine, fir, spruce, larch, Douglas-fir, hemlock), Fagaceae (oak), and Betulaceae (birch, alder) in particular depend on mycorrhiza-forming fungi for the uptake of nutrients. This phenomenon can be traced back some 400 million years to the earliest known fossils of plant rooting structures.

Mycorrhizal fungi absorb nutrients and water from soil and translocate them to the host plant. In turn, the host plant provides sugars from its own photosynthesis to the mycorrhizal fungi. Threadlike fungal hyphae (the "mold" part of the fungus) extend into the soil and serve as extensions of the host's root system and are both physiologically and geometrically more effective for nutrient absorption than the roots themselves.

THE MOST OBVIOUS relationship between the squirrel and the forest occurs on the surface of the ground, when the squirrels forage for food. The nesting and reproductive behavior of these squirrels remains relatively obscure because of their nocturnal habits, as will discussed in a later chapter on rodents. In probing the secrets of the flying squirrel, however, some functionally dynamic, interconnecting cycles emerge, beginning with mycorrhizal fungi.

The host plant provides simple sugars and other metabolites to the mycorrhizal fungi, which lack chlorophyll and are not good at deriving their nutrients from dead or decaying organic material. Fungal hyphae penetrate the tiny, nonwoody rootlets of the host plant to form a balanced, harmless mycorrhizal

Photo 3. The belowground fruiting body (hypogeous sporocarp) or truffle of a mycorrhizal forming fungus. The light tissue (left) is a tough outer coat, whereas the dark tissue (right) contains the reproductive spores, which are surrounded by the tough, relatively impervious outer coat, the white area. (USDA Forest Service photograph by James M. Trappe.)

Mammals of the Pacific Northwest

Photo 4. A northern flying squirrel eating a truffle, Hysterangium *spp. (USDA Forest Service photograph by Jim W. Grace.)*

symbiotic relationship with the roots. The fungus absorbs minerals, other nutrients, and water from the soil and translocates them into the host. Also, nitrogen-fixing bacteria inside the mycorrhiza use a fungal "extract" as food and in turn fix atmospheric nitrogen. (To "fix" nitrogen is to take gaseous, atmospheric nitrogen and alter it in such a way that it becomes available and usable both by the fungus and the host tree.)

In effect, mycorrhiza-forming fungi serve as a highly efficient extension of the host root system. Many of the fungi also produce growth regulators that induce production of new root tips and increase their useful life span. At the same time, host plants prevent mycorrhizal fungi from damaging their roots. Mycorrhizal colonization enhances resistance to attack by pathogens. Some mycorrhizal fungi produce compounds that prevent pathogens from contacting the root system.

Flying squirrels nest and reproduce in the tree canopy and come to the ground at night, where they dig and eat truffles (photo 4). As a truffle matures, it produces a strong odor that attracts the foraging squirrel. Evidence of a squirrel's foraging remains as shallow pits in the forest soil and occasional partially eaten truffles.

Truffles contain nutrients necessary for the small animals that eat them. When flying squirrels eat truffles, they consume

Photo 5. Fecal pellets of a northern flying squirrel act as an ecological pill that helps to inoculate the soil with live fungal spores, nitrogen-fixing bacteria, and yeast. (Photograph by Chris Maser.)

fungal tissue that contains nutrients, water, viable fungal spores, nitrogen-fixing bacteria, and yeast, which are all important in the forest network. Pieces of truffle move to the stomach, where fungal tissue is digested, then through the small intestine, where absorption takes place, and then to the cecum. The cecum is like an eddy along a swift stream; it concentrates, mixes, and retains fungal spores, nitrogen-fixing bacteria, and yeast. Captive deer mice, for example, retained fungal spores in the cecum for more than a month after ingestion. Undigested material, including cecal contents, is formed into fecal pellets in the lower colon; these pellets contain all the viable elements necessary to inoculate the root tips of trees with prolonged life.

A fecal pellet (photo 5) is thus more than a package of waste products; it contains four components of potential importance to the forest: (1) viable spores of mycorrhizal fungi (photo 6); (2) yeast, which is a part of the nutrient base, has the ability to stimulate both growth and nitrogen-fixation in the bacteria, and may also stimulate spore germination; (3) nitrogen-fixing bacteria; and (4) the entire nutrient requirement for nitrogen-fixing bacteria, and an "antifreeze" that protects them during the cold of winter, without which the bacterial cells would rupture and die when feces deposited during winter thawed in spring.

The California red-backed vole and the deer mouse serve similar ecological functions in the same forest as the flying squirrel. When squirrels, mice, and voles dig at the bases of trees, the organisms in their feces can inoculate the trees' rootlets with nitrogen-fixing bacteria, yeast, and spores of mycorrhizal fungi. The deer mouse, however, not only lives within the forest but also is one of the first small mammals to occupy clearings after logging or fire, so it can inoculate the soil with viable nitrogen-fixing bacteria, yeast, and spores of mycorrhizal fungi—even the soil that has been severely altered by a hot fire.

The fate of fecal pellets varies, depending on where they fall. In the forest canopy, the pellets might remain and disintegrate in the tree tops, or drop to a fallen, rotting tree and inoculate the wood. On the ground, a squirrel might defecate on a disturbed area of the forest floor, where a pellet could land near a conifer feeder rootlet that may become inoculated with

Mammals of the Pacific Northwest

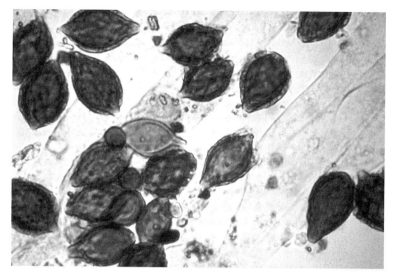

Photo 6. Fungal spore pass unharmed through the digestive tracts of rodents and germinate. (USDA Forest Service photograph by James M. Trappe.)

the mycorrhizal fungus when spores germinate. If environmental conditions are suitable and root tips are available for colonization, a new fungal colony may be established. Otherwise, hyphae of germinated spores may fuse with an existing fungal thallus (the nonreproductive part of the fungus) and thereby contribute and share new genetic material.

These rodents exert a dynamic, functionally diverse influence within their habitat—the forest. The complex of effects ranges from the crown of the tree, down through the surface of the soil into its mantle where, through mycorrhizal fungi, nutrients are conducted through roots, into the trunk, and up to the crown of the tree, perhaps into the squirrel's own nest tree.

Such relationships are by no means confined to the northern flying squirrel or even to the North American continent. Many mammals, such as deer mice, white-footed mice, red-backed voles, chipmunks, mantled ground squirrels, chickarees or Douglas squirrels, western gray squirrels, and even pikas, heather voles, deer, elk, and black bear depend more or less on truffles for food.

Unfortunately, small mammals, especially rodents, can be destructive to young trees and millions of small rodents and other animals have died as a result of poisons and purposeful habitat manipulation. But the more forests are altered by human actions, the greater becomes the need to understand the interaction of all the organisms in the ecosystem. How each component functions is often far more complex than might be anticipated, and the role it plays may be essential in maintaining ecosystem health. The mammals of western Oregon and their habitats must therefore be understood in relation to one another in order to understand either the mammals or their functions within their respective habitats. This understanding is the purpose of studying natural history.

As the statement on the back cover of the book indicates, my field work was carried out in western Oregon, the boundaries of which I define as follows:

•The northern boundary is the Columbia River because it acts as a natural barrier to the dispersal of some mammals.

•The eastern boundary is the interface between the lower edge of the moist subalpine fir forest and the upper edge of the dry ponderosa pine forest, which occurs along the upper to middle elevation of the eastern flank of the High Cascade Mountains. The High Cascade Mountains are the geologically younger eastern portion of the range that is characterized by such peaks as Mt. Hood, Mt. Jefferson, and the Three Sisters, as opposed to the geologically older western portion, which lacks high snow-clad peaks in summer.

I have chosen this line of demarcation for the following reasons: (1) It makes sense ecologically because the height and nature of these mountains, which run north and south, form a "rain shadow" by trapping most of the precipitation west of the mountain's crest; (2) this rain-shadow effect creates wetter habitats west of the crest and drier habitats east of the crest; the boundary between these two extremes is remarkably abrupt at the interface between the lower edge of the moist subalpine fir forest and the upper edge of the dry ponderosa pine forest, which is easily defined in most places in the field within a few hundred yards; (3) this interface of forest types acts as a natural barrier to a number of mammalian species from both sides of the mountains; and (4) you know on the ground when you have crossed from one type of forest into the other.

•The southern boundary of this book is of necessity the California border for two reasons. First, there is a dramatic change in vegetation—habitats—along the southern coastal area that straddles the Oregon-California border. This juxtaposition of habitats also extends eastward from the coast to some extent, but not with the clarity of the forest types on the eastern flank of the High Cascade Mountains. Second, and less satisfactory, I have to draw the line somewhere, and on a map at least people have an idea of where they are.

•The western boundary of a book on land mammals seems intuitively obvious as one stands on the shore of the Pacific Ocean and gazes out over a seemingly endless expanse of water.

I have also devised a simple map, shown on page 16, which divides western Oregon into five physiographical zones, each of which enjoys some ecological integrity. The distribution of each species of mammal is thus designated on the map by the number of the zone or zones in which it is known to occur. Although the map pertains to western Oregon, the physiographical zones, such as the Coast Ranges, the interior valleys, and the Cascade Mountains, have their counterparts into western Washington and northwestern California, and to some extent in British Columbia. Of the 89 species of mammals in western Oregon, 70 occur in western Washington, 67 going all the way into Canada; and 83 occur in northwestern California, some extending to San Francisco and beyond (see Appendix 2).

ACKNOWLEDGMENTS

I am deeply grateful to an anonymous reader for his or her candor in reviewing the manuscript of this book. I have over the years found that the more critical a review is, the more helpful it is, and this review was most helpful.

I extend special thanks, however, to Jo Alexander for a truly excellent job of editing. I say this with sincerity and respect. Hers was, in my estimation, a difficult task given the complexity of the job.

Zone 1. Pacific Coast and Coast Range

Zone 2. Willamette Valley

Zone 3. Klamath and Siskiyou Mountains and the Rogue and Umpqua valleys

Zone 4. Old or Western Cascade Mountain Range

Zone 5. Young or High Cascade Mountain Range

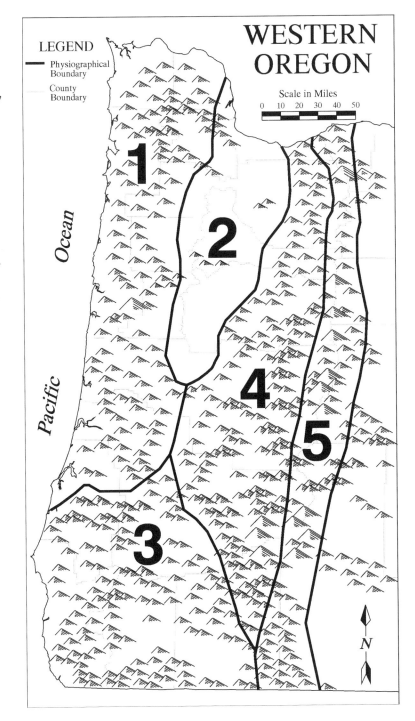

Mammals of the Pacific Northwest

Pouched Mammals
Didelphimorpha

THE POUCHED MAMMALS belong to the order Didelphimorpha, the name of which is derived from the Greek words *di* ("two" or "double"), *delphys* ("womb"), and *morphe* ("form" or "shape") and refers to the structure of the female reproductive organs.

As a group, the marsupials are the oldest, most primitive living mammals of the New World and have survived relatively unchanged for at least 50 million years.

The gestation period in marsupials is short, twelve to thirteen days, and the tiny young are born in a very immature condition. Each baby moves to the mother's pouch, where it finds and grasps a nipple, which then expands in its mouth so it is well attached. When more mature, the young can release the nipple at will but they remain within the pouch for a considerable time before venturing out. Before they are completely independent, they usually return quickly to the safety of the pouch when frightened.

Marsupials are a diverse group of mammals. Some live in trees; others live among rocks or on the ground; still others burrow; and one species is semiaquatic. Their diet is equally varied. Some eat insects and other invertebrates; some eat plants; some eat flesh; and some both plants and animals.

Marsupials occur in the region of Australia that includes New Quinea and Indonesia, as well as North and South America. The opossum is the only marsupial indigenous to the United States.

Opossums
Didelphidae

OPOSSUMS BELONG TO the family Didelphidae, which has the same basic derivation as that of the order, with the Latin suffix *idae* (which designates it as a family).

Opossums are small to medium mammals, ranging in length from about 17 to about 41 inches. They have elongated muzzles and well-developed external ears. The tail is usually long, scaly, and almost naked; it is also prehensile, which means adapted for seizing or grasping an object by wrapping around it. In some species, however, the tail is somewhat hairy. The hind legs are slightly longer than the front legs. All four feet have five toes. The first toe or thumb lacks a claw and is opposable to the others. Although members of several genera have distinct pouches, some have only two longitudinal folds of skin enclosing the teats located near the midline of the abdomen; others have no pouch.

Primarily tropical mammals, most members of this family live in trees or on the ground, but one genus is semiaquatic. Opossums are insectivorous, carnivorous, or omnivorous. They are mainly active in the evening and at night.

The belief that female opossums copulate through their noses and blow their babies into the pouch is false. The idea probably arose from discoveries of newly born young in pouches after mothers were seen investigating their pouches with their noses.

Opossums are restricted to the Western Hemisphere. They occur from southeastern Canada south through the Eastern and Central United States, Mexico, Central America, and into South America to about 47° south latitude in Argentina. There is one genus in western Oregon.

Opossums
Didelphis

THE TWO SPECIES in this genus are the largest of the opossums. Their pelage, composed of underfur and many guard hairs, is unique within the family. Other opossums have few guard hairs. These opossums vary from gray to black, or occasionally white.

The distribution of the genus is the same as that of the family. In North America, however, the species *D.virginianus* (the only species in the United States) has extended its geographical range hundreds of miles north. During the autumn, it accumulates fat and, in the northern part of its distribution, becomes inactive during inclement weather; occasionally, an opossum may lose part of its ears and tail from frostbite.

There is one species in western Oregon.

Opossum
Didelphis virginianus

DISTRIBUTION AND DESCRIPTION

total length 645-1017 mm

tail 255-535 mm

weight 4-6 kg

THE SPECIES WAS named after the state of Virginia, where the first specimen came from that was scientifically described in 1792.

OPOSSUMS, ALTHOUGH INDIGENOUS to the southeastern United States, were introduced into Oregon as pets between 1910 and 1921 and subsequently escaped or were released. Over the years, the species has spread throughout western Oregon, primarily zones 1-4.

The opossum is about the size of a house cat (25 to 40 inches long) but has a heavier body (between 8 and 13 pounds), shorter legs, and a pointed snout. The face is white. The ears are relatively large, naked, very thin, and black, frequently with white tips. The basal one-third to one-half of the long, scaly, prehensile tail is black; the rest is whitish. The first toe on the hind foot is opposable to the others and lacks a claw. Long, whitish guard hairs that overlie the black-tipped, woolly underfur gives the animal a grizzled appearance.

I GREW UP IN Corvallis, Oregon, and no one that I knew of had ever heard of an opossum south of Portland. Then, in 1965, I saw my first opossum. It was caught in a trap set for coyotes in the Coast Range just south of the Lane County line. Today, opossums occupy all habitats in western Oregon except those at high elevations.

Opossums are not social animals. Primarily nocturnal, they are seldom seen except in the headlights of an automobile, where their eyeshine is dull orange, or when investigating strange noises in one's yard, or dead along the road.

Opossums build their nests in a variety of places. They may enlarge abandoned burrows of other animals or take refuge in hollow standing trees or fallen trees, under buildings, or in piles of brush or rock.

Nest cavities are frequently lined with dry grass and leaves. In gathering material for its nest, an opossum collects a mouthful of litter and places it between the forelegs. The tail is drawn forward under the body between the hind legs. Using the forefeet, the animal passes the litter back under the body and places it on the tail. The hind legs are then used to press the litter into the curve of the tail, which wraps tightly around the material. This process is repeated with five or six mouthfuls of litter before the material is carried to the nest by use of the tail, which is held out from the body with as much as three-fourths of the tail holding material. Not all of the material is likely to reach the nest, however.

One aspect of opossum behavior that has become a common phrase in our language refers to their attitude of passive defense or pretending to be dead— "playing 'possum." An opossum's *first* reaction to danger, however, is a bluff—a display of intimidation *not* usually followed by an attack. Such a display

---

I once had a nest under my house, right under the bathtub, which made taking a bath an interesting affair because the scratchings and scrapings from the nest traveled through the floor and the bottom of the tub into my body during what otherwise was a peaceful event.

I did everything I could think of to block the opossums from getting under my house, but I always failed. I eventually trapped eight of them because they were so infested with cat fleas that I could walk where the opossums were going under the porch to their nest and immediately remove anywhere from fifty to sixty fleas from the legs of my pants.

---

consists of crouching, baring the teeth, hissing, growling, screeching, and possibly emitting the foul-smelling secretion of the anal glands.

Anyone who has heard the screeching of an opossum will remember it because it sounds like the shrill screaming of a woman in a horror movie who is slowly being stabbed to death. I have been awakened out of a sound sleep by opossums screeching at each other as they argued over some infraction of 'possum etiquette in my neighbor's backyard, and I can attest that it is a sound most horrible.

If, however, the bluff fails, an opossum goes into a cataleptic state, feigning death. The term "catalepsy" refers to a condition of muscular rigidity, in which the body and limbs keep any position in which they are placed, similar to a temporary paralysis. When feigning death, an opossum lies on its side in a curled position, lolls its tongue from open jaws, and drools. Its eyes may be open or closed, and its respiration is shallow. The pose may last from a few minutes to six hours. When the danger is past, the opossum slowly raises its head, sniffs, and resumes its activities.

The opossum is omnivorous, and nothing edible is bypassed. Its diet consists of such things as fruits, berries, earthworms, insects, frogs, snakes, birds and their eggs, small mammals, and carrion, which means there is usually something in our back yards to feed an opossum.

The breeding season extends from about mid-January to mid-October, and females may have two litters a year. The opossum has the shortest known gestation period of any mammal in Oregon, just twelve to thirteen days. At birth, an opossum is about the size of a small navy bean. A litter of sixteen could easily fit in a tablespoon.

Just before giving birth, a female becomes restless and begins cleaning the pouch in which the young are to be nursed. She licks the genital area as the young are born, and this moisture allows the minute babies to travel the difficult 3 inches from the genital opening to the pouch. Without this help, the moist babies would stick to their mother's hair and perish. During the journey to the pouch, a baby uses an overhand stroke, aided by well-developed forelegs and sharp claws, which drop off when the pouch is reached. Once in the pouch, an embryo must search for and find a teat to which it can attach. There are usually thirteen teats, but may be as few as eleven or as many as seventeen. Litters usually range from five to thirteen, but average nine. Although as many as eighteen newly born young may reach the pouch, those that cannot find teats will starve.

A baby opossum grows rapidly. By the end of the first week, its weight has increased tenfold. It remains attached to a teat for 60 to 70 days and attains the size of a small house mouse, about one ounce. Shortly thereafter, it may occasionally leave the pouch but will return to it for food and protection. It is weaned between 75 and 85 days of age but remains with its mother for three to four months. Once weaned, it spends little time in the pouch but rides on its mother's back, its small tail entwined around hers or partly around her body, clinging tenaciously to her with its feet.

The automobile probably takes the greatest toll of opossums in western Oregon, but people, domestic dogs, and great horned owls also kill them.

Opossums are occasional carriers of the parasitic roundworm *Trichinella spiralis* that causes trichinosis in humans. Those who might be inclined to eat the meat of opossums are thus advised to cook it thoroughly before consuming it.

Insect-eating Mammals
Insectivora

THE NAME INSECTIVORA is derived from the Latin words *insectum* (which refers to insects) and *voro* ("to devour") and means "insect eaters." This is misleading, however, since the food habits of this group vary widely and include fish, amphibians, mammals, green plants, fungi, and invertebrates other than insects.

Insectivores are small mammals. They have long, narrow, flexible snouts and small ears. Their eyes are usually minute and occasionally lack external openings. The body is usually covered with a short, dense pelage sometimes composed of only one kind of hair, or sometimes spines. The teeth are generally primitive in structure.

The group is diverse; its members can be ground dwelling, burrowing, or semiaquatic. With the exception of some shrews, shrew-moles, and semiaquatic species, they are primarily nocturnal. Several species can consume enormous quantities of food, but some of the food has little nutrient value.

Insectivores occupy all continents except Australia, most of South America, Greenland, and Antarctica. There are two families of insectivores in western Oregon.

Shrews
Soricidae

THE FAMILY NAME Soricidae is derived from the Latin word *Soricis* ("shrew-mouse") and the Latin suffix *idae* (which designates it as a family).

Shrews live mainly on the ground. Some are semiaquatic, and a few form burrows. They occupy a wide variety of habitats from the Arctic to the desert and tropical jungle. They are active throughout the year, some day and night, others only at night.

Shrews are small, short-legged mammals with tiny eyes and long, pointed noses. In fact, the smallest mammal in the world is a species of shrew that as an adult weighs about as much as a dime. Shrews cannot see well, but their senses of touch, smell, and hearing are acute. They have a short, thick pelage, usually some shade of gray or brown. Some shrews have scent glands on their sides; these are most active during the breeding season.

Extremely sensitive, shrews have died of fright from loud noises, even thunder. People have tossed hats over shrews running across trails, only to find dead shrews when they remove the hats.

Shrews are mainly insectivorous and carnivorous, but some eat seeds, nut meats, other plant material, and certain fungi. Some shrews have a poisonous substance in their salivary glands that immobilizes small prey; humans have been poisoned by the bites of shrews that, although not fatal, caused considerable pain.

Gestation periods range from 16 to 28 days; there may be one to several litters a year, each consisting of two to ten young. Babies are born naked and blind in a nest loosely constructed of dried grasses or leaves in a secluded place.

Shrews inhabit all major land areas except the arctic islands, Ungava, Greenland, Iceland, the West Indies, Antarctica, Australia, Tasmania, New Zealand, and most of the Pacific islands. Shrews are found only in the most northern part of South America. There is one genus of shrews in western Oregon.

Long-tailed Shrews
Sorex

THE GENERIC NAME of the long-tailed shrews is *Sorex*, which is a Latin word meaning "shrew." Long-tailed shrews have slender, delicate bodies with long, slender, highly flexible snouts. Pelages are light in summer and dark in winter; varying from tan to black, they may have one, two, or three colors. Tails of young animals are hairy, but those of old ones are usually naked. The length of the tail accounts for a third to more than half the total length of an animal. Their eyes are minute but visible, and the ears protrude slightly beyond the fur. The teeth are tipped with a reddish brown, brown, or purplish pigment in young and middle-aged animals, but, because of constant use of the teeth, old individuals frequently lack such pigment. Mature males have visible scent glands on their sides. Young are born in nests—naked, blind, and helpless. Adults lead solitary lives and are active throughout the year.

Shrews of the genus *Sorex* have a lifespan rarely exceeding sixteen months; most probably die before they are twelve months old. Even as adults, these shrews continue to grow slowly until they die.

There are about forty species of *Sorex* inhabiting the Northern Hemisphere, south to Central America in the New World and south to Israel, Asia Minor, Kashmir, and northern Burma in the Old World.

Having spent many years working with shrews in their various habitats, studying their food habits and their behavior, I am most comfortable with the old taxomonic classification because, in my opinion at least, it fits much better the kinds of relationships among species of shrews that I found in the field. For this reason, I shall employ that classification in this book.

To better understand the shrews, let's briefly compare the habitats and food habits of the seven species of western Oregon shrews. Three of the shrews of western Oregon are active on a 24-hour cycle (wandering, marsh, and northern water shrews), whereas the other four are strictly nocturnal (Pacific, dusky, Yaquinae, and Trowbridge shrews). The two aquatic shrews are separated by elevation: the northern water shrew lives in the swift, cold streams at high elevation in the Cascade Mountains while the marsh shrew inhabits the slower, gentler streams at low elevation. The wandering shrew is an inhabitant of grassy areas, such as meadows; whereas the Pacific shrew is mainly an occupant of the marshy areas on the marine terraces along the southwestern coast of Oregon and northern California. The dusky, Yaquinae, and Trowbridge shrews, on the other hand, are mainly forest dwellers; the Trowbridge shrew has the widest geographical distribution and is the greatest generalist in habitat utilization.

In addition to differences in habitat, the seven species of shrews also have different diets. The most common food of the wandering shrew is insect larvae of terrestrial origin while both the marsh and northern water shrews eat larvae of aquatic origin. Common foods of the remaining species are slugs and snails (primarily the medium-sized, thin-shelled, green land snail) for the Pacific shrew; centipedes for the Trowbridge shrew; and internal organs of large insects for the dusky and Yaquina shrews. Truffles were found only in the stomachs of the wandering shrew. Amphibian flesh was found only in the Pacific shrew, beetle larvae in the dusky shrew, and the aquatic stage of the mayfly and earthworms in the marsh shrew.

The shrews' selection of habitats, as well as their partitioning of the available supply of food, allows these small mammals to coexist with a minimum of direct competition, which is an important aspect of their natural history.

Wandering shrew
Sorex vagrans

THE SPECIES NAME *vagrans* is the Latin word for "wandering" or "unsettled." The first specimen of a wandering shrew was caught at Shoalwater Bay, now known as Willapa Bay, Pacific County, Washington, by J. G. Cooper; the date of capture is unknown, but it was cataloged on October 23, 1856 and described scientifically in 1858.

DISTRIBUTION AND DESCRIPTION

Wandering shrews occur in zones 1 through 5. These little shrews are usually somewhere between about $3^1/_2$ and $4^1/_2$ inches long and weigh between a tenth and a quarter of an ounce (photo 7). Wandering shrews have relatively short fur that varies dorsally from a light to dark grayish brown from

Shrews have distinct personalities. I have kept several in captivity over the years and found them all to be unmistakably different. I especially remember two wandering shrews, both males. One was addicted to the exercise wheel in his cage. Once he discovered it and how it worked, he ran it incessantly. As a consequence, his shoulders became so massive that I called him "Mighty Shrew" because he looked so much in physical form like the cartoon character Mighty Mouse.

The other was the antithesis of Mighty Shrew, calm and quiet for a shrew. Although it took about a month, he became so tame that he would climb onto my hand and explore it as well as my forearm. After a while, he would even sit on my hand and eat out of it. He was not frightened of me if I moved slowly, and he was in no hurry to get off my hand or arm once he started exploring.

summer to winter. The summer pelage has light-tipped guard hairs that may make the back appear grizzled. The underside is usually light gray, sometimes with a silvery sheen, sometimes washed with light brown. The tops of the feet vary from tan to brown. The hairy tails of young animals are indistinctly bicolored, brownish above and tannish below, darker at the tips. The naked tails of old animals are brown, indistinctly bicolored, and darkest at the tips. Pigment on the incisors is dark reddish brown.

total length 95-119 mm
tail 34-51 mm
weight 3-8.5 g

Wandering shrews occupy grassy areas, such as meadows and roadside ditches, shrubby areas with grassy groundcover, clearcuts in forested areas, and grassy areas along streams, rivers, and the shores of lakes and reservoirs.

HABITAT AND BEHAVIOR

Because wandering shrews are active both day and night, I have seen them now and then hunting in the runways of "meadow mice," more properly called voles. Although these shrews appear to be intolerant of one another and usually practice mutual avoidance, except when they breed, I have occasionally heard intense squeaking under the cover of meadow grasses only to have a couple of wandering shrews suddenly burst forth as one chased the other along a mouse runway. Normally, mutual avoidance and defense of nests may facilitate spacing of individuals in the wild. When contacts with strangers are frequent, a shrew tends to shift its area of activity to one that is less crowded.

Since the ratio of a wandering shrew's surface area to its body weight is so high, it spends most of its active time searching for food. To equal a shrew's consumption of food, a 150-pound person would have to eat 250 pounds of food per day. The

Photo 7. Adult wandering shrew. (Photograph by Robert M. Storm and Chris Maser)

mutual avoidance behavior of these shrews is therefore beneficial not only in conserving energy that might otherwise be expended in fighting but also in limiting competition for the supply of food in a particular area.

Wandering shrews build their nests in burrows or under fallen trees, boards, or large pieces of bark. Nests are usually simple balls of dry grasses, leaves, or mosses. Nests used as nurseries are usually made of finer materials than those used only as sleeping quarters by adults.

A rickety old hayshed squats like a fort in a cow pasture, its faded boards and tired roof bearing witness to its long years of service. It shudders and moans as the blustery winds of winter batter its tired sides. One day last winter, the howling wind finally succeeded in so loosening a board on the old shed that it fell to the ground, where it now lies bleaching in the summer sun. A meadow mouse found the board first and made a runway under it, but a great horned owl caught the meadow mouse just as spring flowers were beginning to wane.

It is now a month since the disappearance of the meadow mouse, and today a wandering shrew happens along the mouse's abandoned runway and discovers the board. After exploring the cover it provides, the shrew begins building its nest. Searching the pasture within a few feet of the old board, the shrew collects bits of dry grass with its mouth, carrying the material to a little depression scooped out by the mouse during its short tenure. If it drops a piece of grass, it continues to the nest, despite its empty mouth, before searching for more.

Having accumulated a pile of grass of sufficient size, the shrew rapidly arranges the grass around itself by grasping pieces with its mouth and tucking them under and alongside its body. It periodically digs into the floor of the nest, forcing material to the rear or side. By constantly turning as it works, the shrew forms a cup, the sides of which eventually meet and create a roof over its head.

A soft rain begins to fall. Hungry from the exertion of constructing its nest, the shrew begins hunting along the mouse's old runway. Poking its snout here and there, exploring this nook and that cranny, it comes upon an earthworm crossing the runway through the wet grass. The shrew pauses for a second, then begins exploring the worm. Suddenly, it attacks, biting the worm several times along its body. The writing of the injured worm elicits another attack. Again the shrew pauses, but the worm's attempt to escape brings on a third attack.

After several more attacks, the worm's reactions slow and the shrew starts eating it from one end, biting, chewing, and pulling off small pieces. Its immediate hunger quelled, the shrew chews the remainder of the worm into three pieces of about equal length and carries them, one at a time, back to its nest, where they are stored for a future meal. On its last trip, however, the shrew, distracted by a slight movement of the grass, drops the piece of worm it is carrying, but continues nevertheless to its nest, where the rest of the worm is stored. Only then does it retrace its path and retrieve the last piece of the worm.

Three hours later, the last of the worm eaten, the shrew again hunts along the runway. It has not gone far when it encounters a small slug, which it immediately attacks as it did the earthworm. After biting the slug, however, the shrew moves away, pats its forepaws on the soil, and wipes its snout with its forepaws, trying to remove the slimy mucus exuded by the slug in its attempt to ward off the attack. Once the slug has been thoroughly bitten and immobilized, the shrew drags it to a more protected place and eats it.

The slug, being small, does not completely satisfy the shrew's hunger. Detecting a grasshopper eating a meal of juice vegetation along the side of the runway, the shrew swiftly attacks, biting the hapless insect through its neck. The grasshopper is too big to easily carry, so the shrew chews it open along its underside and eats the soft body parts, leaving the hard external skeleton as refuse.

Photo 8. Young wandering shrews huddled together in their nest of grass. (Photograph by Kenneth L. Gordon and Chris Maser.)

Although the shrew eats centipedes, numerous spiders, a snail or two, bugs, and several kinds of beetles over the next few days, as well as a treefrog that was stepped on by a cow searching for hay around the old shed, its real boon comes a week later when it chances across a dead mouse. Biting and chewing as it pushes against the carcass with its forefeet while tossing its head up and down and from side to side, the shrew returns again and again to the carcass until there is nothing edible left.

When the shrew hunts around the old shed in the vicinity of its nest, its diet is mostly non-flying prey, such as earthworms that live in the topsoil. When it hunts where the dairy cattle graze, however, its food is primarily insects that fly into the area, such as grasshoppers. The difference in prey items between the two areas is caused by the presence or absence of the cattle, which are fenced out of that part of the pasture where the shrew has its nest. They are free to graze everywhere else, however, and so compact the soil that numerous organisms, such as beetle larvae, a food favored by the shrew, are excluded from living in the top inch or two of the soil.

Reproductive activity in male wandering shrews occurs primarily from late January through early April in western Oregon, whereas most pregnant females are found in March and April. Litters range from two to nine, but four to six is the usual size, and one litter per year is the norm. The gestation period is about twenty days. At birth, a wandering shrew weighs about 2/100 of an ounce.

During its first month, a baby shrew gains weight rapidly and constantly. At one week, wandering shrews are still naked and their eyes are still closed. In two weeks, their backs are furred and their eyes open, but their teeth have not yet erupted. Females start weaning their young at about sixteen days and finish weaning them about the twenty-fifth day. The young remain in their nest for about one month, during which time their major activity is huddling together, thereby conserving body heat (photo 8).

Owls are probably the main predators of wandering shrews, although I have also found their remains in the droppings of bobcats. Domestic cats kill these shrews but seldom eat them, and snakes, such as the rubber boa, prey on them.

I HAVE PUT THESE two shrews together because I have always found it impossible to tell them apart in the field, though *S. yaquinae* may be a little larger, and even in the laboratory I am unsure that real differences exist. I shall, in writing about these shrews, use the name "dusky" because the dusky shrew was described scientifically in 1900, whereas the Yaquina shrew was not described until 1918.

The specific name *obscurus* is the Latin word for "dusky"; it is sometimes called *S. monticolus*. The first specimen of the dusky shrew was collected at Timber Creek, Lemhi Mountains, Lemhi County, Idaho, on August 26, 1900, by Vernon Bailey and B.H. Hunter. The Yaquina shrew was named after a small tribe of Indians. The individual shrew on which the description of the species is based was captured at Yaquina Bay, Newport, Lincoln County, Oregon, on July 18, 1895, by B. J. Bretherton. Although this shrew is sometimes considered to be the same as the Pacific shrew (see page 30), I have found their habitats and food habits to be very different.

DUSKY SHREWS OCCUR from the northern half of zone 1, through the northern third of zone 2, into zone 4 and hence southward into about the middle of the zone. They are about $4^3/_4$ to $5^1/_2$ inches long and weigh from about one-fifth to half an ounce. In summer, the dorsal pelage is short and reddish brown; the underside is slightly lighter and not as red. The winter pelage is longer and darker brown. The tops of the feet vary from tan to brown. The hairy tails of young animals are indistinctly bicolored: in summer, they are brown above and slightly lighter below, whereas in winter they range from medium brown to dark brown above and slightly lighter below. The naked tails of old animals are brown, indistinctly bicolored, and darkest at the tips. Pigment on the incisors is dark reddish brown.

DUSKY SHREWS, like other shrews inhabiting the coniferous forests of the Pacific Northwest, are usually found in moist areas along streams, in marshy areas with skunk cabbage, under fallen trees, and in the thick vegetation of seepages.

Little is known about the behavior of the dusky shrew, which is nocturnal. The movements of one that I had in captivity for a brief time were rapid. When in motion, it emitted an almost constant high twitter. Its nest in the terrarium was constructed of small pieces of dead fern.

The shrew ate earthworms, sowbugs, pillbugs, ground beetles, native roaches, millipedes, centipedes, and snails. When given a freshly killed deer mouse, it immediately chewed the skull open and ate the brain. It also chewed a hole in the chest

Dusky and Yaquina shrews
Sorex obscurus and Sorex yaquinae

DISTRIBUTION AND DESCRIPTION

Dusky shrew
Sorex obscurus
total length 118-137 mm
tail 48-63 mm
weight 5.5-12.5 g

Yaquina shrew *Sorex yaquinae*
total length 112-143 mm
tail 41-64 mm
weight 5.5-15 g

HABITAT AND BEHAVIOR

of a meadow vole and ate the heart and lungs. These shrews also eat certain fungi that fruit below ground.

Reproductive activity in male dusky shrews appears to begin in late February and to last through August, but most of the males with mature testes were caught in April. Although most reproductively active females were captured in March, one was found in June and another in August. Litters range from three to six, but four is most frequent.

Although owls are probably the main predators of these shrews, domestic cats also kill many, especially in zones one and two, but seldom consume them.

Pacific shrew
Sorex pacificus

THE SPECIFIC NAME *pacificus* is the Latin word for "peace making" or "peaceable" and refers to the proximity of this species to the Pacific Ocean. The first known Pacific shrew to be captured, the one on which the description of the species is based, was taken at the mouth of the Umpqua River in Douglas County, Oregon, about 1858.

DISTRIBUTION AND DESCRIPTION

total length 135-160 mm
tail 52-71 mm
weight 10-18 g

THE PACIFIC SHREW (photo 9) is primarily a coastal mammal and thus occurs only in zone 1, from the vicinity of the Siltcoos River and Siltcoos Lake in Lane County, south into northwestern California. The largest brown shrew in western Oregon, it ranges in length from about $5^1/_2$ to about $6^1/_2$ inches and weighs about half an ounce. In summer, the pelage is short and distinctly reddish brown; in winter the pelage is relatively long and dark reddish brown to dark brown. Pelages are almost uniform in color, both in summer and in winter. The feet are noticeably pale, usually tan or light brown, as are the tails in both young and old animals. Tails are either unicolored or indistinctly bicolored; though usually tan, they may be brown and there may also be a slight darkening toward the tips of the naked tails of old individuals. The pigment on the incisors is light and distinctly reddish brown.

HABITAT AND BEHAVIOR

PACIFIC SHREWS are generally found in wet areas along little forest streams flowing through red alder and salmonberry the banks of which are strewn with fallen trees and interspersed with marshy patches well adorned with skunk cabbage, although they can occasionally be found around fallen trees in the moist conifer forest itself.

Pacific shrews are active at night. They are quick and almost continuously twitch their noses and emit twittering sounds, as though talking to themselves. These twitterings may be a kind of echolocation, like that of bats, which helps them investigate the darkness.

Mammals of the Pacific Northwest

Shelter is important to them, and they become noticeably disturbed when it is not available. Hence, these shrews are seldom found far from protective cover, such as fallen trees or thickets of vegetation.

Although little is know about Pacific shrews in the wild, I have on three different occasions kept one in captivity for several weeks at a time. I kept the shrews on my desk in a terrarium so I could observe them, and quickly found that their personalities were different. One of these shrews jumped and dashed for cover each time my telephone rang, whereas the other two did not—even when the phone startled me.

Pacific shrews make their nests of vegetation, such as grasses, mosses, lichens, or leaves. In building its nest, a shrew collects the material with its mouth, carries it to the selected site, and piles it. When a shrew cannot pull a piece of vegetation free, it tugs at it and, if necessary, bites it in two and carries the free part to the nest site.

After accumulating enough material, the shrew pushes its way into the middle of the pile. Using its mouth to grasp pieces of nest material, it rapidly tucks them around, over, and under itself. The turning and rummaging in the nest shapes the material into a cup and eventually also into a roof as the sides of the nest are pushed up.

A Pacific shrew grooms frequently and at any time, except while hunting. Grooming is performed in a crouched position, except when the anal-genital area is cleaned. The usual grooming behavior is rapid scratching with a hind foot, which is licked clean as soon as the scratching ceases. The shrew bends its body toward the scratching foot, which causes the skin of the body to become taut, thus spreading apart the fur for cleaning. The shrew is even limber enough to twist its body in

Photo 9. Adult Pacific shrew. (Photograph by Bob Smith and Chris Maser.)

order to scratch the mid-line of its back. The face and tail are cleaned orally much in the familiar manner of a mouse. When cleaning the anal-genital area, a shrew lies on its side, turns its head around, and licks, often for a couple of minutes.

Pacific shrews sleep most of the day, waking periodically to eat food cached next to their nests. A shrew enters and arranges its nest when ready to sleep, but only after it has groomed itself. When sleeping, a shrew curls its head underneath its body, nose close to its anus, hind feet resting on its shoulders, its tail either curled around its body or stretched out behind. Such a sleeping position may have survival value because as little surface area as possible is exposed, thus retaining maximum body heat. When disturbed, a shrew can move to another area so swiftly that the movement is difficult to see; thus even a sleeping individual may escape predators.

Urination and defecation are performed during a slight pause in other activity, but not usually while a shrew is hunting for food. The type of feces excreted depends on the source of food. Snails and slugs cause soft feces, whereas earthworms cause soft, muddy feces. After a meal of earthworms, a shrew pauses, lifts its tail, and squirts. When a shrew has eaten insects, centipedes, or millipedes, the feces contain much of the hard, chitinous exoskeleton of the prey, which causes the shrew to drag its anus on the ground after defecation, probably to clean it. The shrews I had in captivity also developed a habit of licking the corner of the cage in which they defecated and urinated. During such activity, some of the feces were undoubtedly reingested, which may have been how they obtained vitamins B and K.

It is rare, in my experience, to observe at close quarters how shrews hunt and kill their prey. The lack of such observation arises in part from the extreme difficulty of keeping shrews alive in captivity, as well as finding enough variety of their natural food on a continual basis throughout the 24-hour cycle to keep them fed. But, as I said earlier, I have been fortunate enough to keep three Pacific shrews alive for some weeks each. Since I was living in a forest at the time, I was able to secure enough food to keep them fed, which allowed me to observe their behavior in a large terrarium that I arranged with natural substrate to mimic their normal streamside habitat. The following account of the hunting and feeding behavior of the Pacific shrew is probably true for shrews in general.

Keep in mind that much of a shrew's prey is potentially dangerous to the shrew itself. To fully appreciate the dangers lurking in a shrew's world, imagine yourself to be only an inch or two in height when confronting an angry wasp or scorpion intent on stinging you to death or an irritated millipede emitting poisonous hydrogen cyanide gas. Now consider that the pygmy shrew, the smallest mammal in the world, weighs about as much as one thin dime. And somewhere in the world a shrew is encountering such danger every few seconds of every day just in order to live.

Although prey is generally immobilized before it is killed, the more dangerous types of prey, such as scorpions, bees, and wasps, are normally killed outright. Some shrews, such as the eastern short-tailed shrew, are known to have poison in their saliva. I do not know, however, if this is true of the Pacific shrew or whether it immobilizes its prey in some other way. A shrew either eats its prey at the site of capture or buries it in a cache next to, within, or under its nest. To bury prey, the shrew pulls nest material over it with its mouth and pushes the material into place with its nose.

When hunting for food, a shrew employs its senses of smell and hearing. Prey that do not fly, such as beetles or ants, are hunted primarily by odor. Coming across the fresh trail of a beetle, for example, the shrew follows every twist and turn the beetle has made—much like a bloodhound on a hot scent, with its nose to the ground. If the beetle has burrowed into the loose, rich litter of the forest floor, the shrew usually forces its way, head first, after the beetle until it is captured. Less often, the shrew digs out the beetle with its forefeet, kicking excess debris out of the way with its hind feet. When no evidence of prey is found on the surface of the ground, the shrew hunts within the litter layer, relentlessly following any prey it locates. Once prey has been found, the shrew attacks immediately.

One young Pacific shrew that I had in captivity encountered a large female flightless tiger beetle, which the shrew attacked immediately. But the attack was generalized—the shrew did not make any apparent effort to bite the beetle on the head or neck, the only place on this beetle that is vulnerable to even an adult Pacific shrew. Instead, the young shrew pursued the beetle, biting it on the back, sides, and rear, and, as a consequence, was severely bitten two or three times by the beetle. Remarkably, the shrew continued the attack for approximately five minutes, resulting finally in the death of the beetle from a bite through the head and neck membrane. The beetle was promptly eaten. I gave these large tiger beetles to the shrew for the next three days, but, after that one experience, the shrew captured and subdued each beetle almost as quickly and expertly as did adult shrews.

Shrews appear to prefer some kinds of beetles over others, but what stimulates the preference I do not know. Medium-sized beetles are usually bitten on the dorsal aspect of the thorax near the attachment of the wing covers. Large beetles with hard external skeletons are bitten at the base of the neck membrane and across the head from on side. Small beetles are simply bitten, killed, and eaten, whereas a medium-sized beetle with a long neck is quickly decapitated and eaten. But these methods of killing must be learned.

Beetles are usually eaten by chewing off the wing covers near the point of attachment to the body and eating out the soft inner portion. Such feeding leaves most of the beetle relatively intact. Sometimes the wing covers are chewed off and softer portions of the external skeleton are eaten, in which case the legs become dismembered and are left as refuse. Occasionally a shrew holds a beetle down on its back, pinning the wing covers to the ground with its forefeet and literally pulls the beetle off the wing covers with its mouth. When eating hard beetles, a shrew's teeth make audible clicking and crunching noises.

Two genera of beetles, however, are exceptions. Some darkling beetles exude an extremely pungent fluid when attacked by a predator. During an attack, a shrew bites a large darkling beetle, bangs it against the ground, and shakes it back and forth. Although these pungent beetles are readily eaten, blister beetles are neither touched nor eaten, and with good reason. The body fluids of the commoner species contain cantharadin, a substance that often causes blisters when applied to the skin.

Compared to beetles, a large millipede is usually killed in a frenzy. A shrew sniffs it, then scratches up the ground with its forefeet much as a dog does. Rapid attacks are interspersed with sniffing, scratching, and running around the millipede. The shrew darts in and bites the millipede across its back from the rear toward the head of the animal. During a biting attack, the shrew sometimes holds the millipede with its mouth and shakes it violently back and forth. At other times the shrew bites the millipede, then rubs its nose and mouth on the ground or thrusts its whole head into the litter on the ground. This behavior may be stimulated by the millipede's defense mechanism—the release of hydrogen cyanide. It often takes a couple of minutes for a shrew to kill a large millipede, after which it is chewed into two, three or more pieces. Those not immediately eaten are stored.

When encountering a large centipede, a shrew simply grabs its with the mouth and bites it just behind the head, which kills the centipede before it can defend itself. This is important because the poison claws of some species of centipedes are strong; the poison is effective against birds as well as large insects and thus may be dangerous to shrews.

Grasshoppers, crickets, and native roaches are all immobilized in a similar manner: they are bitten in the neck membrane between the head and thorax. Those not eaten are stored. Since these insects are relatively soft-bodied, comparatively little is left as refuse, usually only the wings and hind legs.

One shrew that I was studying confronted a scorpion and immediately bit the scorpion on the head, killing it instantly, not giving the scorpion time to sting. The shrew ate the scorpion's abdomen, leaving the rest as refuse.

When a land snail is found, a shrew sniffs it, which causes the snail to withdraw into its shell. The shrew continues to investigate, manipulating the shell with its forefeet, often trying to poke its snout far enough into the shell to pull the snail out with its teeth. When this fails, as it normally does, the shrew begins at the opening and chews the shell apart, until the entire snail is extracted and eaten. These empty shells left as refuse are a sign of Pacific shrews.

Wingless, female mutillid wasps, which are usually referred to as "velvet ants," have extremely long, flexible stingers and potent venom. In addition to bright warning coloration, such as red and black, they produce a warning noise—an audible, high-pitched whining squeal—by extending and contracting their abdomen. The warning noise is usually produced when the wasp is first approached and causes a shrew to appear extremely nervous and jumpy, and to avoid the wasp. But if a wasp fails to make a warning noise, the shrew sniffs it briefly and bites it downward across the top of the thorax in such a way that teeth marks are visible on both sides. At times, a wasp is bitten and tossed with a flip of the shrew's head in what appears to be a single motion. These wasps are immobilized, not killed outright, but they rarely if ever sting the shrew. After the wasp is immobilized, the shrew separates the abdomen between the first and second abdominal segments, eats the contents, and discards the abdominal exoskeleton in one piece. The anterior portion of the eviscerated wasp crawls away. These wasps are not known to be stored by the shrew.

With rare exceptions, the large carpenter ants are simply approached, sniffed, and bitten on the thorax, after which they are eviscerated in the same manner as the female mutillid wasps. But one red, forest-dwelling ant goes through a ritualized defense behavior as soon as a shrew's approach is detected. The ant moves in a circle, swaying its abdomen back and forth, emitting tiny droplets of fluid (presumably formic acid) from the tip of the abdomen onto the ground. The shrew will not attack the ant as long as it continues with this behavior. But when the shrew's presence no longer elicits this behavior, the shrew bites the ant on the thorax, then shakes its head violently, rubs its mouth on the ground, and cleans its face with its forefeet, and then eats the ant.

When flying prey, such as dragonflies, crane flies, moths, or butterflies, are encountered, a shrew apparently locates them through its acute sense of hearing. Except for a bee or wasp, the shrew chases the prey with remarkable quickness, often jumping into the air, and occasionally capturing its prey in flight. When the prey alights above the shrew's head but within reach, the shrew jumps, grabs its victim, and falls to the ground with the prey held securely in its mouth. If the prey lands on the ground, the shrew pounces on it, pinning it down by placing its forefeet on the prey's wings.

If the shrew is hunting a large dragonfly, and fails in its first attempt to pin it down and bite off the head, the shrew retains its grip with its mouth and shoves the dragonfly against the ground before trying to get another hold. This procedure is occasionally repeated several times before a dragonfly is subdued.

Yellowjackets, honey bees, and bumble bees are all treated in a similar manner. Each of these insects produces a buzzing sound when irritated. This initially sends a shrew into "nervous" hiding, running, and exploring behavior, which usually lasts from fifteen to thirty seconds before the shrew begins to investigate the prey. The prey, for its part, tries to keep its stinger toward the assailant; the shrew, on the other hand, approaches the prey in the region of the head. The shrew then plants its feet and stretches its nose as close to the prey as possible without actually touching it. After remaining motionless for a fraction of a second, the shrew snaps at the head with an audible click of its teeth and runs away in what appears to be a single motion. This procedure is repeated until, with astonishing speed, the shrew darts in and bites the prey across the top of the head, between the eyes, killing it instantly. Wasps and bees (with the exception of the wingless mutillid wasps) are killed outright, then either eviscerated or stored. Although the shrew usually eats only the contents of the abdomen, the thorax is occasionally chewed open and the soft, inner portions consumed. Most of the prey is left as refuse.

It is clear that these large shrews have considerable strength. I watched one Pacific shrew that detected a young alligator lizard about 5 inches long and immediately gave chase. The shrew sank its teeth into the end of the lizard's tail; about two-thirds of the tail came off. Dropping the dismembered tail, the shrew again pursued the lizard. Soon caught, the lizard was bitten in mid-body. The shrew released its prey almost instantly and, with startling quickness, secured another grip across the top of the head. No struggle was visible; the lizard appeared to die immediately from a bite across the top of the head over the left eye and across the lower jaw.

I have captured sexually mature male Pacific shrews throughout the year and sexually active females as early as April and as late as November, but most in June, July, and August. Litter size ranges from two to six, but four or five is the usual size.

Although owls are probably the main enemies of Pacific shrews, they are also caught and eaten by the Pacific giant salamander.

Marsh shrew
Sorex bendirei

THIS SHREW WAS named in honor of Army Major Charles E. Bendire who captured the first specimen on August 1, 1882, about a mile from the Williamson River, about 18 miles southeast of Fort Klamath, Klamath County, Oregon.

DISTRIBUTION AND DESCRIPTION

total length 145-185 mm
tail 62-84 mm
weight 14-25 g

THE MARSH SHREW can be found in all five zones. Marsh shrews are the largest species of the genus *Sorex* in North America, ranging from about 6 to $7^1/_2$ inches in length and from about half an ounce to almost an ounce in weight. Dorsally, they are dark brown, blackish brown, or blackish and their undersides are only slightly lighter. There is little difference in color between summer and winter pelages. The tops of their feet vary from brown to dark brown. The hairy tails of young animals are unicolored, almost the same color as the pelage. Although the naked tails of old shrews are also unicolored, they are a lighter shade of brown. The pigment on the incisors is dark reddish brown.

In addition to their large size, marsh shrews can be distinguished from other shrews in Oregon by the small, fleshy projections that line the outside of each nostril. These projections remain conspicuous even in old animals.

Marsh shrews also have a fringe of short, stiff hairs along the margins of the feet, including the toes. This is usually called a "swimming fringe," and is most noticeable on young animals. It appears to sustain much wear during the life of the animal and is not replaced by new hair in old animals.

THESE SHREWS ARE associated most frequently with small streams and marshy areas, but I have found them at considerable distances from their usual habitat during wet winters, especially the young, which seemed to be dispersing to new habitats.

To me, this little-known mammal is one of the most fascinating animals in the forest. Although this shrew is active throughout the 24-hour cycle, it is most active at twilight and during the night and spends much time in the water. Sculling quickly around the surface, it can rapidly submerge, using its long snout and whiskers to explore under water. It swims with its hind legs in almost constant motion, except when it pauses to rest briefly at the surface. Under water, most of the propulsion is provided by its hind legs that are used alternately, not simultaneously like a frog's. Air trapped in its fur gives it not only buoyancy but also a silvery appearance, and a stream of small bubbles rising in its wake, allows its progress under water to be followed. In addition, air trapped by the swimming fringe enables a shrew to run over the surface of the water for a few seconds with its belly just barely above the surface.

Marsh shrews feed primarily on aquatic insects, such as nymphal stoneflies, mayflies, and alderflies. When a shrew's nose or whiskers come in contact with living prey, such as a worm or small fish, normal swimming is greatly accelerated and normal submergence time is often doubled. Not all aquatic animals are eaten, however; small crayfish may be ignored. Other prey taken along the water's edge includes phantom crane flies, ground beetles, snails, slugs, spiders, and harvestmen or daddy-longlegs. When a marsh shrew finds termites in soft, rotten wood, it uses its teeth to tear the wood apart while bracing itself by placing its front feet against the wood. Pushing its long snout into the tunnels, the shrew goes into an excited frenzy extracting and eating the termites.

A marsh shrew attacks an earthworm with a lightning-swift series of bites along its body. Large worms are usually eaten from the rear end to the front; small worms can be consumed from either end. Worms encountered under water are attacked, seized in the teeth, and carried to land to be eaten. A marsh shrew may also immobilize a worm with a rapid series of bites, then carry it to a special area on land to be stored for a later meal, but unlike the Pacific shrew, they neither cover nor conceal their food cache.

Information on the reproduction of marsh shrews is scarce. Sexually active males have been captured as early as February and as late as August. One litter was found in March, and sexually active females have been caught in May, June, and

July. Available data on litter size indicate a range of three to four, with four most frequent.

Although I have no specific data on predation of marsh shrews, some likely fall prey to owls, domestic cats, Pacific giant salamanders, and, since they are aquatic, large fish.

THE SPECIFIC NAME *palustris* is the feminine gender of the Latin word for "marshy" and refers to the "marshy places from Hudson's Bay to the Rocky Mountains," where the species was first captured. It was described scientifically in 1828.

Northern water shrew
Sorex palustris

THE NORTHERN WATER SHREW, which occurs in some places in zone 4 and all of zone 5, is the second largest of the long-tailed shrews in North America, exceeded in size only by the marsh shrew. The northern water shrew is about 6 inches long, of which about half is tail, and weighs about half an ounce. It has a long, pointed, flexible nose, minute eyes, and short, wide, round ears that are almost concealed in the fur. Their teeth are tipped with dark reddish brown. The fur is dense, soft, and velvety. The upper parts are dark gray to blackish, lightly frosted with paler hairs. The underparts, from the throat to the anus, are whitish, tinged with gray or brown. The tail is blackish above and whitish below. Water shrews, like marsh shrews, have swimming fringes of short, stiff, silvery hairs on the margins of their feet, including the toes. The fringe, most noticeable on young animals, sustains much wear and is not replaced by new hair in an old animal.

DISTRIBUTION AND DESCRIPTION

total length 150 mm
weight 8.5-14.5 g

THE WATER SHREW is the most skilled swimmer in the swift, cold, mountain streams of the High Cascades. Its fur is so thick and soft that it traps and holds air, allowing the shrew to sit on the surface of the water and float like a duck. Because the shrew's fur gives it such buoyancy, it must force its way to the bottom of the stream where it literally stands on its long, flexible nose, searching the bottom for food, its hind feet kicking rapidly to maintain this position. To change direction, it twists its body; to come to the surface, it simply stops kicking and rises like a cork, bursting to the surface with dry fur. Wherever the shrew goes under water, it is trailed by a row of bubbles rising out of its fur, and the shining layer of air that clings to the surface of its fur as it swims makes the shrew resemble a silvery fish.

Although a water shrew's fur is remarkably resistant to wetting, water does begin to penetrate after several minutes. After leaving the water, a shrew dries its fur by rapidly and thoroughly working over its body with its hind feet. During this process, which lasts from ten to thirty seconds, fine droplets of water are thrown off. The stiff hairs of the swimming fringe

HABITAT AND BEHAVIOR

along the margins of the hind feet, functioning almost as a comb, facilitate the drying process. The swimming fringe also traps air that allows the shrew to literally walk or run on the surface of the water. The northern water shrew is a truly remarkable mammal; it can swim, dive, float like a duck, and walk both on the surface of the water and on land.

Water shrews feed mainly on invertebrates, such as earthworms and slugs, and the larvae of aquatic and terrestrial insects. In addition, they catch and eat such vertebrates as small trout and sculpins, as well as small larvae, less than one year old, of the Pacific giant salamander. None of these exceed three inches in length.

When hunting fish and salamander larvae, a shrew dives to the bottom and, finding its prey, bites it somewhere around the head, which seems to paralyze it. The shrew then pops to the surface and swims, gripping the fish or salamander larva by the head while its body trails alongside that of the swimming shrew. Once out of the water, the shrew bites its prey through the head with its front teeth to kill it.

When hunting recently hatched trout, called "fry," in the shallow water at the edge of the stream, a shrew rushes about the rocks, stopping frequently to elevate its flexible nose as if trying to detect a scent. It then plunges into the water and swims beneath the surface. If the trout fry are not moving, the shrew has difficulty finding them and frequently bumps into them seemingly by accident, but if the fish are swimming, it has little difficulty in finding and following them. When hunting is good, a shrew will catch more than one fish, carry it ashore, kill it, and store it for later use.

Although water shrews are primarily active at night, they may be abroad at any time during the 24-hour cycle. Their activity is intermittent and only for short periods. After a period of activity, a shrew may retire to its spherical nest, constructed out of moss or other vegetation, or it may simply stop in a crouched position and fall asleep in some safe place. Occasionally a shrew sleeping on a rock under the protection of the overhanging bank along the stream's edge loses its balance and falls into the water—a rude awakening!

Water shrews are active throughout the year, and in winter, as in summer, their major activity is confined to the banks of the stream. A stream, even in the High Cascade Mountains, does not freeze solid under its insulating cover of snow, although a small shelf of ice forms along its banks and, in some places, across the entire stream. The space between the bottom of the shelf of ice and the water's surface becomes the winter domain of many a water shrew.

Male water shrews born in the late spring and summer do not become sexually active until the December or January following their birth, and reproductive activity then continues until August. Females are either pregnant or raising young from February until August, and may produce several litters during the breeding season. Litters normally range from five to eight young, but six is most common.

As far as predation is concerned, one of these shrews is occasionally found in the stomach of a large fish sharing its aquatic habitat.

THE FIRST SPECIMEN of this shrew was caught at Astoria, Clatsop County, Oregon, on June 10, 1855. It was sent to the United States National Museum by W. P. Trowbridge, in whose honor it was named.

TROWBRIDGE SHREWS OCCUR in all five zones, with the exception of the non-forested floor of the Willamette Valley. They range from about 4 to 5^1/$_2$ inches in length and weigh about a quarter of an ounce. They have long, lax fur, which in summer is a relatively uniform brownish gray above and slightly lighter below. Winter pelages vary from dark gray to blackish above and are also slightly lighter below. During spring and autumn molts, distinct areas of lighter and darker fur are readily visible on backs and sides of these small shrews. The tops of the feet range from whitish to very light tan. The hairy tails of young animals are distinctly and sharply bicolored, brownish gray to blackish above and whitish below. The naked tails of old individuals are also distinctly and sharply bicolored; however, they are lighter than the fur, ranging from light brown to brown. The pigment on the incisors is dark reddish brown.

THESE SMALL SHREWS are primarily denizens of the coniferous forest in all its various stages of maturity and are the most common species in the forest. Like other shrews, Trowbridge shrews are usually associated with some type of protective cover, such as fallen trees or thickets of vegetation, but they occasionally venture into unprotected areas. These shrews, when compared with other western Oregon shrews, occupy not only the greatest variety of habitat types but also habitats that range from the wettest to the driest.

Trowbridge shrews become active in late evening as the shadows lengthen in earnest, just before dark. Less aggressive and somewhat more delicate than wandering shrews, Trowbridge shrews seem to be out-matched where the two occur together along the forest's edge, where forest and meadow

Trowbridge shrew
Sorex trowbridgei

DISTRIBUTION AND DESCRIPTION

total length 108-139 mm
tail 47-67 mm
weight 3-9 g

HABITAT AND BEHAVIOR

One evening, as the sun travels westward over the Pacific Ocean, a large bumblebee is knocked off a salmonberry blossom by a passing deer as it walks along a path at the forest's edge. The buzzing of the angry bee reaches deep under the fallen tree bordering the path and arouses a sleeping Trowbridge shrew, which immediately sallies forth to investigate.

Finding the bee buzzing discontentedly on the ground, the shrew attacks without hesitation, but the bee, almost half the size of the shrew, manages to out-maneuver its pursuer. The bee, unable to get a clear flyway in the shrubby ground cover along the path, is closely pursued by the squeaking shrew. The shrew, for its part, is unable or unwilling to secure a firm hold on the loudly buzzing bee during several encounters. The battle continues for perhaps ten minutes, until at last the bee manages to find an opening in the foliage and escapes. The shrew runs frantically about sniffing and squeaking in search of the bee. Feeling suddenly exposed, it disappears under the fallen tree from which it originally ventured, and the evening falls silent once again.

meet, where roads penetrate the forest, and before trees again dominate the vegetation, where forests have been clearcut.

These little shrews are not only the most adaptable in habitat of all Oregon shrews but also in food habits. Compared to the other shrews, they are generalists and eat almost a third again the variety of foods the others do. Their diet includes a wide variety of insects, centipedes, spiders, small snails, fruiting bodies of underground fungi, and the seeds of plants, as well as other plant material. Over several years, I caught hundreds of these shrews in western Oregon in live-traps baited with oatmeal. When I examined the stomach contents of the captured Trowbridge shrews, I found that almost all contained oatmeal. They were the only shrew consistently found to have eaten the bait.

In Oregon, male Trowbridge shrews exhibit most reproductive activity from February through April, whereas females are most active in March and April. Litters range from one to six, but usually contain three to four young.

Some data indicate that Trowbridge shrews may commonly have more than one litter per breeding season. It has been suggested that two or three litters produced successively might compensate for the rather short breeding season and general lack of reproductive activity during summer.

After May, the proportion of adult to immature shrews may drop rapidly, in which case the population turnover is completed by late summer. Perhaps the drastic decrease of adults is because their badly worn teeth make eating difficult. Some students of shrews have suggested that the disappearance of adults is an illusion caused by the dilution of the adult population with young animals, but it is quite apparent that in some places there is a very real and sudden decrease of adults in late spring and early summer.

Although owls are probably the main predators of Trowbridge shrews, rubber boas are likely to take some and others have been found in the stomachs of Pacific giant salamanders.

Moles
Talpidae

THE FAMILIAL NAME Talpidae is derived from the Latin word *talpa* ("mole") and the Latin suffix *idae* (which designates it as a family).

Moles have long, tapering snouts, cylindrical bodies, and minute eyes. Their ear openings are simply holes near the shoulders concealed by dense fur. Keen senses of ground vibrations and of direction compensate for weak sight. They have broad front feet, the toes of which terminate in stout claws adapted for digging. The forelegs and shoulders are anatomically modified in such a way that the front feet, except those of shrew-moles, are permanently turned out.

The pelages of most moles are composed of soft, flexible hairs, all about the same length. The hairs become smaller in diameter near the body, making the pelage much like velvet and allowing the hairs to lie in any direction, enabling an individual to go forward or backward in small burrows. Shrew-moles have pelages more like those of shrews, composed of both guard hairs and underfur directed toward the rear of the animals; they cannot move backward in a burrow.

Moles have white teeth and the first upper incisors of western North American moles are straight, flat, and relatively broad, similar to human teeth. In contrast, the first upper incisors of shrews are pointed and curved sharply down.

Photo 10. Mound of a coast mole. (Photograph by Chris Maser.)

Most moles are well adapted for a burrowing life and normally make two types of burrows; shallow surface-tunnels, marked by visible ridges of soil that are pushed up with their backs, and deeper tunnels that can be located by the cone-shaped mounds of earth on the surface (photo 10). These molehills are formed when soil is pushed out through a tunnel; however, when a mole compresses the soil around a deep tunnel sufficiently to provide the required space for its body, molehills are not created. One mound of primary importance is called the fortress; it is larger than other molehills, and a mole constructs its nest under it.

Although moles can traverse their burrow with surprising speed, they are relatively slow on the surface of the ground. Their digging strength, however, is startling. My late friend Murray Johnson once had a captive Townsend mole in the basement of his home. The mole got out of its container onto the concrete floor. Finding a small crack in the floor, it dug its way straight down through the concrete and escaped.

With this tremendous strength, it is no wonder that some species in the wild can burrow into hard soil in seconds. Other members of the mole family are active primarily in the decaying vegetative litter on the ground and under such protective cover as fallen trees. A few moles (the desmans of Europe) are amphibious or semiaquatic, living in freshwater streams and ponds; these moles have fringes of short, stiff hairs along the margins of the feet, called swimming fringes, as well as partially webbed toes.

Although most moles are insectivorous, the four species occurring in western Oregon are omnivorous. Differences in habitat, depth of burrowing, and size of shrew-moles, coast moles, and Townsend moles suggest a partitioning of the available food resources, which should decrease competition for food among the species where they occur together. However, earthworms comprise between 50 and 75 percent of the diets of both Townsend and coast moles and about 40 to 50 percent of the diet of the shrew-mole. The fourth mole, the broad-handed mole, occurs separately from the other three. It lives in much drier habitat, and what little data there are on its diet in Oregon suggest that snail, slugs, and beetles are among the most important foods. Now let's compare the moles within their respective habitats.

The shrew-mole is active on the surface at any time of the day or night and also is a shallow burrower in the humus of western Oregon forests and the surface soils of meadows. It spends much time around fallen trees rotting on the ground, which probably accounts for the high frequency of centipedes in its diet, including the large *Scolopocryptos sexpinosa*, the same centipede eaten by the Pacific shrew, which shares its habitat with the shrew-mole. Pillbugs, sowbugs, and termites are also readily eaten by shrew-moles. Since the shrew-mole is the smallest mole in North America, its size probably limits the variety of food items it can use, and may explain the relative absence of beetles with hard wing covers.

Coast moles, geographically the most wide-ranging species in Oregon, have a correspondingly varied diet. Primarily inhabitants of western coniferous forests, these moles tend to be shallow borrowers both in forests and along the forest-meadow edge. They are also active on the surface of the forest floor, especially around large, rotten fallen trees. This wide latitude in the use of habitat allows them to exploit a great diversity of food items.

Townsend moles, on the other hand, are the largest of the western moles and are primarily inhabitants of meadows and pastures, a more limited habitat than that utilized by shrew-moles and coast moles, which appears to impose some restrictions on the variety of food items available (23 types of identified food along the Oregon coast) Of the three moles along the Oregon coast, the Townsend mole has the greatest amount of vegetation in its diet, notably roots from pasture grasses.

Many moles (including western Oregon moles) have a strong, musky odor that probably makes them unpalatable to some would-be predators, especially mammals.

The known gestation period ranges from 28 to 42 days, but delayed gestation has been reported in members of one genus. Moles become sexually mature from six to twelve months of age; they have a single litter per year.

Moles are active throughout the year, and numerous methods have been devised to rid lawns and gardens of them. One technique is to put empty bottles into the ground with the bottom in the mole's tunnel and the neck sticking above the surface at an angle. When the wind blows, it makes a piping sound and moles supposedly disappear almost overnight.

Moles are found in Europe and Asia, north of the 63rd parallel and south to the Mediterranean and the Himalayas. In North America, they occur from southern Canada to northern Mexico. There are two genera of moles in western Oregon.

American Shrew-moles
Neürotrichus

THE NAME OF THE genus *Neürotrichus*, is derived from the Latin word *nervus* ("sinew" or "tendon") and the Greek word *trichos* ("hair"); the reference is probably to the little mole's hairy tail.

The shrew-mole, the smallest North American mole, is often mistaken for a shrew (photo 10). The common name is apt since the genus has a combination of shrewlike pelage, composed of both guard hairs and underfur, and the large head and heavy dentition of a mole.

American shrew-moles occur only in the Pacific Northwest from southern British Columbia, south to central California,

Photo 11. American shrew-mole. Note tiny eye, lack of an external ear, and digging claws on front feet. (Photograph by Chris Maser)

Mammals of the Pacific Northwest

and east to the eastern flank of the High Cascade Mountains. There is one species in western Oregon.

THIS SMALL MOLE was named in honor of George Gibbs who caught the first specimen at White River Pass, north of Mt. Rainier, Pierce County, Washington, on July 15, 1854.

SHREW-MOLES OCCUR in all five zones. These tiny moles are from 4 to 5 inches long and weigh less than half an ounce. Except for the tip, which is naked and reddish, the long, tapering nose is sparsely covered with fine hairs. The nostrils are on the sides of the tip. The ears are merely holes located near the shoulders and usually are not visible because of the dense fur. Minute eyes are nearly concealed by fur. The front feet are broad with stout claws adapted for digging. The three middle claws on all feet are longer than the outer claws.

Although definitely constricted at the base, the tail is relatively thick. It is encircled with rows of scales and covered with sparse, long, coarse hairs. In color, the tail resembles the pelage, which is thick, relatively soft, and almost uniform in color, varying from dark brownish gray to blackish gray. In certain lights, the long, glossy guard hairs shine like metal.

THE SHREW-MOLE occupies a wide variety of habitats throughout western Oregon, exhibiting habits that are surprisingly shrewlike. They do not, for example, create burrow systems with the molehills characteristically constructed by larger moles. Instead, they make relatively shallow burrows in the loose, sod-free topsoil.

Today, somewhere in the Western Cascade Mountains of Oregon, a giant Douglas fir, which fell 300 years ago, is gradually becoming part of the forest floor, and is newly discovered by a shrew-mole. The shrew-mole living along the fallen tree makes two types of burrows, shallow and deep. The shallow trough-like burrows are roofed by the decaying vegetation on the floor of the forest and by the fallen tree. This forms a complex, intersecting network through which the shrew-mole regularly travels in search of food, invertebrates that have either crawled or fallen into the shallow burrows.

Most of the shrew-mole's active time is spent in a ceaseless quest for food, and it may eat more than its own weight in a twelve-hour period. When hunting for food, the mole rummages through decaying litter, turns leaves and debris, investigates crevices, and patrols its burrows. It may climb into low vegetation and search the foliage for food, and it may search both the accessible inner and outer areas of the fallen tree. Its

American shrew-mole
Neürotrichus gibbsi

DISTRIBUTION AND DESCRIPTION

total length 100-126 mm
tail 32-50 mm
weight 5.5-12 g

HABITAT AND BEHAVIOR

hunting is aided by an ever-active nose that is thrown high in the air, twisted to one side or the other, rapped on the ground, or hooked under the body. The tapping sound made by a nose-rapping shrew-mole is not altogether beneficial, as it may help a predator to locate it.

If an earthworm is encountered in a burrow, it will be eaten with a degree of individuality because not all shrew-moles handle earthworms alike. Some strip them through the forefeet, presumably to clean off the soil. Some bite worms along their entire length; some merely bite from the portion that is first encountered to the worm's nearest end. Most simply begin eating a worm from one end to the other. But now and then one bites a worm into small pieces, or bites a small piece from the tail of a living worm, then chews and swallows it before pursuing the maimed creature for another bite, giving the worm a chance to escape.

When insect pupae, pillbugs, or sowbugs are encountered in its tunnel, the mole hooks them with its nose, pulling them backward off balance. Pillbugs, which are capable of rolling into tight balls, are bitten open and consumed from their soft undersides. Insects are often struck repeated blows with the mole's forefeet or are covered with earth, whereas centipedes are simply grabbed by the head and eaten. Beetles, ants, and slugs are refused, though they are readily eaten by most western shrews.

While eating, the mole uses its forefeet not only to hold food but also to arrange the food in its mouth. Although the speed of digestion varies with what has been eaten, most food is evident in a mole's feces within forty minutes.

This tiny mole is an excellent digger; it is very strong and can vertically lift almost 7 ounces—twenty times its own weight. When digging, the shrew-mole pushes aside the earth with lateral motions of its forefeet. Using only one foot at a time, it rotates its body at a 45° angle and forms the burrow by pressing aside and packing the loose, damp earth. The mole makes its way through loose litter on the surface of the ground by pushing the litter sideways with its forefeet and pushing its body forward with its hind feet.

The shrew-mole does not construct deep burrows as often as shallow ones, which are used primarily for hunting. The deep, narrow burrows branch, intersect, and cross one another at various levels but are seldom as deep as a foot below the surface of the soil. This burrow system contains a larger sleeping chamber, about 5 inches in diameter, with an arched ceiling about 3 inches high, and close enough to the water table to have a soft, level, mud floor. The chamber is connected to the outside by a tiny ventilation duct in the ceiling.

The shrew-mole moves in a slow walk with momentary pauses. When walking, it bends the claws of the forefeet underneath the feet so that it actually walks on the backs of the elongated foreclaws. It is not only graceful on the ground but is also a capable climber. When frightened, the shrew-mole makes an amazingly swift, scuttling dash for cover, where it remains crouched and, except for rapid breathing, absolutely still. If not too badly frightened, it faces the disturbance with one forefoot raised and its mouth open.

This tiny mole is a good swimmer and sometimes explores along the edge of a stream. Occasionally, while burrowing under the litter on the forest floor above a stream bank, it may fall into the stream. It appears to fly through the water, using the feet on each side of its body alternately to create an undulating motion with its body and tail. Its forward motion is so great that its head and fully two-thirds of its body are above the water.

The shrew-mole's periods of activity are interspersed with periods of rest and sleep that are longest and most frequent when the mole's appetite is satiated and shortest and least frequent when it is hungry. Periods of rest range from one to eight minutes at irregular intervals from two to eighteen minutes apart.

The mole's nest is a simple affair. It grasps small pieces of vegetation with its mouth and takes them to a selected place. When a sufficient pile of material is accumulated, the mole hollows out a cuplike depression by moving around and tucking some of the material around itself with its mouth. To sleep, the mole places its feet under its body fairly close together and tucks its snout and head beneath its body and between its forelegs. Its weight is thus borne by the top of its head and by its hindquarters. The tail is placed around one side of the body. When the mole is resting, its nose is relaxed and flat against the ground. Periods of relaxation are occasionally interrupted when the mole raises its nose and half-heartedly sniffs the air or scratches itself vigorously.

Touch appears to be the most developed of the shrew-mole's senses. Its long, flexible nose guides it much like a blind person's cane and, in almost constant motion, quickly identifies any object it contacts. The gentlest contact of an object by one of the sensitive whiskers on the nose or face causes an immediate response. The mole becomes almost frantic in its efforts to remove with its forefeet any debris adhering to a whisker. The stiff hairs that encircle the openings of the ears are also very sensitive, and the long, stiff hairs on the tail may also be sensitive to touch. The shrew-mole's hearing apparently is so acute that it can detect the sounds made by the invertebrates on which it feeds.

These tiny moles appear to be reproductively active throughout the year, but most reproduction seems to take place between early March and mid May. Although litters range in size from one to five, three or four is probably the norm. Babies, naked and pink, lacking whiskers, and weighing but a fraction of an ounce, are born in a nest excavated about a foot and a half below the surface of the ground and lined with damp leaves. Their eyes are merely prominent black spots under a transparent covering of skin, and their external ear openings are not yet evident. Their broad forefeet are fleshy and paddlelike; the tops of the digits are soft and blunt and there is no indication of claws.

As fur begins to grow when the baby is a few days old, the skin becomes bluish to blackish with developing hair, which upon eruption is dark bluish black, soft, and downlike. The fur is especially short on the upper parts, where it is almost silvery-gray. The feet and nose are still pink, and the skin of the anal region is much folded. The tail is relatively long, thick, and bluish-violet. The forefeet are turned out, with the palms away from the body.

Owls seem to be the shrew-mole's most common enemy. These moles are known to be eaten by Screech Owls, Great Horned Owls, Barn Owls, Long-eared Owls, and Saw-whet Owls. The Pacific giant salamander and the rubber boa undoubtedly capture some shrew-moles, which are also eaten by the northwestern garter snake. Many are killed by domestic cats, but not eaten, and raccoons get one occasionally.

Western American Moles
Scapanus

THE GENERIC NAME *Scapanus* is derived from the Greek word *skapane*, which means "digging tool."

Western American moles are robust, compact animals. The front feet are flat, pale, and scantily haired (photo 12). The palms are nearly as broad as they are long and are permanently turned outward by the anatomical rotation of the shoulders. The claws are broad, flat, and adapted for heavy digging. They have minute but visible eyes and long, tapering snouts. The snouts and lower lips are pinkish and essentially devoid of hair. The tip of the nose is naked with crescent-shaped nostrils opening upward. The short, round, thick tail is slightly constricted at the base and tapers toward the end. There are indistinct circular rows of scales and sparse, coarse hairs on the tail; one species, the broad-handed mole of zone 3, has a

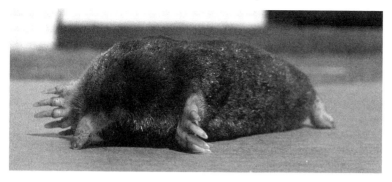

Photo 12. Coast mole. Note lack of external ears, rotated front feet with heavy claws for digging, and flexible snout with crescent- shaped nostrils. (Photograph by Chris Maser)

relatively hairy tail. The dorsal pelage is brownish-black to black and slightly lighter on the underside. When smoothed, the velvety pelage has a metallic luster.

The three species of the genus *Scapanus* are confined to extreme western North America from southern British Columbia, south into the northern part of the Baja Peninsula, Mexico; all three occur in western Oregon.

THIS LARGE MOLE was named in honor of John Kirk Townsend, who captured the first specimen on the banks of the Columbia River, probably at Old Fort Vancouver, now Vancouver, Clark County, Washington, on May 9, 1835. Townsend, primarily a naturalist, was in charge of the hospital at Fort Vancouver during the winter of 1835-36.

Townsend mole
Scapanus townsendi

TOWNSEND MOLES are found in zones 1, 2, the western part of 3, and 4. They range in length from about 8 to about 9 inches and weight from about 2 to 6 ounces. Their physical appearance is essentially the same as that of the genus. Townsend moles and coast moles are similar in appearance, but adult Townsend moles are considerably larger. Because these moles often occur together, it may be necessary to have specimens identified by an expert.

DISTRIBUTION AND DESCRIPTION

total length 198-237 mm
tail 35-61 mm
weight 64-171 g

TOWNSEND MOLES are essentially denizens of meadows, fields, pastures, lawns, and golf courses. They are primarily nocturnal and almost exclusively subterranean in their habits, spending almost all of their time burrowing in the rich soil of the valley floor. Earthworms are the primary food, but the diet also includes centipedes, snails and slugs, larval butterflies, moths, flies, and beetles, and adult beetles, ants, and crickets.

Moles are essentially solitary animals that live in established, interconnecting burrow systems with little or no overlap. Should two moles meet, they frequently try to intimidate one another or fight violently, with much soundless writhing, tumbling, and biting. Such aggressive or defensive behavior may be a

HABITAT AND BEHAVIOR

manifestation of intense efforts to retain homesites. On the other hand, when crowded into restricted areas during times of severe flooding, individuals show some tolerance for one another.

Moles sparsely populating low areas with poorly drained soils move greater distances than do moles inhabiting higher areas with greater densities in population and better drained soils. Poor drainage, coupled with a scarcity of earthworms—the principal food—induces them to extend their movements and to seek higher terrain.

During flooding, low-lying burrows can become inundated with water. Reluctant to travel on the surface of the ground, many moles construct shallow burrows, following the upward contour of the land for great distances, ahead of the rising water. Some make it to higher ground only to be forced to swim when this area also becomes flooded. Although they are good swimmers, they may become disoriented and exhausted, and drown. Some, rather than burrowing, only bury their heads in the soil. When the flood water subsides, one may occasionally find over a hundred dead moles lying about.

The mole's season of reproduction begins in December, and females carry their young from February through April, when they give birth to litters that range from one to four, but usually comprise two to three.

Because of high winter rainfall and the tendency of the soils of the valley to saturate with water during the reproductive season, female moles search for slightly elevated areas in which to construct their nursery nests. Such areas vary from a few inches to several feet above the surrounding land.

Nursery nests are usually situated under, or within 3 feet, of a single, large mound of earth called a fortress; they also may be located near the center of clusters of several normally sized mounds concentrated within a 6- to 9-foot area. Nest cavities are constructed from 3 inches to almost 2 feet below the surface of the ground. They are spherical, and average about 9 inches in diameter and about 6 inches in height. Three to eleven lateral tunnels enter the cavities. Many cavities have a tunnel, called a bolt-hole, leading several inches straight down from the floor of the cavity, then turning up and joining an upper-level burrow. In addition to allowing a quick escape, the bolt-hole also provides drainage.

Nursery nests are usually constructed several days before birth of the young and are composed of two layers. The inner layer, about 25 percent of the nest, is usually made with fine, dry grasses, but occasionally with mosses or leaves. The outer layer is made with coarse grasses that are still green and frequently wet and have their root systems intact. It is thought

When I was boy between the ages of six and twelve (1944-50), I had a friend, Billy Savage, who lived just across the road from the bottom of the golf course of the Corvallis Country Club about 3 miles south of town. Each spring, we waited excitedly for the war on the Townsend mole to begin on the golf course.

We always knew when the war began, because mole traps with long, needle-sharp spikes were set in their mounds to impale the owner as it used its burrow beneath a trap. Because the traps were readily visible, we were admonished by the caretaker of the golf course, Homer Gray, to stay away from them, which, of course, did more to arouse our curiosity than anything else could have done.

Then, one day when I was eight years old, a trap was sprung, and we simply could not contain ourselves. We had to see for ourselves what a mole looked like, so we spent what seemed to us a long time lying in the grasses along the fence peering onto the golf course, cautiously looking for Homer.

"Do you see Homer anywhere?" I asked.

"No," Bill replied.

"Neither do I. Let's go look at the trap."

Taking one last, searching glance, we raced to the stile and crossed the fence. We quickly pulled up the trap, and there was a mole. A mole! A real mole! But try as we might, we couldn't remove it from the silver spikes of death. Then, riddled with guilty excitement, we thought we heard a noise, which caused us to race back across the fence and hide, lest Homer find us.

that a mole may gather the nest material in shallow burrows by grasping the roots of the grass with its mouth and pulling the whole plant underground. The coarse grasses are matted and interwoven to form a compact, protective shell about 2 inches thick around the dry, inner nest. Green grasses are added to the outer layer several times during the months the nest is occupied by the young.

As the wet grasses decay, they generate a considerable amount of heat that is largely retained in the nest cavity, keeping the young warm even when their mother is absent. Nursery nests are kept clean of feces and other waste materials by the mother.

At birth, babies are about 2 inches long and weigh less than $2/10$ of an ounce. They are hairless and their skin is whitish to pink. By the time they are ten days old, their skin is becoming bluish to blackish with developing hair. When 36 days old, they are about 4 1/2 inches long and weigh almost 3 ounces. They are covered with short, soft hair, and at this age or shortly thereafter, they begin to leave the nest.

Young moles disperse at night during the month after weaning; the great majority will stay within about 30 yards of their birthsite. The areas they choose as homesites appear to depend more on the suitability of the habitat than on the number of moles already residing in the area.

The rubber boa is probably the main predator of nestling Townsend moles. Owls are probably the major predator once they leave the nursery nests and disperse over the surface of the ground. Today, domestic dogs and cats catch many of these moles. Dogs dig them out and kill them at all times of the year. Cats primarily kill juveniles that are active on the ground during the summer and occasionally kill adults. But I have seen neither dogs nor cats eat the moles.

Humans probably kill the greatest number of Townsend moles because they are intolerant of the moles' mounds, employing poisons, traps, and "mole-guns." The latter (now illegal) are pipes fitted with shotgun shells and triggered when a mole passes through its tunnel under the gun.

Coast mole
Scapanus orarius

THE SPECIFIC NAME *orarius* is a Latin word meaning "belonging to the coast." The first scientific specimen of the coast mole was caught by J.G. Cooper at Shoalwater Bay, Pacific County, Washington, on August 30, 1855.

DISTRIBUTION AND DESCRIPTION

total length 148-182 mm
tail 34-46 mm
weight 38-67 g

COAST MOLES ARE found in zone 1, 2, the extreme western portion of 3, and the northern half of 4 and 5. Coast moles range from about 6 to about 7 inches in length and weigh from about 1 to 2 ounces. Their general physical appearance is essentially the same as that described for the genus. Coast moles and Townsend moles are similar in appearance, but adult coast moles are considerably smaller. Because these moles often occur together, it may be necessary to have specimens identified by an expert. However, where the coast mole exists in zones 4 and 5 to the exclusion of the Townsend mole, it is easily identified not only because it is the only mole there of the genus *Scapanus* but also because it has a gray to silvery gray pelage.

HABITAT AND BEHAVIOR

THE COAST MOLE has a wide adaptability to different habitats, ranging from the sandy dunes of the Oregon coast and the shores of tideland rivers all the way into the subalpine meadows of the north-central Cascade Mountains eastward into the Wallowa Mountains in the northeastern corner of the state.

Coast moles are primarily nocturnal, but I have occasionally seen them on the surface of the ground in the early morning, late afternoon, and early evening. I have also caught several coast moles in museum snap traps set on the surface of the ground along the sides of large, rotting fallen trees, and I have captured them in the large burrows of the mountain beaver, which I will talk about later.

Although I have often caught coast moles in the same meadows as Townsend moles, I have never captured a Townsend mole cohabiting the forest with coast moles, even where meadow and forest grade gently into one another. Coast moles characteristically inhabit soils that are better drained than those in which Townsend moles occur. In western Oregon, they are usually found in shallower burrows and they construct ridges in the soil by burrowing just under the surface more commonly than do Townsend moles. The molehills created by coast moles are smaller, less conspicuous, and appear to be less numerous than those of Townsend moles.

Although earthworms are their primary food, they also eat centipedes, millipedes, snails and slugs, larval butterflies, moths, flies, and beetles, adult beetles, and ants.

Male coast moles with mature, descended testes have been trapped from December through March, but most records are from February. Three to four young born in late March to early April seem to constitute an average litter, and a single litter per year is probably the norm.

The rubber boa is probably the most important predator of nestling moles. Once the moles have left their nursery nests, owls are most likely their main enemies, especially of the young as they disperse over the surface of the ground. Along the southern Oregon coast, domestic dogs and cats kill many of these moles. Dogs dig out and kill fewer coast moles than Townsend moles because coast moles, being primarily forest dwellers, are not so readily available. Although cats kill primarily juvenile moles active on the surface of the ground during summer, they also kill many adults, in more areas than they do Townsend moles because the coast moles are more readily available in the brushy, forested areas where most cats do their hunting. Neither are eaten, however. Humans probably take a greater toll of Townsend moles than of coast moles since the latter have less direct contact with people.

Broad-handed mole
Scapanus latimanus

THE SPECIFIC NAME *latimanus* is derived from the Latin words *latus* (meaning "side") and *manus* ("a hand") and refers to this mole's out-turned front feet. The first specimen of this mole is thought to have been captured in Santa Clara, Santa Clara County, California. It was described scientifically in 1842.

total length 132.5-
190 mm
tail 45 mm

THE BROAD-HANDED MOLE, or hair-tail mole as it is also known, occurs in western Oregon from the southeastern part of the zone 3 northward into zone 5 to about the vicinity of Mt. Washington in Linn County.

The broad-handed mole is smaller and paler than the Townsend mole, but larger and has heavier feet and claws than the coast mole. It ranges in total length from $5^1/4$ to $7^5/8$ inches. Its tail ranges from almost an inch to $1^3/4$ inches in length. It also has more hair on the feet and tail than either the coast or Townsend moles; the tail of the broad-handed mole is almost covered in silvery hair from the base to the tip, whereas the tails of the Townsend and coast moles are pinkish and all but naked. The fur is soft and velvety and varies from a drab silvery to a somewhat yellowish brown.

HABITAT AND
BEHAVIOR

THE BROAD-HANDED MOLE is found from the sandy flats on the eastern flank of the High Cascade Mountains all the way to the 7,000-foot level along the rim of Crater Lake. In the lowlands, its mounds occur in heavy, moist soils; in the lighter soils all one usually sees are the ridges it makes in burrowing. At high elevations, they can be found in loose pumice soil, which at times seems almost devoid of vegetation.

The broad-handed mole cannot dig well in hard, compacted soil; certain soils, which fail to attract them in the dry season, support them in the wet season. The availability of their food is also a likely limiting factor with respect to their distribution.

In other behavior, the broad-handed is so similar in western Oregon to both the Townsend and coast moles that I refer you to the accounts of these two species for more information.

Winged Mammals
Chiroptera

THE ORDINAL NAME Chiroptera is derived from the Greek words *cheir* ("hand") and *pteron* ("wing") and refers to the four greatly elongated, slender digits or fingers of the forelimbs between which are stretched thin, delicate membranes. The membranes, composed of a double layer of skin, are also attached to the sides of the body, the hind legs, and the tail, forming wings (photos 13 and 14). Bats are unique among mammals in that they have true powers of flight. Other mammals, such as flying squirrels, can only glide in a descending path from a higher to a lower elevation.

Among mammals, only rodents exceed bats in number of species. Their great diversity can be shown in the spectrum of their feeding habits. Insectivorous bats feed mainly on insects that they catch in flight. Fruit-eating bats depend almost exclusively on fruit and some green vegetation for food. Flower-feeding bats, aided by long tongues with brushlike tips, eat primarily pollen and nectar. True vampire bats, which have razor-sharp teeth, make small incisions in the skins of mammals, usually while they are sleeping, and lap the blood that flows from the wounds. Carnivorous bats usually prey on other small mammals, birds, lizards, and frogs. Fish-eating bats catch fish at or near the surface of the water with their large, powerful feet armed with sharply hooked claws.

Photo 13. Long-eared bat showing elongated fingers enclosed by membrane to form wings and free thumb. (Photograph courtesy of Steve Cross.)

Photo 14. Long-eared bat showing wings attach to legs and body and tail membrane attached to legs. (Photograph courtesy of Steve Cross.)

Bats, birds, and airplanes share the same basic aerodynamic characteristics. But there are structural differences among the bats and the resultant aerodynamic characteristics of their wings are directly related to the efficiency with which various sources of food can be exploited. The time, location, and altitude of flight also interact to maximize the efficiency with which bats can use a given source of food, and variations in these factors minimize direct confrontation among species.

Bats are not "blind as a bat." In fact, the predominant senses used by the Old World fruit bats are vision and smell and it has become apparent through research that many echolocating species also depend on vision and smell.

But this in no way detracts from the remarkable feats of bats using echolocation. Bats flying at speeds of up to 11 feet per second can maneuver unhindered through an array of wires and can capture several insects in a swarm in a few seconds. Some bats routinely detect and capture insects or small vertebrates resting on vegetation, the ground, or the surface of

Because bats are primarily nocturnal, little understood, ugly by most human standards, and are known to carry some dangerous diseases, such as rabies, they have probably evoked more superstition, prejudice, and misinformation than any other group of mammals. For example, during the years I studied the mammals of the Oregon coast, I knew a woman who had a small colony of bats in the attic of her house, one of which ended up in her kitchen. Terrified, she called and asked if I would please come and get rid of them. (Here you must understand that I am bald, very bald.) The moment I arrived she said: "Thank God you're here, Chris. It's safe for you to catch the bats. But if I were to try, I'd be in mortal danger because everyone knows that bats lay eggs in your hair and drive you crazy."

the water. In laboratory studies, certain species have even distinguished between real and fake insects.

How does echolocation work? Bats emit sounds—some species through their mouths; others through their noses—that bounce off objects and return to their ears, giving them information on the identity and location of those objects. Bats thus have acute hearing abilities that must be highly directional in both horizontal and vertical planes. Signals emitted by bats are brief and must be perceived and interpreted within milliseconds.

The bat's ear is a specialized organ adapted not only for hearing but also for maintaining balance or equilibrium. The external ears of bats exhibit great differences in size, shape, and complexity. Insectivorous bats, such as the western big-eared bat, that can hunt insects resting on vegetation or on solid objects, usually emit faint signals and have enormously large ears. Fast-flying bats, such as the hoary bat, emit loud signals with high frequencies and have relatively small ears.

Compared with most small mammals, bats have long gestation periods, ranging from slightly less than two to eight months; many rodents, for example, have gestation periods of 21 to 28 days. Nevertheless, bats are helpless and essentially naked at birth and develop slowly; however, they have well-developed thumbs and hind feet and are capable of climbing shortly after birth. Depending on species, bats may be able to fly in three to ten weeks.

The deciduous teeth of most bats are partially developed at birth. They vary in structure by species, but are usually slender, pointed, and directed toward the rear of the mouth. Such hooklike teeth apparently aid young bats in clinging firmly to their mothers' teats. There are fewer deciduous than permanent teeth and they are rapidly replaced by permanent ones.

Bats inhabit most of the temperate and tropical regions of the world but are not found in colder areas beyond the limits of tree growth, nor on certain remote oceanic islands. There are two families of bats in western Oregon.

Bat feces (known as guano) were used for many years as a source of the saltpeter needed to produce gunpowder. They were used for this purpose by the United States during the War of 1812 and the Civil War. During World War II, the U. S. armed forces considered using bats as weapons. It was proposed that large numbers of bats, fitted with small incendiary bombs, be released from airplanes over such enemy installations as ammunition dumps, industrial centers, and fleet concentrations. It was expected that the bats would disperse, seek shelter in numerous obscure places, and produce many fires simultaneously.

Evening Bats Vespertilion- idae

THE FAMILIAL NAME Vespertilionidae is derived from the Latin word *vespertilio* ("animals of the evening") and the Latin suffix *idae* (which designates it as a family).

Most evening bats are blackish, gray, or some shade of brown; members of a few species are red, yellow, or orange. They have small eyes; their ears range from small to disproportionately large. Some members of this family have large glands under the skin on the snout that cause big bulges or folds in the skin, but not all such bulges and folds are caused by these glands.

Most evening bats dwell in caves, but they also take refuge in buildings, mine shafts and tunnels, old wells, storm sewers, culverts, rock crevices, hollow trees, the foliage of trees and shrubs, hollow joints of bamboo, larger tropical flowers, tall grasses, and abandoned bird nests, and under bridges, rocks, and loose bark of tree trunks. Locations where they roost may be found by the whitish stains from their urine, the dark stains from their body secretions (photo 14), accumulations of feces, or any combination of the three. The roosts of some species can occasionally be found by the strong odors emitted from the bats' scent glands. In addition to the ultrasonic sounds undetectable to human ears, bats produce sounds that we can hear; this is particularly true of bats roosting in groups.

Evening bats alight initially with their head up, then quickly shuffle around and hang by their toes with heads down and wings folded along their bodies. Although many hang from vertical surfaces, such as walls, rather than hanging freely suspended from a ceiling, some crawl into small crevices and others hang from limbs of trees amid the foliage.

Members of most species roost in small groups or large colonies, but those species that roost in foliage are solitary by nature. Some species remain in colonies throughout the year; some congregate only in winter, normally returning to the same

Photo 15. Note dark stain from body secretions where long-eared have been entering into the building's attic. (Photograph by Chris Maser)

roosting site annually. Cave dwellers become torpid in winter, periodically changing locations within the cave. Members of a few species migrate south in autumn to milder climates and return in spring. Individuals of some species can find their way back to the home colony after being captured, transported, and released at various distances.

All bats in western Oregon go to the nearest water and drink before they start hunting. They drink on the wing.

Examination of their food habits suggests that bats in western Oregon partition the supply of food among themselves rather than competing for it. They feed at different times, at different heights above the ground, in different areas of a given habitat, and tend to select for certain groups of species within the insect-spider fauna. These behavioral differences allow maximum efficiency in the utilization of both the available habitat and the sources of food.

In western Oregon, for example, eleven species of evening bats eat a variety of insects and their relatives (such as spiders), but the main foods of each species, while greatly influenced by availability, are different. For instance, the little brown bat, Yuma bat, and California bat feed heavily on flies; the western big-eared bat, long-legged bat, long-eared bat, and silver-haired bat feed primarily on moths; and the big brown bat and pallid bat concentrate on beetles.

The following foods are listed in descending order of importance to each species of bat:

Little brown bat—midges, internal organs of large insects.
Yuma bat—midges, termites (always the winged form of the damp-wood termite), moths.
Long-eared bat—moths, flies, spiders.
Fringed bat—moths, harvestmen or daddy-longlegs.
Long-legged bat—moths, termites, spiders.
California bat—midges, flies, craneflies.
Silver-haired bat—moths, termites, flies.
Big brown bat—moths, scarab beetles, other beetles.
Hoary bat—mosquitoes, moths.
Western big-eared bat—moths.
Pallid bat—scarab beetles, long-horned grasshoppers and katydids.

In our studies, we also found that differences in feeding habits among the species of bats allowed them to partition the food supply. The little brown bat, Yuma bat, and California bat had similar food habits, but termites and moths were eaten much more often by the Yuma bat than by the other two. The little brown bat fed more heavily on caddisflies and the internal organs of large insects, while the California bat fed on moths at a rate similar to that of the Yuma bat, but also greatly on craneflies, little on termites, and did not consume the internal organs of large insects.

Some of these differences may reflect differences in availability of insects at the time and place where the bats fed, but others seem definitely to reflect selectivity of prey among the bats and/or adaptability toward capture of certain groups of prey. Consumption of almost 100 percent moths by the western big-eared bat, compared with the complete absence of this food in stomachs of the pallid bat, certainly reflects more than simple feeding on the basis of availability.

The pallid bat is known to regularly take food items from the surface of the ground, but other species of bats also take non-flying prey. Spiders, for example, are often taken by bats, and may reach 10 percent of the diet the the long-eared bat and the California bat. How spiders are obtained is not known.

In addition to partitioning the source of food, bats in western Oregon also use available habitats differently:

Little brown bats exhibit an affinity for forested areas, both coniferous and deciduous. These bats usually emerge about twenty to thirty minutes before full darkness and forage among scattered trees along edges of dense timber. Around human habitation, they are often seen feeding in continuous circular patterns around buildings and small patches of trees, 5 to 10 feet above the ground.

Yuma bats are closely associated with large streams, rivers, ponds, or lakes. They normally emerge twenty to thirty minutes prior to full darkness and often feed just a few inches above the surface of the water, repeatedly flying regular routes. Along streams and rivers, they fly up and down the water course in relatively straight patterns. Over ponds and small lakes, however, they fly in circular patterns.

Long-eared bats, although generally distributed throughout western Oregon, do not seem to be abundant. They are associated with coniferous forest. Long-eared bats emerge from ten to forty minutes after full darkness and feed among the trees.

Fringed bats are rare in western Oregon, except in the area of Ashland. They appear to be associated with coniferous forest, where they emerge late, well after dark, but otherwise little is known about their foraging behavior.

Long-legged bats are generally distributed in the coniferous forest. On warm, overcast evenings, they feed along the edge of the forest and among the trees. On cold, clear evenings, they are not readily seen as the feed among the trees within the forest. Emergence on heavily overcast evenings occurs as much as 45 minutes earlier than on lightly overcast evenings.

California bats are abundant and generally distributed in western Oregon. They are not generally active until after dark, so I do not know where they feed. I have on one occasion, however, shot a California bat while it was foraging along the edge of timber fifty minutes before full darkness.

Silver-haired bats are associated primarily with coniferous forest, but a few also occur in mixed coniferous-deciduous forest. These bats are not abundant, but are widely distributed. They emerge from 15 to 45 minutes prior to full darkness. The slowest-flying bat in western Oregon, they frequently hunt in sweeping circles, often 100 yards or more in diameter. Although they normally forage in and over the forest, they fly 20 to 40 feet above roads through the forest when such are available. Adults generally feed singly, but groups of three to four are occasionally seen.

Big brown bats occupy a wide variety of habitats, but are usually associated with coniferous and deciduous forest. They emerge early, frequently before the swallows have ceased to feed, thirty to forty minutes before full darkness. They usually forage high over the forest, in great sweeping circles, sometimes well over 150 feet above the ground. As dusk deepens, however, they often descend to within 40 or 50 feet of the ground. When feeding along forest roads, big brown bats normally fly only 20 to 30 feet high and tend to fly relatively straight courses.

Hoary bats are large, swift, late-fliers that are seldom captured in western Oregon. They seem to be associated with forested areas, where they forage well after full darkness. I say this based on two that I shot by sound well after dark.

Western big-eared bats are normally associated with abandoned buildings and caves in western Oregon. I do not know where they forage because all of these bats in our studies came from day roosts, which were associated with predominantly coniferous forest.

Although pallid bats occur the full length of western Oregon between the Coast Ranges and the western flank of the Western Cascade Range, I have only collected them from roosts, and thus know nothing of their foraging behavior.

The importance of moths in the diet of so many bats undoubtedly reflects the abundance of the moths. Were it possible to separate the different kinds of moths into their respective taxa, differences in the selection of moths by species of bat would probably also become clear.

There is also a correlation between the habitat in which a bat feeds and the resistance of its wings to punctures. Species

Photo 16. (a) Droppings of the pallid bat. Note shiny insect fragments. Bat droppings crumble easily when gently rubbed with one's finger. (b) Soft and (c) hard droppings of bushy-tailed woodrat. Note the lack of insect remains and the fibrous material protruding from four of the droppings. Rodent droppings do not crumble when gently rubbed with one's finger. (d) Droppings of an insectivorus bird. Note the white "urine" component that covers part of the dropping. (e) Droppings of a lizard. Note that although the white "urine" component is present, as in the bird droppings, that it is barely attached to the fecal component. (Photographs by Chris Maser)

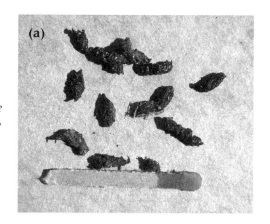

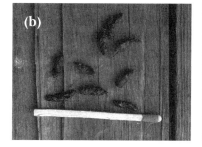

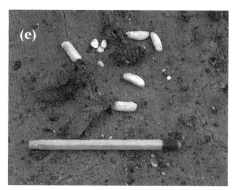

that feed on or near the ground or in areas of dense or thorny vegetation usually have wings that are more resistant to punctures and are less elastic than the wings of species that feed in more open areas.

Although a bat may capture some prey directly with its mouth, most is caught by "netting" it with a wingtip and immediately transferring it into the cupped tail membrane, formed by curling the tail forward under the body. The prey is then grasped with the teeth. Small prey are eaten in flight, but large prey are consumed at a resting site, where their remains may occasionally be found. Many evening bats also pause during feeding to rest at night roosts, specific places, separate from day roosts, where bats hang up to allow their food to digest, which accounts for the accumulation of feces under such resting stations.

The feces of evening bats found under their roosts are usually segmented and crumble easily, revealing bits of shiny exoskeleton of insects (photo 16). The droppings of rodents, in comparison, which may also be found under bat roosts, are much harder and more fibrous. The feces of bats do not contain the white, calcareous material characteristic of the droppings of birds, lizards, snakes, or some frogs and toads, which I have on occasion found under bat roosts. This urine component does not show up in mammalian droppings because mammals have separated bodily canals for reproduction, urination, and defecation, whereas the three systems are combined into one common opening—termed a cloaca—in birds, reptiles, and amphibians.

In temperate climates, species that become torpid in winter breed from August through October and frequently again in the spring. Sperm deposited in the females' reproductive tracts in the autumn are stored through the winter. The combined autumn and spring breeding periods produce a litter in spring, because ovulation and fertilization occur only in spring. (In tropical regions, however, breeding is immediately followed by fertilization and development of embryos.) The gestation period of most species ranges from 40 to 70 days but is 100 or more days for some species. Litters range from one to four young.

In many colonial species, the females segregate into maternity colonies to bear and raise young. As the time of birth approaches, an expectant mother that has been hanging quietly from a vertical surface becomes restless and changes position frequently. She is nervous, irritable, and reluctant to eat. She periodically grooms her underside, genitalia, and tail membrane; this behavior lasts from a few minutes to an hour.

Just before onset of labor, a fetus changes from a horizontal to a vertical position. An expectant mother also normally reverses position until, head up, she is suspended from the vertical surface by her thumbs and feet. Her hind legs are spread, and her tail is curled forward over her vaginal opening, forming a "cup" into which the young are born. During labor, a bat cries, closes her eyes, bares her teeth, and makes chewing motions, indicating that labor is probably painful.

In normal labor, there may be as few as ten muscular contractions. The babies are usually born in a breech position. Unlike most mammalian offspring, a bat starts to grope with its feet as soon as they emerge; the feet grasp whatever they encounter, usually the fur or leg of the mother.

After securing a foothold, the baby pulls and in a few seconds, the body is freed up to the head, which remains momentarily in the vagina. Pushing or pulling by the baby, coupled with the muscular contractions of the mother, frees the head suddenly. The mother then tears the birth membrane off her baby with her teeth. A single birth is usually completed within thirty minutes; double births often take longer. During or shortly after birth, the mother resumes her normal head-down position. The mother then grooms her baby until it is clean and dry, after which she cleans herself, using her wings to shift the baby around.

After delivery, the naked, blind baby remains attached to the placenta by the remarkably elastic umbilical cord. In members of some species, the expulsion of the placenta with its attached umbilical cord is delayed, and the cord functions as a safety line after the birth. Should a baby fall off its mother, it remains suspended close enough to her to be able to secure a firm grip on her with its well-developed thumbs and feet.

Blood continues to circulate through the umbilical cord for three to ten minutes until it blanches and circulation ceases. The delicate cord dries rapidly. After its expulsion, the placenta may be eaten by the mother. If the umbilical cord ruptures too early, the baby bat bleeds to death in minutes.

Most bats probably live from four to eight years in the wild, but banding records show that some individuals have exceeded 21 years of age, and captives have lived over twenty years.

Evening bats are distributed throughout the temperate and tropical regions of the world, except some islands in the South Pacific. They range from tropical forests to deserts and from sea level to timberline.

There are six genera of evening bats in western Oregon—all of which must be handled with care because they are capable of carrying rabies.

Mammals of the Pacific Northwest

THE GENERIC NAME *Myotis* is derived from the Greek words *myos* ("mouse") and *otos* ("ear").

Mouse-eared bats are usually some shade of brown, darker dorsally and lighter ventrally. Dark and light color phases are exhibited by members of certain species. Members of some species roost singly or in small groups; others are colonial. In cold regions, mouse-eared bats become torpid in winter, some flying long distances to find acceptable sites for hibernation.

Females of some species are gregarious, and their maternity colonies may number into the tens of thousands. Although females have two teats, one by each forelimb, members of most species have a single young, but some may have two. By late summer, maternity colonies disintegrate. The young become sexually mature in about one year. Mouse-eared bats are long lived; individuals have been known to live to eighteen years in the wild.

Mouse-eared bats are the most widely distributed of the bats. Except for humans, they probably have the most widespread natural geographical distribution of any genus of terrestrial mammal. They are not, however, found in the arctic or antarctic regions nor on many of the oceanic islands. There are six species of mouse-eared bats in western Oregon.

Mouse-eared Bats
Myotis

The specific name *lucifugus* is derived from the Latin word *lucis* ("light") and *fugio* ("to flee"). The first little brown bat was caught in Georgia, probably on the LeConte Plantation near Riceboro, Liberty County, and described scientifically in 1831.

Little brown bat
Myotis lucifugus

LITTLE BROWN BATS occur in all zones. They weigh about a quarter of an ounce, and range in length from $3^3/8$ to almost 4 inches. The tail ranges from $1^1/4$ to $1^5/8$ inches.

These small bats have long, soft, thick pelages that are reputed to have among the finest hair of any mammals, as well as the greatest number of hairs per unit area of the skin's surface. The hair on their backs varies from yellowish brown, to brown, to dark brown, with shiny tips that create a characteristic glossy sheen. The undersides are pale brown with little or no sheen. Ears, when laid forward, reach to the nostrils. Flight membranes and ears are usually dark brown and almost devoid of hair. There are no keels on the calcars. Little brown bats may be confused with Yuma bats.

DISTRIBUTION AND DESCRIPTION

total length 84-98 mm
tail 33-41 mm
weight 5-9.5 g

LITTLE BROWN BATS appear to be generally distributed in virtually all habitats, but they exhibit an affinity for forested areas. The earliest I have observed these small bats to become active in

HABITAT AND BEHAVIOR

the spring was along the Oregon coast, where they were feeding by the middle of March.

Little brown bats usually emerge about twenty to thirty minutes before dark and forage among scattered trees, along the edges of dense timber, and over lawns and meadows. They can occasionally be seen feeding in continuous circular patterns, 3 to 6 feet above the ground, around buildings and small patches of trees. Although these bats in Oregon eat a fairly wide variety of foods, they concentrate primarily on flies, including midges, but also eat moths, beetles, and the internal organs of large insects.

By August, adult bats have accumulated layers of fat for winter. Females accumulate more fat than males. In western Oregon, these bats usually disappear sometime in October. Where or if they hibernate in Oregon, I do not know. They are known, however, to migrate as far as 290 miles between their summer feeding areas and their winter places of hibernation.

When hibernating in northern regions, these bats form tightly packed clusters that aid in conserving body heat, but they do not exhibit this habit in southern regions. During hibernation, they select areas that will remain several degrees above freezing and also have high humidity (85-100 percent), often high enough to form droplets of condensation on their fur. Bats arouse at about two-week intervals and, if conditions have changed, move to more favorable areas.

Photo 17. Female little brown bat with her baby attached to a teat. (Photograph by Chris Maser)

When wind blows into the place of hibernation during an extreme cold spell, exposed bats increase their metabolism to maintain a body temperature a few degrees above that of the surrounding air. But prolonged exposure to subfreezing temperatures is fatal, especially to young bats.

Little brown bats breed in the autumn. Sexually active adults have been captured along the Oregon coast as early as August 1; in large hibernating colonies, however, some breeding occurs throughout the winter. Delayed fertilization takes place in the spring when the bats emerge from hibernation.

The sexes segregate during April when the females form maternity colonies, often in the attics of houses, where they can frequently be heard during the day as they squabble among themselves. Some of the attics with bats in them that I have examined reached more than 91° F during the day. Males are frequently found under loose flashing around chimneys, where temperatures also get very high. Although I have occasionally found males in attics, they are usually under shingles, or under wooden or tar-paper siding on houses, where temperatures are cooler than in the hot attics favored by the females.

Photo 18. Close up of baby little brown bat. (Photograph by Chris Maser)

The gestation period is fifty to sixty days. In some areas of North America, maternity colonies are large, frequently three to eight hundred individuals but occasionally thirty thousand or more, consisting primarily of mothers and their offspring. One maternity colony that I examined under the loose flashing around a chimney at the beginning of May in Bandon, Coos County, contained seven pregnant little brown bats and one pregnant Yuma bat. A second colony, found in Lincoln City, Lincoln County, in the middle of August, consisted of twenty to thirty females and young (photos 17 and 18).

Little brown bats usually have a single young each year, occasionally two. Old females tend to give birth earlier than young females; hence, late births may be from young females with their first litters. There is considerable variation in the dates of birth within the species and within a given colony; the time of birth may be correlated with latitude to some extent.

At birth, a baby is blind, almost devoid of hair, and weighs about 6/100 of an ounce, about one-fourth the mother's weight after she has given birth. The young have disproportionately large thumbs, feet, and head. After the mother has licked her baby dry, it climbs about on her body until it finds one of her two teats on her chest near her forearms; the suckling baby is then wrapped in its mother's wing. Although the baby's eyes open on the second day, it remains wrapped in its mother's wing, firmly attached to a teat throughout its first few days of life, after which the mother probably leaves her baby hanging in the roost at night when she leaves to feed. When about half grown, the youngster begins to hang beside its mother in the roost during the day.

Young bats are capable of flight at about three weeks of age. They learn to fly in the home shelter. They venture outside when about a month old, at which time they weigh nearly as much as adults. Young bats fly in circular patterns near the exit of the roost. Their flight tends to be slow and steady in contrast to the zigzag pattern characteristic of adults. Once outside, they must learn to detect and capture prey, which apparently requires time and practice. Therefore, they are probably not weaned until they become competent at capturing food. Even though young bats continue to feed after the adults are in hibernation, they are not able to accumulate as much fat as the adults, which may cause some mortality during their first winter. Little brown bats are not sexually mature until the year following their birth. Several have lived twenty years.

The only predators that I have seen capture little brown bats have been domestic cats, which leap into the air and snag them in and around their roosting areas.

Yuma bat
*Myotis
yumanensis*

THE FIRST SPECIMEN of this species was captured at Old Fort Yuma, Imperial County, California, on the right bank of the Colorado River opposite the present town of Yuma, Arizona; hence the proper name "Yuma" plus the Latin suffix *ensis* ("belonging to"). It was described scientifically in 1864.

DISTRIBUTION AND DESCRIPTION

total length 84-99 mm
tail 32-45 mm
weight 5-7 g

YUMA BATS OCCUR in all zones, wherever there is appropriate habitat. They range in length from $3^3/8$ to 4 inches. Their tails range in length from $1^1/4$ to $1^3/4$ inches, and they weigh about a quarter of an ounce.

Yuma bats have dull fur that varies dorsally from brown to dark brown. Their undersides are pale brown to almost tan; their throats are often whitish. When laid forward, the ears reach to the nostrils. Flight membranes are usually dark brown. There are no keels on the calcars. Yuma bats may be confused with little brown bats.

I have found night roosts, but not day roosts, under bridges in western Oregon. Ten miles east of Brookings, Curry County, these bats were using an exposed chimney of a house as a night roost. They would fly in and rest just under the overhang of the roof at about fifteen- to twenty-minute intervals. At Bandon, in Coos County, on the other hand, numerous Yuma bats used a low, permanently open garage for a night roost. The bats flew into the garage and rested on the beams near the ceiling. There were occasionally as many as twelve bats in the garage at one time. The garage had been used as a night roost for several years even though it was constructed under the house. The noise and movement of the human inhabitants had no apparent effect on the bats.

Mammals of the Pacific Northwest

YUMA BATS ARE closely associated with water. They can be found along large streams, rivers, ponds, and lakes, but since they feed close to the surface of the water, they need ample room to maneuver.

The bats emerge to feed twenty to thirty minutes before full darkness. They normally feed just a few inches over the surface of the water, where they often have regular routes that they fly again and again. Along streams and rivers, they fly back and forth in relatively straight patterns; over ponds and lakes, they fly in circular patterns. I have also seen these bats flying in continuous circular patterns 3 to 6 feet above the ground around a house near a small river 10 miles east of Brookings in southern Curry County.

Yuma bats are efficient feeders. Taken from resting sites fifteen minutes after dusk, they already had full stomachs. In western Oregon, they eat a modest variety of food items, of which the four most important seem to be midges (which are flies), other flies, the winged form of termites, and moths.

Yuma bats have been observed feeding as early as March along the Oregon coast. The sexes segregate in April and the females form maternity colonies in buildings, mines and caves, and under bridges. One colony in Nevada contained an estimated five thousand individuals. Maternity colonies do not tolerate human disturbances; once disturbed, a colony may partially or wholly abandon the original site.

In western Oregon, Yuma bats can be found roosting under loose tar-paper on the sides of buildings, wooden siding, and cedar shakes, as well as in attics and under loose flashing around chimneys. They occasionally share roosting sites with little brown bats, long-eared bats, or California bats.

Maternity colonies may be formed in buildings, where temperatures reach 122° F in the afternoon. Yuma bats roosting in a barn in California tended to congregate in tight groups at the tops of the beams adjacent to the ceiling during the cool morning. As the temperature reached about 104° F, the bats spread out and moved down the beams away from the hot ceiling. When the temperature exceeded 104° F on all portions of the beams, most of the bats flew to cooler areas of the barn, where the temperature stayed below 86° F. The remaining bats withstood a temperature of 106° F. As the temperature decreased in the late afternoon, the bats that had remained on the beams moved closer to the ceiling, but those that had left the beams did not return to them until they had finished foraging during the night.

Evening Bats • Vespertilionidae

Yuma bats, having substantial accumulations of fat by September, disappear in late October. Although these bats have been studied in numerous areas, no hibernation sites have been found that I am aware of. The bats simply disappear in the autumn after the maternity colonies are abandoned.

Little seems to be known about the reproduction of Yuma bats. Each female gives birth to a single young between late May and July. Breeding probably begins in the autumn.

The only animal that I have known to kill Yuma bats was a domestic cat that was extremely agile and adept at snagging them as the flew into a garage. Although the cat caught and killed three bats in thirty minutes, it did not eat them, though it undoubtedly killed many during a summer.

Long-eared bat
Myotis evotis

THE SPECIFIC NAME *evotis* is derived from the Greek words *ev* ("good") and *otos* ("ear"); it refers the species' large ears. The first specimen was collected at Monterey, Monterey County, California. The scientific description was published in 1864.

DISTRIBUTION AND DESCRIPTION

total length 89-96 mm
tail 37-43 mm
weight 5-7 g

LONG-EARED BATS occur in appropriate habitat in all zones (see photos 13 and 14 on pages 57 and 58). They range in length from $3^5/8$ to $3^3/4$ inches, with a wingspan of about 10 inches. Their tails range from $1^1/2$ to $1^3/4$ inches, and they weigh about a quarter of an ounce.

Long-eared bats have dull, yellowish brown to light brown backs and tan to light tan undersides. The hairs are dark gray at the base, much darker than the tips. The large, black ears, when laid forward, reach noticeably beyond the end of the nose. Flight membranes are dark brown. Long-eared bats are sometimes confused with fringed bats, but *lack* the conspicuous fringe of stiff hairs along the margins of the tail membrane that are characteristic of the fringed bat. The calcars do not have keels.

HABITAT AND BEHAVIOR

THESE BATS ARE primarily inhabitants of coniferous forests. Although generally distributed throughout Oregon, long-eared bats are not abundant anywhere, and there is a corresponding lack of knowledge about their life history.

Long-eared bats usually emerge late in the evening, somewhere between ten to forty minutes after full darkness, and feed among the trees. They eat a moderate variety of foods, including flies, moths, beetles, and spiders.

Their day roosts throughout their geographical distribution are most frequently in buildings, although they use slabs of loose bark attached to dead trees as day roosts in the coniferous forests of western Oregon (photo 19). I once found a long-eared bat roosting during the day in the fold of an old rag in an abandoned house in the Willamette Valley. I have also found these bats to use caves as night—but not day—roosts in the Columbia River Gorge. However, long-eared bats have in the past been the most common species using the cave at Oregon Caves National Monument, especially in early August. Most bats were captured as they returned to the cave between 10:00 p.m. and 2:00 a.m. Of the 381 bats caught and banded, 213 were long-eared bats—185 males and 28 females.

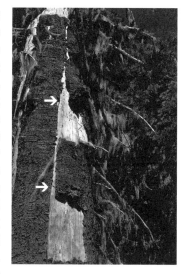

I have found none of these bats later than October 31, and know nothing of their hibernation.

Small maternity colonies form in late spring or early summer; maternity colonies of twelve to thirty individuals have been found in buildings (see photo 14 on page 60) and hollow trees. Each pregnant female bears a single baby born around the middle of July. Babies weigh about $4/100$ of an ounce and are almost 2 inches long, with a wingspan of about 4 inches. They are devoid of hair, but their milk teeth are sharp.

Fringed bat
Myotis thysanodes

THE SPECIFIC NAME derives from the Greek word *thysanos* ("fringe"), and refers to the short, stiff hairs along the margin of the tail membrane. The first specimen of this species was caught on October 16, 1897, at Old Fort Tejon, Kern County, California.

DISTRIBUTION AND DESCRIPTION

total length 91-98 mm
tail 40-41 mm
weight 7-9 g

FRINGED BATS ARE rarely seen in Oregon, but should potentially occur in all zones (photo 20). The first record from the state was a subadult male secured in Tillamook, Tillamook County, on September 13, 1928; a second specimen was captured at Tillamook on August 3, 1940. In addition, I caught three males about one-half mile northeast of Otis Junction, Lincoln County, on August 11, 12, and 14, 1971. Although fringed bats seem rare over much of Oregon, a few appear to regularly use the cave at Oregon Caves National Monument.

Fringed bats range in length from $3^1/2$ to almost 4 inches, and their tails are about $1^5/8$ inches long. They weigh about a quarter of an ounce.

Fringed bats are the only species with a conspicuous fringe of short, stiff hairs along the margin of the tail membrane (photo 21). Dorsally, they vary from dull to moderately glossy yellowish brown, slightly reddish brown, or brown. Their undersides are

Photo 20. Fringed bat. Note large ears. (Photograph by Chris Maser)

light tan or light grayish tan. They have black ears and dark brown flight membranes. When laid forward, the ears extend slightly beyond the nose. There are no keels on the calcars.

ESSENTIALLY NOTHING IS known about habitat use by fringed bats in Oregon; three that I captured in August 1971 were associated with young coniferous forest. They are known, however, to use caves, mines, rock crevices, and buildings as both day and night roosts. They emerge well after dark. The three fringed bats I caught were roosting under loose flashing around a chimney. To my knowledge, nothing is known about their habits in winter.

The stomachs of the fringed bats caught in August 1971 had the following food items in them, listed in order of importance: moths, harvestmen or daddy-longlegs, crickets, craneflies, and spiders.

Little detail appears to be known about reproduction in this species, except that a single young is produced annually. Fringed bats are colonial, and maternity colonies of several hundred individuals may be common in some areas. Females containing embryos have been captured in June; young are born in July. Maternity colonies apparently break up in the autumn, but little or nothing is known about their subsequent movements.

Photo 21. Tail membrane of fringed bat showing conspicuous fringe of stiff hairs along the margin. (Photograph by Chris Maser)

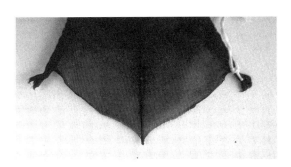

THE SPECIFIC NAME *volans* is the Latin word for "flight" and refers to the excellent flying capabilities of this bat. The first specimen came from Cabo San Lucas, Baja, Mexico, and was described scientifically in 1866.

Long-legged bat
Myotis volans

LONG-LEGGED BATS occur in appropriate habitat in all zones (photo 22) and throughout the state. They range in length from 3$^5/_8$ to 4$^1/_4$ inches, and their tails vary from 1$^1/_2$ to 2 inches in length. They weigh about a quarter of an ounce.

DISTRIBUTION AND DESCRIPTION

total length 90-106 mm
tail 38-49 mm
weight 7-9.5 g

Long-legged bats have long, soft, dark hair that is slightly glossy in some individuals. Dorsally, they vary from reddish brown, to brown, to nearly black. Their undersides are slightly paler than their backs and vary from yellowish brown, to reddish brown, to nearly black. The rounded ears are attached low on the head, making them appear smaller than they are. When laid forward, the ears barely reach the nostrils. The ears and flight membranes appear black. These are the only brown bats with belly hair extending onto the undersides of the wings, covering the membranes as far as the knees and elbows and forming a line that parallels the body. They have prominent, long, wide keels on the calcars.

LONG-LEGGED BATS are associated primarily with coniferous forest. Even though these bats are widely distributed in Oregon, little is known about their habits. The only roosting long-legged bat I ever found in Oregon was a male discovered under loose tar paper on the side of a building 10 miles east of Brookings in Curry County on April 24, 1972. A few have also been caught in nets in the cave at Oregon Caves National Monument. In

HABITAT AND BEHAVIOR

Photo 22. Long-legged bat. (Photograph by Chris Maser)

other parts of their geographical range, they roost in crevices in the face of cliffs by day, and use caves and mines as night roosts. Almost nothing is known about their winter habits; however, two of these bats were beginning to accumulate body fat when they were captured in western Oregon in August.

I have found long-legged bats flying only on warm evenings, emerging as early as 8:00 p.m. along the coast in August when the overcast was heavy; when the overcast was light, they emerged as late as 8:45 p.m. On cold, clear nights, however, the only *Myotis* captured feeding along the edge of the forest were little brown bats. Perhaps long-legged bats fed among the trees on such nights, because occasionally a large *Myotis* would fly out of the forest and directly back in again without paralleling the edge of the forest.

Long-legged bats do not appear to establish well-defined territories for feeding. An area might be occupied by a single foraging long-legged bat one night, three the next, and none the next. On August 16, 1972, I watched two bats flying at considerable speed along a road through a forest; one bat appeared to be chasing the other for they were only inches apart. From the size and style of flight, both bats appeared to be of the same species. The second bat made every twist and turn made by the leader. I shot the pursuing individual, which proved to be a long-legged bat. Thus, if my assumption that they were both of the same species is correct, there may be some territoriality among foraging long-legged bats. In western Oregon, these bats eat a moderate variety of prey, but moths seem to be their most important food.

Breeding probably begins in the autumn. Long-legged bats form maternity colonies, often containing several hundred individuals. Pregnant females, which annually bear a single young, have been captured from May 31 to August 1.

California bat
Myotis californicus

THIS SPECIES WAS named after the state of California with the Latin suffix *cus*, denoting possession, hence "California's bat." The first specimen was collected in Monterey, Monterey County, California and was described scientifically in 1842.

DISTRIBUTION AND DESCRIPTION

total length 78-88 mm
tail 33-42 mm
weight 3-7 g

THE CALIFORNIA BAT not only occupies all zones but also occurs in appropriate habitat throughout the state (photo 23). It is from $3^1/_4$ to $3^1/_2$ inches long and weighs less than a quarter of an ounce.

California bats are the smallest bats in western Oregon; they also have the smallest hind feet. Dorsally, their soft hair ranges from reddish brown to brownish red; their undersides are slightly lighter with a little less red. The bases of the hairs are much darker than the tips. Ears and flight membranes are dark brown. There are definite, although small, rounded keels on the calcars.

HABITAT AND BEHAVIOR

THESE SMALL BATS are common and generally distributed, occupying a wide variety of habitats. They are perhaps best characterized as crevice-dwellers and probably use structures made by people as night roosts more than any other species of bat. These bats roost in buildings made of wood, adobe, and concrete, as well as in mines, caves, crevices in the faces of cliffs, hollow trees, and beneath flakes of rock and loose bark.

In western Oregon, they also seem to use structures made by people as day roosts more frequently than other species of bats. Tar paper on the outside of buildings is commonly used if it is loose, torn, or has a large enough hole for entry. I have found some of these bats under loose tar paper that had been in place less than two weeks. I have also found them roosting in crevices in the roofs of attics, under shakes on roofs, under window shades, in cracks around exterior moulding, and in the rough-texture, craggy bark of old-growth Douglas fir trees within the forest. I have even found them occasionally hanging exposed on the ends of rafters that extended to the exterior of a building but were protected by the slight overhang of the roof.

Photo 23. California bat. Note small feet. (Photograph by Chris Maser)

With the exception of occasional maternity colonies, California bats seem to move around without any habitually used hiding place, apparently roosting in the most readily available site when their evening foraging is complete. They have been found, for example, in the fold of a temporary canvas shelter less than 24 hours after it was erected, and two females, with their young, were found within the folds of an old rag hanging on the outside of a building. Since most roosts are used only occasionally, there are seldom evident accumulations of feces or stains from urine, but I have found that when the bats roost under tar paper that is fastened securely at the bottom, the accumulations of feces can become great enough to cause the tar paper to bulge outward. Other

than that I have found such evidence of continued use of particular roosting sites only rarely.

So far as I know, the California bat is the only species within Oregon in which healthy individuals occasionally fly during the day. On May 9, 1950, for example, a California bat flying in bright sunlight in a ponderosa pine forest was shot at 9:00 a.m. as it fed among the trees, six miles northwest of Sisters in Deschutes County. Another was collected as it fed on mayflies along Fishhawk Creek near Jewell in Clatsop County, at 11:00 a.m. on April 30, 1955. And I collected a third on April 10, 1971, near Otis Junction, in Lincoln County, as it fed over a lawn at 4:55 p.m., just as the sun emerged from the clouds on a predominantly overcast day.

I found three hibernating California bats in western Oregon on December 3, 1969, between the inner and outer walls of an old, abandoned house in a coniferous forest in the vicinity of Alsea, in Benton County. The bats, two females and a male, showed no evidence of having eaten recently. Two males captured four miles southeast of Bandon, in Coos County, on November 15 and 17, 1971, both had food in their intestines, which indicated that they had been feeding recently.

I found thirty California bats under tar paper on the outside of a shed in a coniferous forest 9 miles north of Gardiner, in Lane County, between February 26 and March 23, 1972 (photo 24). Because all showed signs of recent feeding, it appears that California bats emerge from dormancy in late February or early March, at least along the Oregon coast. When they enter dormancy in western Oregon is not known, but none that I

Photo 24. Shed nine miles north of Gardiner, in Lane County, where I collected Caifornia bats during February and March as they roosted during the day under the tar-paper siding. (Photograph by Chris Maser)

Mammals of the Pacific Northwest

have collected in August had any sign of body fat. The California bat in western Oregon thus appears to emerge from dormancy earlier and to enter it later than other species.

California bats emerge to feed at different times, depending on location, from well before dark to well after dark. Regardless of when they begin, the bats feed rapidly, then retire for a period to a night roost. After resting a while, they feed again, resting periodically throughout the night. These bats may forage within 10 to 12 feet of the ground, often in proximity to vegetation, where they secure a relatively wide variety of foods. The four most important are midges, unidentified flies, craneflies, and moths.

Breeding probably commences in the autumn because males with enlarged testes have been captured from August through October. Female California bats form maternity colonies in the spring, where each female gives birth to a single young. There is considerable variation in dates of birth. I have found pregnant females along the Oregon coast as early as February 26 to as late as May 6. Recently born babies have been found as late as June 20, whereas others were nearly as large as their mothers by that time and were well able to fly.

Silver-haired Bats *Lasionycteris*

THE GENERIC NAME *Lasionycteris* is derived from the Greek words *lasios* ("hairy") and *nykteros* ("nocturnal") and refers to the species' hairy tail membrane and its time of activity.

Silver-haired bats dwell primarily in trees, and are usually associated with wooded or forested areas. In some parts of their geographical distribution, they are permanent residents and hibernate throughout the winter. Over most of their geographical range, however, including Oregon, they migrate south in the autumn and north in the spring. In the mountainous regions of the western United States, their migration is primarily elevational.

Except for the Hawaiian Islands, silver-haired bats occur throughout most of the United States, including southeastern Alaska, and southern Canada. They have also been found in the Bermuda Islands, but most oceanic records are from the autumn and spring, the time of migration.

There is a single species of silver-haired bat in western Oregon.

Silver-haired bat *Lasionycteris noctivagans*

THE SPECIFIC NAME *noctivagans* is derived from the Latin words *noctis* ("night") and *vagans* ("wandering"). The silver-haired bat was first collected in the eastern United States, and was described scientifically in 1831.

SILVER-HAIRED BATS occur throughout western Oregon in all zones. They range in length from $3^3/4$ to $4^1/2$ inches, and their tails vary from $1^1/2$ to 2 inches in length. They weigh between a quarter and a half of an ounce.

Silver-haired bats are distinctive (photo 25). Their backs vary from dark brown to black; their undersides are slightly paler. Numerous white-tipped hairs give the pelage a silvery appearance, especially on the back. The black ears are about as broad as they are long. Wing membranes are dark brown to almost black. Tail membranes are dark brown and well furred on the topsides from their bases to about two-thirds of their lengths.

The only other bat in western Oregon with a furred tail is the hoary bat. Hoary bats have small clusters of stiff, light tan hairs on the wrists and elbows and soft, yellowish fur extending to the wrist on the undersides of the wings; silver-haired bats do not.

HABITAT AND
BEHAVIOR

USUALLY SOLITARY BY nature, silver-haired bats are erratic in numbers and are scarce throughout much of their geographical distribution. Although the species is generally distributed throughout the forested portions of Oregon in low numbers during the summer, it may be locally abundant anywhere in the state during its brief migration north in spring but less so during its more prolonged migration south in autumn.

The spaces between pieces of loose bark and the trunks of trees are probably their most typical roosting sites during the day. They also are known to roost in hollow trees, abandoned woodpecker cavities, and birds' nests. Because they encounter a scarcity of normal roosting sites during migration, they may be found in a wide variety of places, such as open sheds, garages, and outbuildings. They have even been discovered in piles of lumber, railroad ties, and fenceposts.

The migratory habits of silver-haired bats are not well understood. At times, they migrate in groups; there are records of weary bats descending on ships at sea off the Atlantic coast of North America and being found on oceanic islands, where they are normally absent.

Silver-haired bats hibernate in hollow trees, buildings, and crevices in rock faces and have been found hibernating under the loose bark of western redcedar trees.

Silver-haired bats emerge to feed at various times in western Oregon, anywhere from fifteen to forty-five minutes before full darkness. Such variability may be due to the prevailing conditions of weather and/or the location of the roosting site—when roosting in trees, some bats may be in areas where daylight lasts longer (such as an open forest edge) than it does in others (such as dense forest interior).

I find silver-haired bats to be the slowest flying bats in western Oregon. Although their wings are moved with a characteristic fluttery motion, this is apparent only to an experienced observer.

Silver-haired bats frequently hunt in sweeping circles, often more than 100 yards in diameter. Although they usually forage in and over the forest, they also fly along roads through the forest, feeding 20 to 40 feet above the road's surface. Adults generally appear singly, but they may be found in pairs or in groups of three or four. These bats eat a wide variety of foods in western Oregon, the most important of which are moths, the winged form of damp-wood termites, and flies.

Photo 25. Silver-haired bat. Note broad ears and silvery hair. (Photograph by Chris Maser)

Little is known about the breeding habits of silver-haired bats. Two young (occasionally one) are born to each pregnant female in June or July. Males with enlarged testes have been captured in the Pacific Northwest from July through September.

Silver-haired bats occasionally fall prey to owls, such as the Northern Spotted Owl, which hunts in the coniferous forests, where the bats roost.

Big Brown Bats
Eptesicus

THE GENERIC NAME *Eptesicus* is derived from the Greek words *epien* ("to fly") and *oikos* ("house") and refers to the affinity members of this genus have for occupying human habitations. The genus is distributed nearly worldwide and contains about thirty species. They occur from southern Canada to Argentina and Uruguay in South America, the Antilles, Africa, Madagascar, Europe, Asia, and Australia. There is a single species in western Oregon.

Big brown bats are reddish brown, dark brown, or nearly black above and usually paler below. Some species in Africa have white or translucent membranes. Members of this genus weigh from about a quarter to almost an ounce. During the summer, they often roost in hollow trees or in and around buildings. In the Antilles (West Indies), these bats roost throughout the year in the "twilight zone" near the entrances of caves and do not hibernate. One species in Africa is associated with fertility by some natives and is used in religious rites.

Big brown bat

Eptesicus fuscus

DISTRIBUTION AND
DESCRIPTION

total length 108-139
mm
tail 43-59 mm
weight 12-28 g

*HABITAT AND
BEHAVIOR*

THE SPECIFIC NAME *fuscus* is the Latin word for "brown." The first specimen of this bat was captured in Philadelphia, Pennsylvania, and was described scientifically in 1796.

BIG BROWN BATS occur throughout western Oregon in all zones. They range in length from $4^{1}/4$ to $5^{5}/8$ inches, and their tails vary from $1^{3}/4$ to $2^{3}/8$ inches in length. They weigh from less than half to nearly 1 ounce.

Big brown bats are large (photo 26). Dorsally, they vary from light reddish brown to dark reddish brown and may have glossy hairs. The undersides are paler and lack gloss. They have black faces, which at first glance appear devoid of hair. Ear and flight membranes vary from dark brown to black. There are well-developed keels on the calcars.

THOUGH USUALLY ASSOCIATED with forests, these bats occupy a wide variety of habitats; hence they are not only abundant but also generally distributed throughout their geographical range. Their wintering and summering areas are almost the same. Because they are closely associated with humans, they are often called "house bats" or "barn bats," and are probably the best known of the North American bats.

These bats seek roosts in dark places and are intolerant of disturbance. If disturbed while roosting in an exposed area of an attic, for example, they immediately retreat to some inaccessible crevice and may not use the exposed area again that season. They may elect to abandon the roost altogether. On the other hand, some may become tame with repeated handling.

*Photo 26. Big brown bat. Note black face.
(Photograph by Chris Maser)*

Mammals of the Pacific Northwest

Even though they roost in attics, they are not as tolerant of high temperatures as are little brown bats and Yuma bats. When temperatures exceed 90° F, they may seek cooler places, perhaps bringing them into direct contact with people, or even abandon their roost altogether.

Along the Oregon coast, temperatures seldom get high enough to force these bats into direct contact with people inside houses. Temperatures in the Willamette Valley, on the other hand, may be getting hot enough in summer to bring these bats into more frequent contact with people in buildings, especially in rural areas.

Their summer roosts are usually in buildings, hollow trees, or under bridges, but they hibernate in buildings, caves, mines, tunnels, and similar shelters. The places I have found them roosting most often over the years have been under loose flashing around chimneys, in crevices under the exterior siding of houses, and occasionally in attics, where they have been few in numbers. I have also found them in barns in the Willamette Valley. They are known to roost in hollow trees in the forest. I have never found a big brown bat under tar paper, even when the building hosted other species of bats and was situated within a forest occupied by big brown bats.

In late summer and early autumn, big brown bats accumulate thick layers of fat. In western Oregon, the accumulation of fat begins in July, and by the end of August, many bats have thick layers. Big brown bats captured in September may weigh over an ounce, and a third of this weight is due to the enormous accumulation of fat.

Big brown bats are remarkably hardy. They retreat into underground shelter, such as caves and mines, only during the coldest weather, and their stay in such retreats may be short. Even though some individuals are always found in caves during hibernation, the winter quarters of the majority of these bats are not known. In the northern part of their range, most of them apparently hibernate in buildings, which is where I have found all the hibernating big brown bats that I have encountered in western Oregon over the years.

Big brown bats have a strong, steady flight, which can be readily seen because they often emerge early to forage, before the swallows have stopped feeding, thirty to forty minutes before full darkness. These bats often feed over lighted streets and in city parks. When feeding in the vicinity of their forested habitat, they usually forage high over the canopy of trees; at times they appear to be more than 150 feet above the ground. As dusk deepens, they sometimes descend to with 40 feet of the ground. When feeding along roads through the forest, on the other hand, they may be only 20 feet above the ground.

When hunting over treetops, they often fly in great, sweeping circles; along roads through the forest, however, they normally fly in a relatively straight course with frequent sallies after insects. They are relatively solitary while hunting and appear to have definitely established foraging territories in which they fly again and again. Although they will contest the intrusion of another big brown bat into their feeding territory, they readily tolerate the presence of other species of bats.

Big brown bats eat a wide variety of foods over the breadth of their geographical distribution and in western Oregon, but moths, scarab beetles, winged damp-wood termites, and unidentified beetles are the most important items, collectively comprising just over 60 percent by volume of their diet, with a heavy concentrate on beetles.

The production of sperm occurs during the summer, a little earlier in adults than in young of the year, and ceases by the middle of October. Breeding pairs have been observed from November through March. Ovulation takes place about the first week of April, and the eggs are fertilized by the sperm previously stored in the females' reproductive tracts.

From the eastern slope of the Rocky Mountains west to the Pacific coast, females nearly always produce a single young per year, though two are the rule in the rest of the U.S. Maternity colonies, ranging from twenty to three hundred individuals, are formed about the middle of May, two to three weeks before the young are born.

As the time of birth approaches, the females become reluctant to fly and spend more time in the roost at night. In the northern portion of the United States, including western Oregon, pregnant females are found in late June, and the young are born near the end of June or the beginning of July, a full month later than in the southern U.S. I have found nursing females as late as the second week of August.

Newborn big brown bats are large, averaging one-tenth of an ounce. They cling so tenaciously to the mother's teats that they are difficult to remove without injury. Even the tiniest baby is left in the roost when the mothers leave at night to forage, but how a mother induces her young to release its grip is unknown. Mothers with very young babies feed early in the evening, close to the maternity colony, and return within a hour to nurse. On returning, each mother crawls around among the babies as though looking for her own, while each baby tries to grab any adult that comes within reach. Before a mother allows her baby to nurse, she licks it about the face and lips.

Babies occasionally fall to the floor of the roosting site; although some die, others are retrieved by their mothers and live. When bats less than two weeks old fall from the roost or otherwise appear to be lost, they squeak continually. Squeaking may have survival value in aiding the mother to locate and retrieve her offspring.

Young bats grow rapidly, gaining as much as 1/100 of an ounce per day. As the young mature, their mothers spend less time at the maternity colony during the night and start using outside night roosts. The young can fly when they are three weeks old and, shortly thereafter, begin joining their mothers at the outside night roosts. Maternity colonies either disband when the young are weaned or remain intact throughout the summer and break up with the arrival of cold weather.

Long-tailed weasels sometimes hunt big brown bats in maternity colonies. Great Horned Owls are also known to catch some.

Hairy-tailed Bats
Lasiurus

THE GENERIC NAME *Lasiurus* is derived from the Greek words *lasios* ("hairy") and *oura* ("tail"); it refers to the characteristically hairy tail membrane of members of this genus.

These bats are strong, fast fliers, usually feeding from 20 to 50 feet above the ground. They roost singly or in small groups among the foliage or on the trunks of trees and shrubs. Their coloration resembles the trunk of a tree or a dead leaf and, when disturbed while hanging, they often sway from side to side like a dead leaf.

Members of this genus almost invariably select roosts in trees along the edges of forests or other treed habitats, usually on the south to southwest side of trees. The roosts provide dense shade and cover above and at the sides with an open space below, allowing the bats to be undetected except by inspection from below, while simultaneously permitting them to drop downward to initiate flight. The roosts lack lower perches from which birds or other animals can detect the hanging bats, while the dark ground cover provides a minimum of reflected light from the sun. In addition, the roost interrupts the currents of air, retards the distribution of dust, and contributes to the heat and humidity.

Females have four teats and have from two to four young, usually two or three. *Lasiurus* is the only genus of bats known to commonly have more than two young per litter. In northern latitudes, young are born from late May through early July. The mother leaves them in the roost when they become too heavy to carry, though occasionally a mother carries two or three young whose combined weights exceed her own.

In Canada and the United States, red bats (*Lasiurus borealis*) and hoary bats are migratory, flying south in the autumn and north in the spring.

There are about twelve species within the genus *Lasiurus*. They normally occur from southern Canada south to Uruguay and Argentina and from the Hawaiian and Galapagos Islands east to the Greater Antilles, Cuba, and the Bahamas. There is a single species in western Oregon.

Hoary bat
Lasiurus cinereus

THE SPECIFIC NAME *cinereus* is the Latin world for "ash colored." The hoary bat was first collected in Philadelphia, Pennsylvania and was described scientifically in 1796.

DISTRIBUTION AND DESCRIPTION

total length 114-150 mm
tail 42-65 mm
weight 17.5-33 g

HOARY BATS OCCUR in all zones. They are large, ranging in length from $4^5/8$ to 6 inches, with tails that vary from $1^3/4$ to $2^5/8$ inches. Their weight ranges from just over half an ounce to a little over 1 ounce.

The hoary bat is the most colorful and distinctive bat in Oregon (photo 27). Its fur is long, thick, and very fine; the hairs are longer on the neck than on the back, forming a ruff. Most of the hairs on the back have four bands of color. The back is yellowish tan to light orangish brown, occasionally grayish, with many white-tipped hairs that produce a heavily frosted appearance. The face and cheeks are more yellowish than the back, but with little evidence of white-tipped hairs. The chin and forethroat are particularly yellowish and lack white-tipped hairs. The underside is paler, more yellowish than the back and only slightly frosted. The rounded, relatively short ears are partially covered with stiff, yellowish tan hairs, but the edges of the ears are naked and black.

The tops of the wings have small clusters of stiff, light-tan hairs on the elbows and wrists (see photo 27) and soft, yellowish fur on the undersides, extending from the body to the wrists. The wing membranes are conspicuously bicolored, brown to slightly reddish brown along the forearm and partway down the finger from the wrists; the rest of the wing membranes are dark brown (see photo 27). The tail membrane and feet are completely covered with thick hair on the topside; they have a moderately frosted appearance. The underside of the tail has soft yellowish fur extending about half an inch onto the membrane. There are no keels on the calcars.

HOARY BATS ARE associated primarily with coniferous or mixed coniferous-deciduous forests and are the most widely distributed bat in North America. Records are scarce in the northern Rocky Mountains but more numerous in the Pacific Northwest. I found it to be reasonably abundant in some forested areas of northeastern Oregon, where I netted a goodly number of the species as they came to drink during the night. These bats are considered to be common, though seasonal, in California, Arizona, and New Mexico.

Hoary bats spend summer days sleeping singly or in small family groups concealed in the foliage of trees. They are found occasionally in trees along busy city streets, and once in a while in cavities abandoned by woodpeckers in tall dead standing trees. Although hoary bats usually roost 10 to 17 feet above the ground, females with young as well as solitary youngsters normally roost higher in trees than do solitary adult males.

The hoary bat is one of the most accomplished migrants, and data suggest that they migrate in waves as do birds. Migration waves have been observed in spring and autumn. Spring migration appears to begin in April in the southern U.S., and autumn migration begins in August in the northern regions of the United States and southern Canada. Pregnant females migrate north earlier than males and are more or less well distributed throughout the southern United States in April and May. By June, they are mostly concentrated in the northcentral and northeastern U.S., where they give birth to their young. Males, on the other hand, concentrate in the western United States.

Photo 27. Specimen of a hoary bat. Note small clusters of hairs at elbows and wrists, pointed out by arrows. (Photograph by Chris Maser)

Evening Bats • Vespertilionidae

The earliest record of a hoary bat in Oregon that I know of is a male I shot as it foraged 10 miles east of Brookings, along the Oregon coast in Curry County, just north of the California border, on April 19, 1972. Otherwise, records point to the arrival of hoary bats, all males, in eastern Oregon in June. One study, by Findley and Jones in 1964, indicated that only males had been taken within the western portion of Oregon in July, and a single female, near the coast, in August.

Late records include four hoary bats found in northwestern Oregon in October, two females and two of unrecorded sex. According to the above-mentioned study by Findley and Jones, all hoary bats had left Oregon by November. On February 24, 1972, however, I found a female hoary bat just north of Gardiner, along the Oregon coast in Douglas County. The bat, dead no more than a week, was lying on the ground under the overhang of a roof and was completely wet. The weather had been cold and rainy for several days, and since the bat had neither fat reserve nor food in its stomach or intestine, the adverse weather was likely the cause of death.

The hoary bat is a strong flier, often making straight forays for a mile or more before returning at speeds that may be close to 60 miles per hour. Most hoary bats shot in Oregon, however, were foraging 20 to 40 feet above roads through forested areas, where patrolling up and down the road for insects appears to be a common practice among these bats.

Its large size and swift, rather straight, flight pattern makes the hoary bat relatively easy to identify as it forages, provided the individual is flying early enough to be seen. Identification is further aided by the frequent, though not constant audible vocalization uttered during flight. Hoary bats are usually considered late foragers, emerging after full darkness, though they fly earlier during migration. The bat I shot on August 12, 1971, near Otis Junction, was secured at 9:15 p.m., when it was too dark to see. I located the bat by its audible chattering and shot it, but I did not see it until, with the aid of a flashlight, I found it dead on the ground. Vernon Bailey found several hoary bats in eastern Oregon, where he observed them roosting by day in large cottonwood trees. "They did not leave their leafy retreats," he said, "until so late in the evening that it was too dark for successful bat shooting...."

Although hoary bats eat a moderate variety of prey items, their principal diet seems to be moths. One of these bats from western Oregon, however, had eaten nothing but mosquitoes, which, in my experience, is not a common food item for any species in Oregon.

As far as I know, little information exists on the reproductive behavior of hoary bats. Breeding probably takes place in the autumn. Females are known to eat prodigious amounts of food before and shortly after giving birth, often eating an amount equal to a quarter of their normal body weight. Thus rapid weight gain, both before and after giving birth, is not surprising. Pregnant bats can weigh more than three times their normal weight.

The normal size of a litter for hoary bats is two young per female, although a single young is occasionally born. Youngsters are born in May and June, and possibly early July.

Although females can carry their babies in spontaneous, well-controlled flight within 24 hours of birth, and continue to do so until the young are six to seven days old, there are numerous instances cited in the literature of mothers and their babies being found helpless on the ground because the offspring became too heavy for their mothers to carry. Thus, when the babies are about a week old, they are usually left hanging in the roost while the mothers forage. A mother may move her youngsters, however, if she experiences disturbance at the original roost.

Newly born babies are covered with fine, silvery gray hair on the back of the head (including the ears), shoulders, tail membrane, and feet, but are otherwise naked. When they are twelve days old, their eyes open, low-frequency clicking is audible, and hair is present in the armpit. The entire underside is covered in hair by eighteen days of age, and patches of stiff, white hairs are present on the thumbs and elbows. At 27 days, the young spread their wings when launched, but do not flap them, and landings are uncontrolled. By day 31, the wings are fluttered on launching, and landing consists of a gentle glide. Day flight is purposeful by day 33, and gently controlled turns are possible. When 34 days old, the young can sustain directed flight for one minute. Weaning is complete. A 44-day-old youngster can launch itself and make deft, vertical swoops in flight. At this time, the young are about 5 inches long.

Big-eared Bats
Plecotus

THE GENERIC NAME *Plecotus* is derived from the Greek words *pleko* ("twist") and *otos* ("ear"). The name refers the bat's habit of resting with its large ears coiled into a shape reminiscent of a ram's horn (photos 28 and 29).

Big-eared bats may be distinguished by their enormous ears, just over $1^1/_2$ inches in length, and by the glandular masses on their muzzles. In members of some species, known as "lump-nosed bats," these masses rise above the nose as flaplike lumps.

Big-eared bats forage well after dark. Their flight is slow, and they are able to hover at a point that interests them. They pick insects off foliage, as well as capturing some in flight. All big-eared bats hibernate; none are known to make extensive migrations.

These bats occur in Eurasia, northern Africa, and in North America, where they are generally distributed in the western half of the United States from southern British Columbia, to southern Mexico. They are also present in the southeastern United States. A single species occurs in western Oregon.

Western big-eared bat
Plecotus townsendi

THIS BAT WAS NAMED in honor of John Kirk Townsend who first recorded the species while at old Fort Vancouver, Washington, in 1835. Townsend, primarily a naturalist, was in charge of the hospital at Fort Vancouver during the winter of 1835-36. The bat's scientific description was published in 1837.

DISTRIBUTION AND DESCRIPTION

total length 80-111 mm
tail 32-55 mm
weight 5-22 g

WESTERN BIG-EARED BATS occur in all zones. They are medium-sized bats, ranging in total length from $3^1/4$ to $4^3/8$ inches. Their tails vary from $1^1/4$ to $2^1/4$ inches in length. They weigh from less than a quarter to just over three-quarters of an ounce.

Western big-eared bats have larger ears than any bat in western Oregon except for the pallid bat, which is easily distinguished by other features. One or two fringes of short, stiff hairs extend two-thirds to three-quarters the length of the ears along the top edges. In addition to large ears, they have dull, rather long, soft hair. Their backs vary from brown, grayish brown, to black, with paler undersides. There are two prominent glandular lumps, one on each side of the nose. Their flight membranes are dark brown and very thin.

HABITAT AND BEHAVIOR

ALTHOUGH WIDELY DISTRIBUTED throughout western Oregon, these bats are seldom abundant. I have found them most often in buildings in western Oregon, but more than any other species in the western United States, they are characteristic dwellers of caves and abandoned mine tunnels, where they can be found at any season. They are the only species of bat that regularly hibernates in fairly large numbers in western caves and mines. In the higher-elevation forested regions of the West and along the west coast, on the other hand, they regularly reside in buildings. They probably do not often use buildings as day roosts in the hot, dry regions of the western United States, which could include the southern parts of zone 3 in western Oregon, because dehydration during the day would be a serious problem for them.

Photo 29. Western big-eared bat at rest; note the coiled ears and the lump by each nostril. (Photograph by Chris Maser)

Photo 28. Western big-eared bat at rest; note the coiled ear. (Photograph by Chris Maser)

Western big-eared bats seem to prefer dim light in their retreats, near total darkness. They apparently hang at the spot where they land, which means clusters of suspended bats are in areas easily reached by flight, namely the open ceiling.

These bats are extremely sensitive to disturbance, especially by humans. It is thus a sad fact that the growing popularity of cave exploration can cause them to permanently abandon a cave. Even conservative scientific collecting among colonies of bats in caves can cause serious disturbance to the colony. In one study, for example, 75 young bats were banded after the adults had left to forage; on returning, the mothers picked up their babies and moved to another roost well over a mile away.

Western big-eared bats living in occupied buildings sometimes become accustomed to the presence of people as long as they are not unduly disturbed. In fact J.K. Townsend, after whom the bat is named, wrote that they inhabited the "storehouses attached to the forts" and seldom left them even at night. They were protected by the "gentlemen of the Hudson Bay Company for their services in destroying the *dermestes* [a species of beetle whose larvae ruin furs by eating them] which abound in their fur establishments." More often in contemporary western Oregon, however, as opposed to the early days at Fort Vancouver, a single bat seems to use a human abode at any one time.

In flight, the western big-eared bat is one of the most versatile of the North American insectivorous bats. Its movements vary from swift, easy flight with deep, smooth wingbeats to slow, deliberate flight, to hovering. Wingbeats are often alternated with set wings and short glides. During a glide, they noticeably lose altitude, which is smoothly recovered with the next series of wingbeats. When released in a room, western big-eared bats, unlike most North American bats, which merely circle repeatedly close to the ceiling, inspect the room from ceiling to floor. The flight patterns of such tours of inspection are usually narrow figure 8's varied occasionally with ovals and circles.

Western big-eared bats readily move from one roost to another and are known, throughout their geographical distribution, to hibernate in caves and mine tunnels. They probably do not undertake major migrations, although banded individuals have been known to move from 20 to 40 miles. Where these bats hibernate in western Oregon is, to my knowledge, unknown. Western big-eared bats can become dormant at any time of year. Bats of either sex roosting in places warmer than 62.5° F are normally awake, and those at lower temperatures are dormant. Dormancy appears to be voluntary because the bats seek different temperatures at different seasons. There is a stronger tendency to hibernate among females than males, and in midwinter females are usually dormant both day and night. Females tend to arrive earlier and remain longer in the hibernation caves than males. Males, choosing warmer places than females, may awaken at night and fly around in midwinter hibernation caves. Some occasionally even fly outside the caves and use night roosts in buildings. Although bats begin arriving at hibernacula in late October, and females may arrive earlier, these winter colonies do not reach maximum size until January. Juveniles accumulate substantially greater stores of fat before hibernation than adults, whereas adults of other species of bats have greater accumulations of fat than their offspring.

Western big-eared bats are by and large late fliers and, with rare exceptions, emerge to forage only after full darkness. Like several other species of bats, western big-eared bats exhibit light sampling behavior: as twilight deepens, individuals fly to the entrance of the roost, then return to its interior and rest a few moments before again testing the light. These bats primarily feed on moths, although they will occasionally take such other insects as bugs and small flies. They are efficient and rapid feeders, retiring to night roosts when their stomachs are full, although they do not seem to chew their food as finely as other species of western Oregon bats.

Females breed when they are four months old, but data suggest that males probably do not breed until their second year. Although some females have bred before they arrive at the winter roost, most breeding occurs there. Data indicate that females are largely dormant when they are bred. This passivity of the females allows each to be mated many times during the course of the winter.

The duration of pregnancy varies widely because of delayed fertilization. It appears to depend on the body temperature of the female and the length of dormancy as well as on the temperature in the roosting site and the general climatic conditions. Nursery colonies of up to two hundred in the western United States and as many as a thousand in the eastern U.S. are formed in early summer. Here, pregnancy proceeds at a more regular rate because higher metabolism is maintained, and therefore, higher body temperatures. Gestation periods may vary from 56 to 100 days in different colonies in different years and birth dates probably extend over a long period. Most young in California are born in late May and early June, while in both Oregon and Washington, most are born in July.

Female western big-eared bats produce a single young per year. Babies weigh about 1/10 of an ounce, are pink and naked at birth, with large ears that flop over unopened eyes; the disproportionately large hind feet and thumbs give them a spiderlike appearance. Within a few hours of birth, a youngster can utter a characteristically sharp, metallic chirp. The dried umbilical cord may remain attached for a day or two. A baby becomes covered with short, gray hair within the first four days. At about seven days, the ears become erect and the youngster can make the squawking noise of a disturbed adult. The eyes open at about nine days.

The young do not fly until they are about three weeks old. They are instead left hanging in the roost, where they cluster together, when their mothers leave during their nightly foraging flights. On returning to the colony, each mother finds her own baby, perhaps by odor and covers it with her body and wings, as if encouraging it to attach itself to her, before she flies off with her offspring clinging to her body.

The youngsters grow rapidly and are almost fully grown at one month of age. They can fly at about three weeks, and by six weeks some fly out of the roost at night as the adults do. They are not weaned, however, until they are about two months old. One banded 42-day-old bat, as large as its mother and a skillful flyer, had its stomach full of milk when it was examined, but no fragments of insects.

Nursery colonies disband about the time weaning is complete, in August or perhaps early September. Adults that have lost their young leave the nursery colonies earlier than the lactating females. Young males depart before young females.

On another occasion, Dalquest tried for nearly half an hour to catch one of these bats, which was trapped in a low-ceiling room about 20 feet square. When placed on a flat surface, such as a floor, the bats stretch their wings, and with a single violent motion, are in flight; they neither walk nor scramble over the surface.

Pallid bats
Antrozous

THE GENERIC NAME *Antrozous* is derived from Greek words *antron* ("cave") and *zoon* ("an animal," "a living being").

Pallid bats vary in length from $2^3/8$ to $3^3/8$ inches, with tails that range from $1^3/8$ to $2^5/8$ inches in length. Adults weigh from about half an ounce to an ounce. The woolly fur varies from yellowish white to dark brown on the uppersides to nearly white on the undersides.

Distinctive features of the genus are large ears and the presence of a small horseshoe-shaped ridge on the squarely truncated muzzle. The nostrils are located on the front of the muzzle beneath the ridge. Although the ears are large and separated, they are not as large, proportionately, as in the genus *Plecotus*, discussed immediately above.

Bats in the genus *Antrozous* roost in caves, crevices in rock faces, cavities in trees, and buildings. Three species occur in western North America from southwestern Canada to central Mexico, in the Tres Marias Island, Nayarit, Mexico, and Cuba. There is one species in western Oregon.

Pallid bat
Antrozous pallidus

THE SPECIFIC NAME *pallidus* is the Latin word for "pale." The specimen on which the species description is based was caught in 1856 at El Paso, El Paso County, Texas, and was described scientifically the same year.

DISTRIBUTION AND DESCRIPTION

total length 100-135 mm

THE PALLID BAT occurs in zones 1 through 3, and in zone 4 near the Oregon-California border. Although these large bats are found sporadically in western Oregon, their main geographical distribution is south through western California, west of the Sierra and San Bernardino Mountains into Mexico.

The pallid bat is big and pale with large, broad, widely separated ears about an inch long, large eyes, and broad wings. It ranges in total length from 4 to $5^3/8$ inches. It has a wingspread of about 14 inches.

The upper parts are light yellow heavily washed with light to dark brown. The underparts are light to pale yellowish. Ears and membranes are naked and dull dark brown. The muzzle is squarely truncated, with low, flattened swellings (scent glands) on each side from which the bats emit a characteristic faintly skunky odor, which can sometimes be detected as much as 15 feet from the entrance to a roosting site. Numerous tiny droplets exude within seconds from the glands when a bat is disturbed, which increases the strength of the odor. Since the odor is strongest when the bats are alarmed, it may be used to repel predators.

Pallid bats usually pass the day in caves, secluded spots under concrete bridges, cracks in rocks, hollow trees, inside old buildings, hollow walls, under roofs, and so on. I once found a colony a few miles west of Junction City in Lane County, Oregon, living in a hollow between the outer chimney and the flue. The main requirements for a daytime roosting site is that it must be dark, well concealed, and protected. Pallid bats are intolerant of disturbance and may abandon a roost, not to return for years. In addition, they shift about among daytime roosts for no apparent reason; this may be a result of seasonal changes in temperature, because they do not tolerate temperatures over $100°$-$104°$ F.

Pallid bats are silent and motionless on cool days unless they are provoked, in which case they generally open their mouths and make a buzzing sound. Although their movements are slow at first, if a disturbance lasts for several minutes, they warm up enough to fly. On warm days, however, they can often be heard squabbling from as much as 100 feet away. During their bouts of squabbling, should vocal warnings be ineffective in dissuading an annoyer, they readily bite one another. If an intruder invades the daytime roosts, the bats give numerous intimidation and squabble notes as they retreat into the deepest recesses of the roost. As such times, they are alert with heads raised and mouths often open. If the disturbance continues, some individuals may take flight. Should one succeed in capturing a pallid bat, one would soon discover that they are vicious biters when handled.

Pallid bats begin emerging at early dusk and usually fly 3 or 4 feet above the ground as they head of a source of water, where they drink repeatedly on the wing. After drinking, they go hunting for food. Because these bats are efficient hunters, some individuals arrive at night roosts within a half hour after taking wing, and mothers begin returning to their young.

Night roosts are open shelters readily accessible by flight and include rock shelters, open buildings (such as garages), porches, concrete bridges, and mine tunnels. Years ago, I netted pallid bats in an abandoned railroad tunnel along the Columbia River between Portland and Hood River.

The feeding habits of the pallid bat are unlike those of most North American bats. Little of their food is captured on the wing; it is taken instead primarily from the ground, as well as from trees. Perhaps their low flight and rather large eyes adapt them for this type of feeding. That insects are taken on the wing is evidenced by the bat's zigzag hunting flight, though their flight is slower and less erratic than that of other species of western Oregon bats. Their prey includes such large flightless insects as Jerusalem crickets, grasshoppers, June beetles, and

Vernon Bailey wrote in 1936 that a Mr. Elmer Williams knew of and collected two specimens of Pacific pale bats on September 16, 1923, from a colony in a large hollow ponderosa pine on a ridge between Salt Creek and Evans Creek in the northwest corner of Jackson County in southwestern Oregon. The tree was located abut 40 miles north of the Rogue River post office at about 4,000 feet elevation.

When Bailey and Williams visited the tree 1927, they found it had been burned. Williams told Bailey, however, that bat droppings or guano had been at least a foot deep on the ground at the base of the tree in years past, indicating that the tree had been a regular residence for the colony.

scorpions, and their sharp teeth are advantageous in dealing with such large prey. Although prey may be eaten on the ground, I have found pallid bats to bring their food to a particular night roost and eat it at their leisure, as evidenced by the telltale heads, legs, wings, and scorpion stingers on the ground below the roost.

The tendency of the pallid bat to fly close to the ground and alight to capture and devour food exposes them to dangers escaped by other bats that fly higher and capture their food on the wing. This may account for the fact that about 80 percent of pallid bats have two youngsters.

Testes reach their maximum size in late August and September. Mating commences in late October, after the summer colonies have dispersed, and probably occurs sporadically in hibernacula throughout the winter, at least until late February. As with other species of bats in western Oregon, females retain live sperm throughout the winter and fertilize ova as they are released in spring.

Unlike many bats, summer colonies, consisting of thirty to a hundred pallid bats, may comprise adults of both sexes and the young. Maternity colonies are also established and begin forming in early April in California. Males may separate from the females before the young are born and remain by themselves in other places or scattered throughout the landscape, though some can be found in maternity colonies. In western Oregon, young are probably born sometime in June, after a gestation period of about nine weeks.

Pallid bats appear to mature more slowly than most other species. Their eyes are closed at birth, their ears are tightly folded against their heads, and their pink skin appears devoid of hair. Although scattered hairs can barely be seen by the fourth day, scanty fur is clearly evident over much of the upper parts of their bodies by the tenth day. Their eyes open between eight and ten days of age, and their ears become erect at about the same time. Youngsters also begin showing signs of fear at ten days, emitting the intimidation buzz as they withdraw from a perceived disturbance.

When they are eighteen days old, a scanty short pelage is present over all of the back and sides, except between the shoulders. The skin is darkened wherever fur is present, and the flight membranes are somewhat grayish. By 24 days of age, the fur is dense, darker than that of the adults, and long enough to hide the skin, except between the shoulders. By day 34, the light basal parts of the hairs become evident on the back.

The young can fly well by six weeks of age, but nurse for several more weeks, until their mothers discourage them by threatening them whenever they approach. By the time the young have reached the size of their parents in late summer, their pelages are indistinguishable from those of the adults.

Summer colonies break up into smaller groups in the autumn. Pallid bats are migratory, at least in the northern portions of their geographical distribution, which may include northern Oregon, although in some regions single individuals and small groups hibernate.

Although I have no information on predation of pallid bats, it is likely that owls catch some. Others are caught in mouse traps set by mammalogists trapping on the ground for species other than bats.

Free-tailed Bats, Mastiff Bats Molossidae

THE FAMILIAL NAME Molossidae is derived from the Greek word *Molossos* ("the Molossus hound or mastiff"), plus the Latin suffix *idae* (which designates it as a family), and undoubtedly refers to the facial features of members of this family of bats.

The length of the head and body range from $1^5/8$ to approximately $5^1/4$ inches. The length of the tail varies from $^5/8$ to $3^1/4$ inches and projects far beyond the free edge of the narrow tail membrane, hence the alternate common name of "free-tailed bats."

The short hair of the body has a velvet-like texture. In one genus, the hair is so short that the animals appear to be naked. The coloration is usually brown, tannish, gray, or black. Some species have two color phases.

The head is rather thick. The muzzle is broad and truncated in a downward sloping manner, and usually has a scattering of short hairs with spoon-shaped tips. The eyes are small. The ears are thick, directed forward, and rather leathery; they vary in size and shape, and often unite across the forehead. The nostrils usually open on a pad, the upper surface of which is often adorned with small, horn-like projections. The lips are large; the upper lip is frequently furrowed by vertical wrinkles. The wings are long and narrow, and the flight membranes are thick and leathery. The legs are short and strong, each terminating in a broad foot. There are curved, spoon-shaped bristles on the outer toes of each foot, which are used by the bat in cleaning and grooming its fur.

Members of this family of bats roost in caves, tunnels, buildings, hollow trees, foliage, crevices in rock cliffs, and under bark and rocks. A characteristic strong musky odor generally permeates their roosts. Some species live in colonies of hundreds of thousands or even millions; others associate in small groups; still others are solitary.

Although bats within this family tend to be active throughout the year, the northernmost species may be inactive for short periods during winter, but there is no definite evidence of true hibernation. Whereas some species move about locally rather than hibernate, the Brazilian free-tailed bat of southwestern Oregon makes seasonal migrations north in spring and south in autumn.

These bats are swift and relatively straight in flight as compared to many other insectivorous bats. Their diet often consists of hard-shelled forms of insects.

Free-tailed bats usually have a single young per year, although two are born on occasion. Breeding generally takes place just before ovulation in late winter or spring. There may be partial or complete segregation of the sexes in some species.

This family of bats consists of twelve genera and approximately eighty species, which are found in the warmer parts of the world. In the Old World, they occur from southern Europe and southern Asia south through Africa and Malaysia and east to the Solomon Islands. In New World, they occur from the central United States south through the West Indies, Mexico, and Central America to the southern half of South America. There is one genus in western Oregon.

THE GENERIC NAME *Tadarida* is a New Latin word created by natural historian Rafinesque to designate a this genus of bats, but he left us no clue as to the origin and meaning of the name.

Free-tailed Bats
Tadarida

The length of the head and body is about $1^3/4$ to 4 inches, and the length of the tail varies from three-quarters of an inch to $2^3/8$ inches. Most of the New World species weigh about half an ounce.

Coloration is from reddish brown to almost black. Several species have tufts of glandular hairs arising from the crown of the head behind the ears, and others in the Old World have a heavy crest of long, straight hairs on the back of the membrane uniting the ears. These patches of specialized hairs are often restricted to the males.

Some species roost singly or in small groups, though most seem to seek shelter in groups of hundreds or thousands. The Brazilian free-tailed bat, the species found in southwestern Oregon, shelters by the millions in the Carlsbad Caverns of New Mexico and in certain other caves. These bats roost in cracks and crevices of caves, rock formations, hollow trees, and human-made structures. A species in the southern Cameroons, in central Africa, often resides in the holes bored into the trunks of dead trees by barbets, where the bats and birds occupy the holes together.

The flight of members of this genus is swift and high. Free-tailed bats tend to be active throughout the year, but some apparently become torpid for at least short periods. Moths and beetles seem to be their preferred foods.

Breeding generally occurs in late winter and early spring, just prior to ovulation. The young are born about twelve weeks later. Most births in North America occur in June and July. Both sexes are usually together throughout the year. A single species occurs in southwestern Oregon.

THIS SPECIES OF BAT was named after the country of Brazil in which it was first captured, to which the Latin adjectival suffix *ensis* is added, meaning "belonging to." Its scientific description dates back to 1824.

Brazilian free-tailed bat
Tadarida brasiliensis

AS THE NAME IMPLIES, the Brazilian free-tailed bat has a wide geographical distribution that extends from northern South America and the Caribbean Islands through Central America, Mexico, the southern tier of states north of the Mexican-U.S. border into the vicinity of Ashland and Medford in zone 3 of western Oregon.

This rather small bat ranges to total length from $3^1/2$ to $4^1/2$ inches, and its tail ranges from $1^1/4$ to $1^5/8$ inches. Its weight ranges from just over a quarter to a half ounce. The long, narrow wings have a span of $11^5/8$ to 13 inches, and the lower half of the tail is free of the tail membrane. The ears almost meet in the middle of the forehead but are not joined. These bats have hairs protruding from their toes that are as long as their feet. The fur is a nearly uniform dark brown or dark gray, although there may be some white hairs or occasional patches of white fur on some individuals.

THE HABITAT OF THE Brazilian free-tailed bat differs in the various parts of the United States in which it occurs. In zone 3, they are associated primarily with the tile roofs of buildings.

The Brazilian free-tailed bat is one of the most gregarious of bats, forming colonies at all seasons, some of them numbering into the millions of individuals. The daily activity of a colony seems to be influenced by fluctuations in temperature.

Although activity begins in the morning on warm days, the bats remain quiescent when it is cool. They squeak and chatter incessantly when awake and seem to spend much time moving about, becoming quiet in late afternoon, only to arouse again shortly before sunset. On foggy or cloudy evenings, they take wing just before sunset; on clear evenings they leave the roosts shortly after sunset.

These bats are known to fly at 40 miles per hour while leaving some of the large roosts in caves, with top speeds estimated at 60 miles per hour. The line of flight of Brazilian free-tailed bats when foraging is relatively straight, with deviations in flight mostly downward rather than upward or sideways.

The Brazilian free-tailed bat eats moths almost exclusively. It captures its prey on the wing like the other insectivorous bats of western Oregon, which use their tail membranes to "net" the prey; but the short tail membrane of the Brazilian free-tailed bat apparently poses a problem. To overcome it, the bat points its heels backward in flight, which extends the membrane almost to the tip of the tail, forming a usable net in which to catch its food. Unlike the other bats in western Oregon, the Brazilian free-tailed bat does not seem to use night roosts, which means it may be on the wing all night.

Although 60 to 64 million Brazilian free-tailed bats migrate many hundred of miles between their summer colonies in Texas, Oklahoma, New Mexico, and Arizona and their winter habitat in Mexico, at least some of this species overwinter in Oregon, where they have been found hibernating in buildings.

Males are sexually active from February until April; breeding occurs in February and March. Ovulation takes place in the latter part of March, and the gestation period ranges from 77 to 84 days in the southeastern United States but is estimated to be 100 days in the west. Babies are born from the latter part of May through mid July.

A female usually produces one young, rarely two, which is naked and blind at birth. Mothers produce a great deal of milk; nursing females returning to a maternity colony in the morning have such a quantity that it oozes out with the slightest handling by a human researcher. The other bats of western Oregon will suckle only their own babies, but lactating Brazilian free-tailed bats seem to nurse the first two babies to find their two nipples. If this is true, a mother's tremendous supply of milk may be important for the survival of the young.

Where these bats occur in large colonies, they have a variety of enemies, including hawks, owls, raccoons, opossums, and striped skunks. Even some snakes enter the caves where these bats roost by the millions and eat their fill.

Hopping Mammals
Lagomorpha

THE ORDINAL NAME Lagomorpha is derived from the Greek words *lagos* ("a hare") and *morphe* ("form" or "shape").

For many years, pikas (rock rabbits), rabbits, and hares were considered a suborder of the rodents, Rodentia. But several characteristics that differentiate them from rodents have been identified, and they are now considered a distinct order. Three easily distinguishable differences are: (1) Lagomorphs have four upper front incisors—two large teeth in front and two small peglike teeth, lacking cutting edges, directly behind them—whereas rodents have only two upper incisors. (2) The testes and scrota of lagomorphs are in front of the penis, whereas those of rodents are behind the penis. (3) Depending on species, lagomorphs have 26 to 28 teeth; rodents have 16 to 22 teeth.

Lagomorphs usually have a thick, soft pelage, but some species exhibit a coarse pelage. They are all ground dwellers. A few are diurnal, but most are crepuscular and nocturnal. Although a few species are distinctly colonial, most are not. The species that are poor runners generally stay close to thickets, burrows, or rock crevices. Species that are good runners normally use a shallow nest or "form" situated under cover on the surface of the ground, usually in relatively open areas; their inhabitants depend on speed for protection. The white-sided jackrabbit (hare) also shifts its skin from side to side as it runs, thus revealing alternating flashes of white ventral and dark lateral pelage—such maneuvering presumably confuses pursuers. Another escape maneuver is to freeze after making a short dash, thus seeming to disappear.

If captured, rabbits and hares can deliver vicious, raking kicks with their powerful hind legs and sharp claws, as well as a singularly loud, piercing distress cry or scream. Except for pikas, which communicate vocally, lagomorphs are otherwise essentially nonvocal as adults.

Strictly vegetarians, lagomorphs normally eat grasses and other herbaceous plants. Some species also feed on the leaves and bark of trees, and a few are known to eat the fruiting bodies of subterranean fungi. Lagomorphs have a remarkable method of obtaining maximum nutrition. They produce two types of fecal pellets: dry pellets that are expelled as waste material, and moist, cecal pellets, which, after they have been expelled, are reingested almost without being chewed; hence, most of the food is passed through the digestive tract twice.

Lagomorphs do not hibernate. Some forms change color from brown during summer to white during winter. In addition, rabbits and hares are noted for periodic fluctuations in population. These are particularly well marked in northern latitudes, where peak populations tend to occur at approximately ten-year intervals.

Lagomorphs occupy a wide variety of habitats on most continents and many large islands, except Antarctica, Madagascar, and most islands of southeast Asia. They have been introduced into Australia and New Zealand and other islands.

There are two families of lagomorphs in western Oregon.

Pikas or Rock Rabbits Ochotonidae

THE FAMILIAL NAME Ochotonidae is derived from the Mongol name for the pika *ochotona* combined with the Latin suffix *idae* (which designates it as a family). This family contains a single recent genus, *Ochotona*. Members of the genus vary in length from 5 to 12 inches, with most averaging about 8 inches in length. They have no visible tail. Weights range from just under $4^1/2$ to 14 ounces.

Pikas (correctly pronounced *peekas*) have long, dense, soft, fine pelages. In most species, the general color is grayish brown, usually darker above than below; one species, however, is reddish. Some species molt twice a year; the summer pelage is brighter and more yellowish red, and the winter pelage is grayer. The head is short and blunt. The ears are very short, rounded, approximately as wide as high, and often rimmed with whitish hairs. The nostrils can be completely closed. The legs are short, the hind limbs scarcely longer than the forelimbs, and the soles of the feet are covered with dense woolly fur.

The warning call is a sharp bark or whistle and is reflected in some of the common names of this small mammal, such as whistling hare and piping hare. Species that occur in talus, live among the rocks. Those that occupy plains or desert areas dig burrows.

The remarkable custom of curing hay seems to exist throughout members of this genus. Green vegetation is gathered in late summer and dried in the sun. The resulting hay is then stored for use during the winter.

The breeding season appears to be late spring and summer. Gestation takes about thirty days. Litters vary in size from two to six, although three to four is the usual number. Two or three litters are born to a female each year in a cozy, well-protected nest of plant material. The young at birth are naked, helpless, and weigh about a quarter of an ounce. Youngsters are weaned when they are one-fourth to one-third the size of their mothers.

Members of this family are represented in Asia by twelve species and in North America by two. In the Old World, the geographical range is from southeastern Russia, near the Volga and Urals, through Siberia to Kamchatka and Korea, and southward, east from Iran, Afghanistan, and Turkestan to Mongolia, Tibet, and into the Himalayas. Although they do not occur south of the Himalayas in eastern and southeastern China or in southeast Asia, one species does inhabit the northern Japanese island of Hokkaido.

In North America, the species *O. collaris* occurs in the region of southeastern Alaska and the Yukon. The species *O. princeps* occurs from the mountains in southwestern Canada, down the Rocky Mountains to New Mexico, throughout the scattered high mountain peaks of the Great Basin, and down the Cascade-Sierra Nevada Mountains of Washington, Oregon, and California. Consequently, there is a single genus in Oregon.

Pikas
Ochotona

Ochotona HAS THE SAME derivation as that of the family. Since there is but a single genus within the family Ochotonidae, the generic description is identical with that of the family. There is a single species of *Ochotona* in western Oregon.

Pika
Ochotona princeps

THE SPECIFIC NAME *princeps* is a Latin word meaning "first, in front, chief," hence one of the common names of the pika is "little chief hare." The first specimen of this species came from the headwaters of the Athabaska River, near Athabaska Pass, in Alberta, Canada. Its scientific description was published in 1828.

Photo 30. Pika. Note the tell-tale white margin of the short, rounded ear. (Oregon Department of Fish and Wildlife Photograph.)

Mammals of the Pacific Northwest

Pikas occur in zones 4 and 5 (photo 30). They measures about 8 inches in length, including the tiny inconspicuous tail, and the broad, rounded ears are a little less than an inch in width. Their weights range from 5¹/₄ to 6¹/₂ ounces. The soles of the feet are covered with rather stiff, woolly hair. The soft fur on the upper parts of the body is a uniform reddish brown that is slightly darkened over the back by black-tipped hairs. The ears are blackish with a distinct white margin. The belly is tannish, and the throat is a clear reddish brown.

DISTRIBUTION AND DESCRIPTION

total length 200 mm
weight 150-181 g

The pika usually makes its home high in the Cascade Mountains in the rocky rubble at the base of cliffs called talus. If you were to sit quietly and listen and watch, you would soon hear the nasal *waaa*, *waaa*, *waaa*, *oink*, *waaa* of a pika and then another as here and there they magically appear in the midst of sun-drenched boulders. The call is usually made from the top of a rock or from the doorway of a cavern between rocks, and it may run into a trill of alarm as a pika dives into the protection of the nearest crevice. Occasionally, however, a call is faintly heard from deep within the pika's rocky fortress.

    To see these small creatures, who scamper silently and deftly over the roughest rocks on fur-cushioned feet, you either must catch the flicker of a movement out of the corner of your eye or be able to pick out the white margin of a small, round ear that is somehow out of place in this high mountain world of jags and angles.

HABITAT AND BEHAVIOR

Photo 31. Note the pika in the center of the picture and the parsley fern at the base of the two small stems of the shrub just beyond it. Parsley fern is an important component of the pika's diet in the Cascade Mountains of Oregon. (Photograph by Chris Maser)

Pikas • Ochotonidae

Photo 32. A pika's hay pile exposed to the sun for drying as food for the coming winter. (Photograph by Chris Maser)

Pikas keep cool deep in their rocky caverns during the hottest of days and warm in their snug nests under the deep snows during the coldest of nights. They are alert, keen of sight and hearing, and quick to dive into the depths of the talus at the first sign of danger.

The pikas have two main periods of feeding during the summer and early fall, between 4:30 and 10:00 in the morning and between 3:00 and 8:00 in the late afternoon and evening, but some activity is ongoing during all daylight hours. They have lookout stations between their home areas and the meadow where they gather vegetation to carry back to their particular portion of the talus. These lookouts are invariably used on their way back from a sojourn into the meadow.

Beginning in early June, adult pikas gather vegetation from the meadow and transport it to their own territory within the talus, where, once it is dried by the sun, they store it under the protective cover of the boulders (photo 32). This activity, called haying, continues until the beginning of November, unless a heavy, early snow makes the vegetation temporarily inaccessible. Pikas carry one mouth full of vegetation after another from the meadow to their home area and store it in piles deep within the talus. Among their foods, which include grasses, herbs, and shrubs, subalpine lupine is especially favored.

An individual's storage area often forms a complex of hay piles, some of which may grow so large that they spill out of the confines of the talus and become visible in the openings between the boulders. Although pikas usually store more hay than they need in a given winter, they do not use the same storage areas every year. Their social organization, however, is centered around their hay piles.

During the haying season, they spend much time in gathering food. An adult might make more than a hundred trips to the meadow in a day to feed, eating like a rabbit, seizing a large leaf at the tip and drawing it into its mouth with rapid chewing motions without the assistance of its forefeet. In addition to feeding, it might make twice as many trips to gather vegetation for its hay pile. The trips to the meadow are periodically interrupted to chase other pikas for trespassing in its territory, being chased for trespassing in another's territory, calling, periods of observation, grooming, and heeding predator alerts sounded by pikas elsewhere in the talus. A pika defends its own specific area against its own species by vocalizing, consistent spacing of hay piles, chasing trespassers, fighting, and marking its territory by rubbing secretions of well-developed chin-glands on rocks.

Although pikas are visibly active during daylight hours, they are also active to some extent at night, but their nighttime activity is confined to the safety of the talus and consists mostly of calling. Nighttime disturbances are instantly noted and immediately challenged vocally. During the breeding season, territorial calls are given on bright moonlit nights throughout the talus.

In winter, where the snow drifts to depths of 8 feet or more around the edge of the talus, pikas dig tunnels into the snow from about 2 feet below its surface down to the buried vegetation of the meadow. By spring, the vegetation of the meadow will be severely grazed immediately adjacent to the talus but only where the protective cover of the snow is sufficient to give the pikas a sense of security.

During the winter, the pikas' piles of droppings and urination spots mark the exits of the tunnels they use to gain access to the surface of the snow. Sometimes their activity results in accumulations of droppings, called fecal towers, in tunnels beneath the surface of the snow. When they feed on dry vegetation, their droppings are round, like the familiar rabbit pellet, but when they feed on green vegetation, the feces are soft and irregular in shape (photo 33).

The pikas' reproductive activities begin in March, peak in May through July, and are largely completed in August. The onset of the breeding season is announced by the long territorial calls of the males and the short, answering calls of the females. Although breeding males visit females in adjacent territories, they tolerate one another only during the mating season and again practice mutual intolerance soon after the youngsters emerge from the inner protection of the talus. Litters range from one to six young, but two to three is the norm. The gestation period is about thirty days, and there may be two litters per year.

Young pikas begin to appear on the surface of the talus by the end of June, and display their first territorial behavior with high-

Photo 33. When feeding on dry vegetation, pika's droppings are round, like the familiar rabbit pellet, but when feeding on green vegetation, the feces are soft and irregular in shape. (Photograph by Michael Castellano and Chris Maser.)

pitched calls about one week after they emerge. Within two weeks the calls of the young are indistinguishable from those of the adults. The adults become intolerant of their young soon after the youngsters become independent of parental care. The young lack alertness when they first emerge from the talus, and are vulnerable to predators, such as marten and the indigenous red fox. But weasels, both long-tailed and short-tailed, are the pikas' main predators because they can penetrate the rocky fortress and either kill the adults or carry off whole litters of young from their nests. Because martens and red foxes are too large to follow them into their talus, pikas keep marten and fox in view when either approaches the talus and warn one another, even though they will be seen by the predator. But pikas are markedly silent when a weasel is near, because, with its snakelike body, a weasel can follow a pika anywhere, leaving little chance of escaping the sharp fangs and deadly jaws.

THE FAMILIAL NAME Leporidae is derived from the Latin word *leporis* ("a hare") and the Latin suffix *idae* (which designates it as a family).

The vernacular names "rabbit" and "hare" are used as though they are synonyms, and they are often applied to the wrong group. The names "jack rabbit" and "snowshoe rabbit" are wrongly applied to hares, and the name "Belgian hare" to a rabbit.

Although the physical differences between hares and rabbits are not well defined (the major differences are in the structures of the skulls), behaviorally they are quite distinct. Young hares are born under cover, but not in a constructed nest. They are fully haired at birth, their eyes and ears are open, and they are capable of running around within a few minutes after delivery. Rabbits, on the other hand, are born naked and blind in a nest constructed by their mothers.

Unlike most mammals, female leporids are usually larger than the males. Members of this family range in total length from about 10 to 30 inches. Hares, which are larger than rabbits, may weigh as much as 15 pounds. The pelage is usually thick and soft, and if small patches are torn out, they are quickly replaced with new hair.

A sensory pad, normally hidden by hairy folds of skin, is located at the entrance of each nostril, and a naked Y-shaped groove extends from the upper lip to and around the nose. The term "hare-lip," an unusual condition in humans, is named after this groove.

Leporids produce from two to eight young, often more, to a maximum of fifteen. Gestation periods range from 28 to 47 days; hares have longer gestation periods than rabbits. Most species produce several litters per year.

Leporids inhabit most of the major land masses and some islands. Their natural distribution does not include the antarctic region, Madagascar, parts of the Middle East, southern South America, Australia, or most oceanic islands. Among the most widely introduced groups of mammals, leporids have been introduced into Australia, New Zealand, and other islands. There are two genera of leporids in western Oregon.

THE GENERIC NAME *Lepus* is the Latin word for "hare."

The genus contains the largest of the lagomorphs. They range in head and body length from 16 to 28 inches and weigh 3 to 15 pounds. All members of this genus have long ears and large hind feet; the latter are well haired regardless of the climate in which they live. Most species—such as the big, long-eared,

Hares and Rabbits Leporidae

Hares
Lepus

slender-bodied jack rabbits—prefer open, grassy areas. A few, however, are associated with forested areas.

Hares do not dig burrows but spend their inactive periods hidden among vegetation in shallow depressions called forms.

Northern species of hares exhibit drastic fluctuations in numbers that appear to be in cycles of nine to ten years. They increase to great abundance, then suddenly decline. Although the reasons for the fluctuations are not definitely known, various diseases, as well as disruption of the endocrine gland system, may be responsible. The most regular fluctuations occur in the northern half of North America. When hares are abundant, the populations of foxes, lynxes, weasels, mink, and other predators increase because of the abundant supply of food; when the populations of hares decrease, so do the populations of predators that depend on them as food.

Various species of hares are used as food by people. Although the pelts of hares are neither durable nor valuable, they have been used extensively in the manufacture of felt; they are also used as trimming and lining for garments and gloves.

The combined, original distribution of the approximately 26 species of hares include most of Eurasia, as far south as Sumatra, Java, Taiwan, and Japan; most of Africa, except the rain forest of the Congo and the region along the Gulf of Guinea on the Atlantic coast; and most of North America south to the end of the Mexican plateau. In the eastern half of the United States, however, their distribution is limited. After their introduction, they became common in parts of South America, Australia, New Zealand, islands off the northwest coast of Africa, and in some portions of the northeastern United States. There are two species of hares in western Oregon.

Snowshoe hare
Lepus americanus

THE SPECIFIC NAME *americanus* is a proper name. The specimen on which the 1777 scientific description is based came from Fort Severn, Ontario, Canada.

DISTRIBUTION AND DESCRIPTION

total length 382-430 mm
tail 35-56 mm
weight 950-1,416 g

SNOWSHOE HARES OCCUR in all five zones, but not on the floor of the Willamette Valley proper in zone 2 or in the hot interior of the Rogue Valley in zone 3. Snowshoe hares range in length from 15 to 17 inches from the tip of their noses to the tip of their tails and weigh from 2 to 3 pounds.

In summer, a snowshoe hare's back and sides are covered with a long, thick, soft, relatively light reddish brown coat with a few intermixed black-tipped hairs. The throat is also reddish brown, as are the ears, with their blackish tips and white margins. The reddish brown gives way to clear white on the chin, belly, the insides of the long legs, and the tops of the large hind feet. The inconspicuous tail is blackish on top and light whitish gray underneath.

In winter, the hare's pelage remains essentially the same in zones 1 through 3, but changes to pure white (except for the black tips of the ear) at high elevations in zones 4 and 5 (photo 34).

SNOWSHOE HARES LIVE in the coniferous forest from almost sea level along the northern Oregon coast up to timberline at around 7,000 feet in the High Cascade Mountains of zone 5. They are normally shy and secretive, seldom venturing far from protective cover; their dark, rich color blends into the deep, shadowy forests and dark fern undergrowth. Often, the only evidence of the hares' presence is their trails and droppings or pellets. In western Oregon, and particularly in the dense forests and vegetation of the coast, their trails and pellets are inconspicuous. Although these hares may have lived close to humans for years, many local residents are unaware of their presence.

Snowshoe hares are primarily active during the evening and throughout the night into the early morning. On foggy or rainy days, however, it is not unusual to see a hare feeding or just sitting along the edge of a road in the forest. During the day, snowshoes normally retire to their forms, a shallow depression located among clumps of swordfern, in thickets of salal or other suitable vegetation, or under jumbled piles of windthrown

Photo 34. Winter-white snowshoe hare in a protective thicket out of winter's storm. (Oregon Department of Fish and Wildlife Photograph by Mark Henjum.)

timber. I have even trapped hares that were using the large burrows of mountain beaver as daytime retreats.

When disturbed, adult hares usually make a few short hops, freeze (remain motionless) for a few seconds, hop a few more feet, then freeze again; in this way, a hare may slip quietly away without attracting attention. Juveniles, on the other hand, freeze immediately. Juveniles in particular depend so much on freezing that, by walking slowly and quietly, I have occasionally been able to get close enough to a youngster to catch it by hand.

During the breeding season, snowshoes seem to be moving most of the time, rendering them more conspicuous than at other times of year. Along the northern Oregon coast, the juveniles appear more active and less cautious than the adults.

Although snowshoe hares are not gregarious, they often play with one another, but only during the breeding season do males and females seem to tolerate one another for an extended length of time. Males are generally intolerant of one another, particularly during the breeding season, and fights are frequent.

Adult showshoes are relatively sedentary and may remain in a small area for a long time. A hare's home range may be on the order of 16 acres.

In addition to the characteristic distress cry, snowshoes make other vocalizations: a grunt, a chirp or click resembling a human sound sometimes written "tch," and a birdlike warble that is difficult to describe. Both sexes frequently thump their hind feet on the ground. The thumping is surprisingly loud and can be heard for some distance. It probably denotes an intercommunication or warning signal.

The drastic fluctuations in population for which the snowshoe hare is famous are neither readily apparent nor, to my knowledge, have they been studied in the Pacific Northwest.

In western Oregon, during the spring, summer, and autumn, snowshoes feed on a wide variety of herbaceous plants (such as grasses, clover, false dandelions, woolly everlasting) and some woody plants (such as young sprigs of spruce and fir, and young leaves and twigs of salal). During winter, their diet consists mainly of the needles and tender bark of such conifers as spruce, fir, and hemlock. They also eat the leaves and green twigs of salal; twigs, buds, and bark of willow; and some herbaceous vegetation that remains green.

The testes of male snowshoe hares begin to enlarge in December and reach maximum size in May. In June, they start to shrink and reach minimum size in November. Sperm was found in about 30 percent of the males in February, about 70 percent in March, 100 percent from April through July, and 50 percent in August. Snowshoes are normally not reproductively active until their second summer. The gestation period is 35 to 37 days. They have two litters per year, occasionally three. Litters range from one to seven; the young are commonly called leverets. Although five young seem to constitute the usual litter in western Washington, the little data I was able to gather in western Oregon, especially the coastal region, indicated that the usual size of a litter there may be three.

Young are born from about the middle of April through the middle of August. Snowshoe hares are not fully grown until they are about five months old, and young born in August would be only four months old in December, during some of the worst winter weather. Young born earlier in the summer may therefore have a better chance of survival.

Snowshoe hares have a long list of predators, including bobcats (probably the main predator in western Oregon), mink, long-tailed weasels, foxes, coyotes, domestic dogs, domestic cats, Great Horned Owls, and Northern Spotted Owls. With this many predators, life for a snowshoe hare can be both complicated and difficult.

Black-tailed jackrabbit
Lepus californicus

THE SPECIES WAS NAMED for the state of California, plus the Latin suffix *cus*, "belonging to." The first specimen was collected by David Douglas in 1831, probably on the coastal slope of the mountains near the Mission of San Antoine, Jolon, Monterey County, California, and was described scientifically in 1837.

DISTRIBUTION AND DESCRIPTION

total length 604 mm
tail 95 mm
weight 2.3-3.2 kg

BLACK-TAILED JACKRABBITS occur in zones 2 and 3 (photo 35). They are large, averaging about 24 inches in length, with tails almost 4 inches long, and they weigh between 5 and 7 pounds.

These hares have very long ears. Winter pelages are dark yellowish brown, darkened by long black outer hairs. The top of the tail and the backs of the ears near the tips are black; the underside and flanks are dark yellowish brown to slightly orangish. Summer pelages are paler and grayer than in winter. Young are heavily furred at birth, dark, coarse gray, becoming paler and less grizzled when half grown.

HABITAT AND BEHAVIOR

THESE LARGE HARES extend from California northward into the open country of the Rogue, Umpqua, and Willamette valleys, where they formerly reached the vicinity of Salem, Oregon. They generally inhabit open spaces, such as prairies like that once covering the floor of the Willamette Valley, clearings along the edges of the floors of valleys, old fields, and pastures, but more than most other jackrabbits, they readily enter thickets and shrubby areas.

Photo 35. Black-tailed jackrabbit enjoying the morning sun. (Photograph by Chris Maser)

Although they were not numerous in western Oregon even in the early days, and are much less so now, especially in the Willamette Valley, they used to be fairly common in particularly favorable habitat, such as the hill south of Corvallis, Oregon, where the Corvallis Country Club is situated and I grew up. I remember the black-tailed jackrabbit as quite common from the early 1940s until the mid 1950s, when housing developments began to rapidly destroy their habitat. Vernon Bailey wrote of counting nine individuals killed by automobiles along Highway 99 West between Eugene and Salem on November 26, 1930. I have not seen a black-tailed jackrabbit in the Willamette Valley in many years, which is not so say that a few might not linger in secluded places. Many, or most, of them disappeared, however, as the vegetated fencerows that once separated the fields of small family farms gave way to huge open fields devoid of appropriate habitat when corporate-style farming took over the valley, although they are still fairly common in the valleys and foothills of zone 3.

During the day, these large, brown hares lie concealed in shallow forms, which are really no more than depressions in the protective cover of overhanging vegetation, where their brownish coloration blends in well with their surroundings. Depending on the quality of their protective coloration, they often lie still until almost stepped on before bounding away in

Mammals of the Pacific Northwest

great leaps, with a dazzling display of legs and ears. Once out of their forms, they depend on their tremendous speed for survival.

They begin to emerge of their own volition toward evening, however, seeking succulent green vegetation for both sustenance and water, and when I was young I would occasionally see one abroad as it foraged at night, bathed in the quiet light of a summer moon. I remember seeing them often as they fed along the edge of the golf course as the sun rose. Not long after full sunrise they sought protected areas in which to sun themselves for a while before retiring to their forms for the remainder of the day.

Although they seem to eat anything succulent and green, they are especially fond of clover and alfalfa. But no meal is entirely relaxed, because they frequently stop eating to look and listen, their sensitive ears forever searching this way and that for signals of danger. They are most likely to visit fields, gardens, pastures, and golf courses in late summer and autumn when most unirrigated herbaceous vegetation has dried out for the year. In winter, when green vegetation is scarce, they eat buds, twigs, and tender bark.

The young, usually two to six in number, are born fully furred, with their eyes and ears open, their incisor teeth well developed, and the ability to move about. Born from late spring throughout the summer into early autumn, they are well concealed, both through the effect of their cryptic coloration and by their mother's skill at hiding them in or near her form.

In western Oregon, coyotes, Golden Eagles, and Great Horned Owls prey on black-tailed jackrabbits.

Rabbits, *Sylvilagus*

THE GENERIC NAME *Sylvilagus* is derived from the Latin word *silva* ("wood," as in "woods" or "forest") and the Greek word *lagos* ("hare").

The head and body of rabbits range in length from about 10 to 18 inches, and they weigh from about 14 ounces to about 5 pounds. Their pelages usually vary from grayish brown to reddish brown above and are normally whitish, tannish, or grayish underneath. There usually is a patch of bright reddish brown hair on the nape of the neck. Unlike hares, rabbits do *not* turn white in winter.

Rabbits are active in the evening, throughout the night, and into the early morning. Although most members of this genus live in burrows, only one is known to dig its own. A few do not inhabit burrows. All tend to have relatively small home ranges and usually maintain definite trails within them.

Herbaceous vegetation is the major food, but during winter, twigs and bark of woody plants are also eaten.

Gestation periods range from 26 to 30 days, and litters vary from two to seven young. Youngsters are born in nests constructed in shallow depressions by the mother, occasionally under protective structures made by humans, and they are generally composed of soft fibers from plants lined with fur from the mother's underside.

These rabbits occupy a variety of habitats; although most species seem to prefer open or brushy areas or clearings in forested areas, a few inhabit forests, swamps, marshes, sandy beaches, or deserts. Their ability to survive in areas of dense human population makes some species popular as game animals for hunting.

Rabbits of the genus *Sylvilagus* are restricted to the Western Hemisphere. They occur from southern Canada throughout most of the United States, south to Argentina and Paraguay in South America.

There are two species in western Oregon, one indigenous and one introduced.

Brush rabbit
Sylvilagus bachmani

THIS SPECIES WAS NAMED in honor of Dr. John Bachman. The first specimen of the brush rabbit is thought to have been collected between Monterey, Monterey County, and Santa Barbara, Santa Barbara County, California. It was described scientifically in 1839.

DISTRIBUTION AND DESCRIPTION

total length 280-363 mm
tail 25-43 mm
weight 450-965 g

BRUSH RABBITS OCCUR in zones 1 through 3. They are relatively small, ranging in total length from $11^1/4$ to $14^1/2$ inches, with tails that vary from 1 to $1^3/4$ inches. They weigh from 1 to almost $2^1/2$ pounds.

Brush rabbits are small and compact, with short, dark ears that are distinguished by neither a black tip nor a white margin. Their legs are also short and their tails small. Their pelages are fine and soft with few white hairs. Summer pelages are lighter than those of winter. Dorsally, summer pelages are reddish brown and heavily mottled because of numerous black-tipped hairs. Winter pelages are also mottled but are brown dorsally. The napes of brush rabbits' necks have a patch of bright reddish brown fur. The sides vary from slightly more grayish brown to slightly more yellowish brown than the back. The undersides are gray in winter and light gray washed with light tan or tan in summer. The tops of the feet are grayish to almost gray. The undersides of the small, inconspicuous tails are usually gray but may be almost white.

THE BRUSH RABBIT is well named since it inhabits the thick, brushy edges of various habitats in western Oregon. The only time it penetrates particularly dense habitat, such as along the Oregon coast, is when roads or unusually wide trails exist.

Humans have had a profound effect on the distribution of these rabbits. Removal of forests by logging, construction of roadways, and subsequent severe grazing by livestock have created brush rabbit habitat of almost impenetrable brush, especially in the Coast Range and along the coast. Further, stabilization of moving sand dunes along the coast through the planting of Scotch broom, lodgepole pine, and European beachgrass has extended the habitat of brush rabbits onto the sand dunes. In some cases, these rabbits have followed favorable habitats across a mile or more of sand almost to the open beach.

Brush rabbits are seldom seen more than about 3 feet from their brushy retreats. Adult rabbits are most active in the evening, throughout the night, and into the early hours of morning. On foggy days or days of constant, drizzling rain, some activity occurs throughout daylight hours. During the summer and early autumn, young rabbits are often active during the day—even clear, sunny days. Adults, on the other hand, usually retire to some secluded place.

In my experience, brush rabbits do not dig burrows and seldom use those available. When an individual does use a burrow as a retreat, it is often the large burrow of a mountain beaver. Use of burrows by brush rabbits occurs, as far as I can determine, only during the hottest days in summer and the coldest and stormiest days in winter. The rabbits normally have several secluded refuges within protective thickets. Because these thickets are almost impenetrable to larger animals, a rabbit can usually zip along its runways to safety. Although adult rabbits may watch the approach of an animal for a second or two, their move into cover is immediate when danger seems imminent. Youngsters, on the other hand, frequently freeze in place, which has allowed me, by moving slowly and quietly, to capture several by hand.

Brush rabbits live in small areas, which largely conform to the size and shape of the thickets in which they live. The home ranges of females are about 70 feet in diameter and those of males about 115 feet. Males move around more freely than females and have overlapping home ranges, whereas females tend to have discrete areas of occupation. A rabbit of either sex maintains a good network of trails within its home range, allowing it easy access from one part of its living area to another, to choice areas for feeding or resting, and for quick escape from predators.

These small rabbits eat a variety of vegetation (photo 36). During spring, summer, and autumn, they feed primarily on herbaceous plants, such as false dandelion, plantain, pearly everlasting, grasses, clover, and the fruits of blackberries and salmonberries. During winter, they consume whatever green herbaceous vegetation is available, but woody plants form the bulk of their diet, primarily the leaves and green twigs of salal and, to some extent, the

Photo 36. Brush rabbit feeding. (Photograph by Chris Maser)

needles and small twigs of Douglas fir, which are gleaned from boughs that have been broken off by the wind and blown down to within the rabbit's reach (photo 37).

Along the coast, where the vegetation is particularly dense, the rabbits' feeding areas parallel the thickets they inhabit and are usually restricted to within 3 feet of the edge of protective cover. The feeding areas are kept open because the rabbits constantly crop the vegetation to within an inch or two of the ground. After some practice, these feeding areas can be easily located.

Along the coast, the most reingestion of cecal pellets (see pages 102-3) by brush rabbits appears to take place during winter, when as much as a fourth to a third of their stomachs was found to be filled with these pellets. In summer, many of the rabbits did not have any cecal pellets in their stomachs, but one lactating female, captured in mid-July, had her stomach filled with them. (If you remember, cecal pellets are the moist pellets that are reingested be a rabbit as they are expelled, almost without being chewed, which allows the rabbit to get the maximum nutrition our of its food be passing it through the body twice.)

The breeding season of brush rabbits in Oregon begins by mid-February and lasts through mid-August. The testes of males begin to enlarge in December, and contain mature sperm by January. Development of the testes culminates in March, then declines gradually until September or October, when they reach their minimum size. By July, however, there is so little sperm remaining within the shrinking testes that males are effectively sterile. Neither males nor females appear to breed until the year after their birth.

In the Willamette Valley, 13 percent of the adult female brush rabbits examined in one study were reproductively active in February, 70 percent in March, 100 percent from April through July, 50 percent in August, but none thereafter. I encountered no pregnant females along the Oregon coast before March, but during the beginning of March females exhibited the onset of

reproductive activity, which seemed to terminate no later than mid-August.

Photo 37. Because brush rabbits cannot manipulate with their front feet, they can only eat those needles of Douglas fir that are readily exposed to being bitten off. (Photograph by Chris Maser)

Although the reported range in size of litters is from one to six, along the Oregon coast three females each gave birth to seven young, and one female had 10 fresh placental scars. But the usual number of offspring appears to be three. Considering that the gestation period is about 27 days and that females average about five litters per year, one can still produce around fifteen offspring in a single breeding season.

Young are born in a hidden nest constructed by the mother. One such nest, which I found in a cranberry bog in the vicinity of Bandon, in Coos County along the Oregon coast, on July 27, 1970, was composed of an outer layer of dried grasses and an inner layer of fur that the female had plucked from her own underside that was thick enough to form a roof over the four youngsters, which were born naked and blind. The nest was 6 inches in diameter and 4 inches deep.

The bobcat is probably the major predator of brush rabbits in western Oregon (photo 38), though mink and long-tailed weasels are also important predators. A mink or a long-tailed weasel in pursuit of a brush rabbit in daylight hours is so intent on catching the rabbit that it may almost run into a person observing the chase. When pursued by a mink or a weasel, both of which can easily negotiate its runways, a rabbit will

Photo 38. The molar teeth of a brush rabbit in the dropping of a bobcat—the fate of many an individual. (Photograph by Chris Maser)

often break cover in apparent panic and run down the middle of a forest road, where it is soon overtaken and swiftly dispatched by a deft bite through the back of the skull.

Spotted skunks also are adept predators of brush rabbits. Mink, weasels, and spotted skunks probably kill more nestling brush rabbits than any other predator. Striped skunks also occasionally kill nestlings. Foxes and coyotes, as well as domestic dogs and cats, kill a considerable number of rabbits each year, especially juveniles. Great Horned Owls and Cooper Hawks prey on a few of these rabbits along the Oregon coast. Red-tailed Hawks undoubtedly also capture some.

Eastern cottontail
Sylvilagus floridanus

DISTRIBUTION AND DESCRIPTION

total length 450 mm
tail 65 mm
weight 1.4-1.5 kg

THIS RABBIT IS NAMED after the state of Florida, where it was first taken along the Sebastian River in Brevard County, Florida, and described scientifically in 1890.

EASTERN COTTONTAIL RABBITS were introduced into the Willamette Valley, zone 2, as a game animal. They average about 18 inches in length, and have tails that are between 2 and just over $2^1/2$ inches long. They weigh between 3 and almost $3^1/2$ pounds.

Eastern cottontail rabbits are small and relatively dark, varying in color from dark grayish mixed with pale yellowish to reddish brown mixed with pale yellowish. The nape of the neck and the legs are reddish brown; the ears are short, rounded, and darker than the back. The top of the head and back are dark yellowish brown interspersed with reddish and dark yellowish hairs. The rump and sides of the body are dark yellowish gray, washed with black. The top of the tail and the front of the forelegs are reddish to orangeish brown. The underside of the tail, which is readily visible when a rabbit hops, is white, hence the name "cottontail." The lower flanks have a clearer pale yellowish than the back. The dark grayish ears are heavily bordered and washed with black, especially the terminal half.

HABITAT AND BEHAVIOR

THE EASTERN COTTONTAIL was much more abundant when I was a boy. Today, in comparison, it is relatively rare in most areas. It used to live along the fencerows of the valley floor and in the patches of blackberry brambles, from which it seldom ventured very far.

Cottontails are timid, and rely on their speed to zip into cover at the first hint of danger. When not eating or sunning themselves, they remain tucked away, often in the impenetrable mass of a Himalaya blackberry bramble, where they sleep away the day, ever alert to explode into flight should danger threaten.

Mammals of the Pacific Northwest

Although these rabbits are known to use burrows in some portions of the their geographical distribution, even digging their own, I have not found a burrow in Oregon.

These rabbits are active chiefly at night, venturing beyond the confines of protective cover with the approaching dusk to forage for food. I used to follow their nightly activities in the winter snows that periodically blanketed the valley floor. I was always amazed to see how often their tracks crisscrossed back and forth, making their nightly activities a real challenge to unravel.

During the summer, cottontails eat succulent herbaceous plants, such as grasses and clover, and can become regular visitors in vegetable gardens, where these are close enough to their hideaway. In winter, their food consists of buds and tender twigs of shrubs, and they also nibble on the canes of their blackberry fortresses.

Like the brush rabbit, cottontails reingest cecal pellets. Green food is rapidly chewed and swallowed, after which the rabbit returns to its shelter, where the soft, green pellets, which consist of undigested vegetation, are defecated and eaten again at a more leisurely pace.

The cottontail is prolific, producing several litters of one to seven, but usually of three to six, youngsters over a long breeding season, which may commence in January, with the first youngsters being born in March after a gestation period of 28 to 32 days. The babies are born into a warm nest, which the mother creates by scraping out a shallow depression that she lines with shredded leaves and grasses, finally adding a goodly amount of her own fur plucked from her breast and belly. The fur adds substance and warmth to the birth chamber, which the mother hides in a safe place and which she carefully covers with vegetation when she leaves it.

The young are born blind, essentially naked, and quite helpless, weighing slightly less than an ounce. The mother returns to her nest each dawn and dusk for about sixteen days. She opens the top of the nest, and lies over it to let her babies nurse. After having fed them and licked them clean, she carefully closes the nest and remains nearby feeding and resting.

Cottontails in the Willamette Valley have many enemies, including long-tailed weasels, mink, foxes, coyotes, bobcats, domestic dogs and cats, owls, hawks, and people—especially hunters. Gopher snakes will also readily eat the young.

Gnawing Mammals
Rodentia

THE ORDINAL NAME Rodentia is derived from the Latin word *rodentis* ("gnawing") and refers to the gnawing habits of this group of mammals. Rodents comprise over one-third of the known species of mammals. In many areas of the world, they are the most abundant animals, in species as well as in numbers. Members of this order usually have high birth rates.

Rodents are unique among the mammals of North America in having four incisors, two above and two below. They lack canine teeth (commonly thought of as "fangs"). The incisors of rodents grow throughout an animal's life. Growth is from the base, the only portion that contains nerves, and replaces the top portion worn away by chewing hard materials. The outer surface (enamel) is harder than the underlying surface (dentine) and is somewhat self-sharpening because the dentine wears down faster than the enamel. Rodents often grind their front teeth together when they are irritated and also at other times; this may help to keep the teeth in proper condition, because if the tips are not constantly worn down, they can grow past one another in spiraling form (photo 39). The result may be upper incisors that curl around into the mouth, grow upward, and possibly pierce the roof of the mouth. Sometimes, however, they grow upward and outward, forming a spiral on each side of the mouth. The lower incisors may grow upward in front of the nose or face. Such abnormal growth of incisors may cause death by starvation.

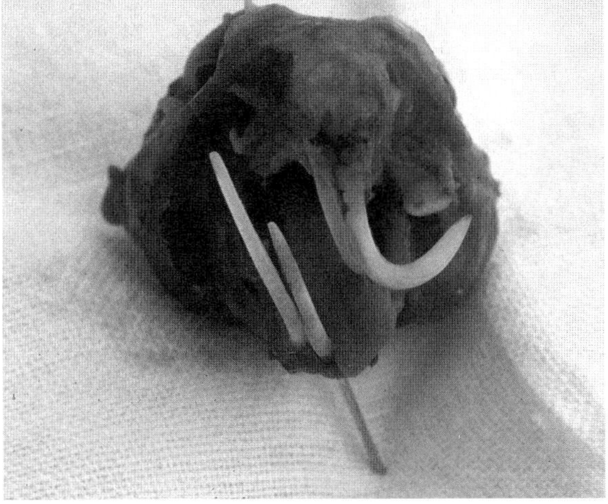

Rodents walk on the entire surface of their feet. The hind feet have from three to five toes; the front feet have five toes, although the "thumb" may be small or absent. Rodents have evolved considerable ability to manipulate objects with their front feet.

Several members of this order possess either internal or external cheek pouches that open near the angle of the mouth and serve

Photo 39. The incisor teeth of a Townsend ground squirrel from southeastern Oregon that have grown faster than they could be worn down. (Photos by Chris Maser and Gerald Strickler.)

as storage areas during the gathering and transportation of food. External cheek pouches are lined with hair and can be turned inside out for cleaning. Internal cheek pouches lack hair and are attached firmly to the cheek, preventing inverting of their position.

The tails of some rodents are adapted for swimming, some for gliding, and a few for the storage of excess fat. Most are important in the maintenance of balance. The tails of spiny mice, *Acomys*, break off readily when they are grabbed, enabling the animals to escape. Other rodents have skin on their tails that breaks or tears readily and slips off, leaving flesh and bone exposed. The exposed flesh and bone dry and fall off. The tail then heals but does not grow.

Generally speaking, the testes of male rodents remain small and are retained within the body cavity except during the breeding season, when they enlarge and descend into the scrotum.

Rodents are extremely diverse in form and highly adaptable. They have a wide geographical distribution and are usually abundant in most land areas. Some rodents are specialized for digging or burrowing; some are primarily arboreal; some are primarily aquatic; most are terrestrial. Of the latter, some are adapted to arctic regions, others to desert areas. Members of this order are adapted for digging, running, leaping, climbing, gliding, or swimming. Many species use a combination of methods of locomotion.

Rodents are extremely important to humans. Some species destroy insects and weeds; others, such as beaver, muskrat, nutria, and chinchilla, are valuable for their fur. Members of this order are used extensively for laboratory research. Some are considered pests because of the extensive damage they cause to agricultural crops or because they carry parasites that transmit diseases to which people are susceptible or transmit diseases directly to people.

Rodents are cosmopolitan. They are indigenous to most land areas except some arctic and oceanic islands, such as New Zealand and Antarctica, and almost everywhere people have traveled, they have introduced rodents.

There are eleven families of rodents in western Oregon.

Mountain Beaver Aplodontidae

THE FAMILIAL NAME Aplodontidae is derived from the Greek words *haploos* ("single" or "simple") and *odontos* ("tooth") combined with the Latin suffix *idae* (which designates it as a family).

The Sierra Nevada miners of California named this unique rodent mountain beaver because it occasionally gnaws bark and cuts off limbs in a manner similar to the true beaver, *Castor*.

Mountain beaver is really a misnomer because this rodent, which belongs to the oldest known family of living rodents, is more closely related to the squirrels than it is to the true beaver.

Since there is only one living genus within the family Aplodontidae, refer to the genus for the general description.

Mountain beaver inhabit the humid regions of western North America from southern British Columbia, Canada, south to San Francisco Bay, California, and east to the Cascade Mountains and Sierra Nevada.

Mountain Beaver
Aplodontia

THE DERIVATION OF the generic name *Aplodontia* is the same as that of the familial name.

Mountain beaver are chunky, short limbed, long whiskered, and cantankerous. They have five toes on each foot. The toes of the forefeet are fairly long and are used for digging and grasping. The pelage is composed of sparse guard hairs and thicker underfur. Mountain beaver can be distinguished from all other mammals within their geographical distribution by their size, shape, uniform coloration, and the apparent absence of a tail. They are active throughout the year.

The geographical distribution of the genus is the same as that given for the family. Mountain beaver are represented in western Oregon by a single species.

Mountain beaver
Aplodontia rufa

THE SPECIFIC NAME *rufa* is the Latin word for "reddish." The original description of the species was based on a description by Lewis and Clark, who, in 1806, had obtained skins from the Indians near the Columbia River, Oregon.

DISTRIBUTION AND DESCRIPTION

total length 238-370 mm
tail 19-55 mm
weight 502-1,419 g

MOUNTAIN BEAVER (photo 40) can be found in zones 1 through 5, but today are probably absent from the floor of the Willamette Valley proper (zone 2) and the Rogue Valley proper (zone 3) due to the thorough alteration of their habitat with the advent of agriculture and urban sprawl.

Mountain beaver range in length from $9^{1}/_{2}$ to about $14^{1}/_{2}$ inches and weigh from about 1 to 3 pounds. They are chunky with large, wide, flat heads; long, stiff whiskers; and small eyes and ears. There is a little patch of whitish hair at the base of each ear. They have small, inconspicuous tails and short, stout legs.

Their forefeet are relatively small. The first finger is thumblike and has a small, blunt nail, whereas the other four digits are armed with long, sharp, slightly curved claws adapted for digging. The hind feet are larger, but the toes are shorter with more sharply curved claws.

Mammals of the Pacific Northwest

Photo 40. Mountain beaver. (Oregon Department of Fish and Wildlife Photograph by Ron Rohweder.)

The fur is dark reddish brown with numerous black-tipped hairs on the back, but the number of black-tipped hairs varies greatly by individual. Ventrally, pelages are grayish brown to tan, often with patches of white hair. Mature females have dark brown to black hairs encircling each of six teats. The pelages of young animals are woolly and dark brown but lack the shiny guard hairs.

They have well-developed senses of smell, touch, and apparently taste. Their long, stiff whiskers are well adapted for detecting the sides of their extensive tunnels through which they usually make their way in total darkness. Their small eyes may not be very efficient, but they can detect light and movement. A thick, sticky, whitish substance often covers the eyes and probably aids an animal in keeping its eyes clean of the earth through which it burrows.

HABITAT AND BEHAVIOR

ALTHOUGH MOUNTAIN BEAVER are primarily animals of the coniferous forest and riparian habitats, they can survive in clearcut areas provided enough large wood remains on the ground for protective cover while the animals forage on the surface of the ground. Along the coast, I have trapped them in people's gardens and even under their houses, where the forest was close enough to provide habitat for dispersing young.

The burrow system of a mountain beaver is often conspicuous. It is characterized by large holes ranging from about 6 to 8 inches in diameter and occasionally as large as 10 inches. The holes are surrounded by large piles of earth, rocks, and other debris (photo 41). The burrows are frequently near the surface of the ground and are easily broken through. Caved-in burrow roofs are not repaired; the debris is merely removed

Photo 41. Burrow of a mountain beaver. (Photograph by Chris Maser)

from the tunnel, leaving an open trench. Burrow systems are normally located in or near cover. Most of the entrances and exits, as well as the short trails on the surface that connect entrances of burrows, are usually well hidden.

A mountain beaver's nest chamber is circular and is situated from 14 inches to 5 feet below the surface of the ground. The nest of an adult may contain a bushel of vegetation and is composed of an outer layer constructed of coarse vegetation and an inner layer of soft, dry vegetation. Fern and occasionally Douglas fir and hemlock sprigs are used for the outer layer, but any readily available vegetation may be used. When available, the leaves of salal are used for the inner layer. Although nests may be moist on the outside, they are dry inside.

Dead end tunnels, located near the nests, are used as fecal and food refuse chambers. Fecal pellets are deposited at the rear of the tunnel and when the tunnel is full, a new one is constructed. New nest sites, particularly those of young animals, have only one fecal chamber. But long-established nest sites, which may be used for as long a three and a half years, have several. Feeding and food-storage chambers are also located near the nest.

These chunky rodents also have chambers that provide storage for their "mountain beaver baseballs," found while they excavate their burrows. The baseballs, composed of heavy clay or friable rock, may be spherical or lopsided, about 3 inches in diameter and about 7 ounces in weight. The function seems twofold: (1) because the animals' diets furnish little abrasive material, gnawing the balls keeps their ever-growing front teeth in proper trim; (2) they are used to close the nest-feeding chamber complex during an animal's absence, preventing trespass.

Mammals of the Pacific Northwest

Although these burrowing rodents are active primarily from late evening into the early morning, they may also be active during daylight hours. Mountain beaver do not travel particularly far in their daily wanderings. Most adults stay within about 25 yards of their nest. The size and shape of a mountain beaver's home range is influenced by the arrangement and quality of the habitat, as well as by the territorial behavior of the individual animal. Nest sites, defended against trespass by these generally solitary rodents, are located in such a way that advantage is taken of both good drainage and available cover; the same is true for the burrows that radiate from the nest, although much-used tunnels away from the nest site may have permanently flowing rivulets coursing through them.

Mountain beaver have various vocalizations, ranging from soft whining and sobbing to a kind of booming noise, hence another name for them—"boomer." The most frequent sound, however, is a harsh chattering or grating produced by gnashing the tips of the lower front teeth across the tips of the upper front teeth. Gnashing of teeth indicates irritation and is best heeded because mountain beaver are normally cantankerous and are swift, vicious biters.

Strictly vegetarian, a mountain beaver consumes a wide variety of plants, but swordfern is the most important food on a yearly basis. There are changes in diet throughout the year, however, related to the protein content of the food. Adult males, for example, eat principally ferns, but shift to red alder for a short period in the early autumn when these leaves are amassing

Photo 42. Burrow of a mountain beaver; note the wilting Oregon oxalis piled at the entrance. (Photograph by Chris Maser)

their greatest protein content of the year. Because milk production in mammals depends on a high-protein diet, in spring the nursing females shift their diet from ferns to the new growth of coniferous trees, which is then high in protein. Later, when the protein content of grasses and herbs reaches its peak, the females again shift their diet.

There is a definite difference between the harvesting of food and the gathering of nest material. Nest material is always dry, whereas food plants are never allowed to dry out. For nest material, the animals gather from the ground the plentiful already-dry vegetation from a previous year. Food, on the other hand, is cut while it is fresh and piled next to a burrow, under a fallen tree, or on top of one. There it is allowed to wilt before a mountain beaver transports it by mouth into the feeding chamber (photo 42). Since the relative humidity of the feeding chamber is 100 percent, the food does not dry out. So by wilting some vegetation prior to eating it and mixing it with a certain amount of fresh plants, a mountain beaver ingests the desired volume of water.

Mountain beaver climb hardwood trees, such as red alder and vine maple, to get the leaves. The animals climb to heights of about 15 feet in alder and about 6 feet in vine maple to lop off living branches that are sometimes nearly an inch in diameter. They cut the branches off as they climb, but leave stubs for their descent.

Like rabbits and hares, mountain beaver produce two kinds of pellets—hard and soft. The hard fecal pellets of waste material are discarded, whereas the soft pellets produced in the cecum (the large, blind pouch that forms the beginning of the large intestine) are reingested as soon as they are expelled, allowing maximum use of the nutrients and vitamins contained in the food.

Mating occurs in January and February, and litters, usually comprising four or five young, are born after a gestation period of about thirty days. Youngsters are born naked and blind with disproportionately large heads. Although their front feet serve as hands for grasping, as they do in the adults, there are two elongated processes on the heel of the palms, which together with the thumb, oppose the four fingers and thus assist in grasping while they are babies. Youngsters are evident outside their mother's burrow in early June. Although some young animals become sexually mature the year after their birth, most do not mature until the second year. One litter per year seems normal.

During dispersal, the young may travel along the surface of the ground or follow existing burrow systems. In the latter case, they may attempt to establish several nest sites before finding one that suits them. Nest sites are established either by enlarging or extending burrows or by occupying vacant nest sites. After a nest site has been established, movements of the young are similar to those of the adults.

Mountain beaver burrows are used by a variety of mammals: shrew-moles, coast moles, snowshoe hares, deer mice, western red-backed voles, short-tailed weasels, long-tailed weasels, mink, and spotted skunks. Most of these mammals are innocuous to the mountain beaver, but the long-tailed weasels and mink (primarily large males) prey on the young. These unique rodents are also prized by bobcats and coyotes, who seem to catch them whenever possible. Even pumas take one now and then. Their main enemy, however, is the human being, who all too often tries to eradicate them from large areas in tree farms.

In times past, the Indians of the Northwest coast used the skins of mountain beaver to make fur robes and blankets, as well as eating their meat. I, too, have eaten mountain beaver and think they have a pleasant flavor, provided the fatty scent glands are removed from the flanks of the carcass before it is cooked.

Chipmunks and Squirrels Sciuridae

THE FAMILIAL NAME Sciuridae is derived from the Greek words *skia* ("shadow") and *oura* ("tail") combined with the Latin suffix *idae* (which designates it as a family). The name alludes to the shadow or shade cast when a squirrel holds its bushy tail over its back.

The family Sciuridae includes chipmunks, tree squirrels, ground squirrels, flying squirrels, marmots, woodchucks, and prairie dogs. All members of this family are diurnal, except the flying squirrels, which are nocturnal.

Tree squirrels and flying squirrels nest in trees. They live in hollow trees or in the cavities of woodpecker nests, but when these are not available, they construct nests made of twigs, shredded bark, leaves, or mosses on limbs.

Tree squirrels make long leaps. When leaping, they extend their legs widely, broaden and flatten their bodies, and stiffen and slightly curve their tails. Such a position presents the broadest possible surface to the air, providing lift. Flying squirrels, on the other hand, have furred membranes extending along the sides of the body from the forelimbs to the hind limbs. In members of some genera, the membrane extends from the neck to the tail. At the outer edge of the wrist, the gliding membrane is extended by a cartilaginous projection. These squirrels are noted for their ability to glide.

Ground squirrels dwell in burrows. They normally come out of hibernation early in the year and feed on green plants in the spring and early summer. Many species retire to their burrows during the hottest part of the summer and become dormant until early the next spring.

Squirrels occupy a wide variety of habitats, ranging from the Arctic to the Tropics and from the coast to the mountains to the desert. Accordingly, some are active throughout the year, while some are dormant during the hottest part of the year, and others are dormant during the coldest part of the year. They have a correspondingly varied diet that includes such items as seeds, nuts, green vegetation, insects, meat, birds' eggs, and fungi.

Squirrels are cosmopolitan, except for the Australian region, Madagascar, southern South America (Patagonia, Chile, most of Argentina), the polar regions, certain desert regions (such as Arabia and Egypt), and many oceanic islands.

There are eleven species of squirrels in western Oregon, nine indigenous and two introduced, but nowhere do all nine indigenous species occur together. Where several species do occur together, their behavior is very different. Here is an example.

In zone 1, there are five species of squirrels representing five genera: northern flying squirrel; chickaree; western gray squirrel; Townsend chipmunk; and Beechey ground squirrel.

•The northern flying squirrel is associated primarily with the Douglas fir forest and is active at night, so does not interact with any of the other four squirrels, which are active during the day. It nests in the tree tops, feeds on truffles in season, and lichens in the forest canopy in winter. It is active all year.

•The chickaree is also associated with the Douglas fir forest, where it, too, nests in the tree tops and feeds on truffles in season, but in winter it feeds on the seeds of coniferous cones, which it has cut and stored in autumn. It is also active all year.

•The western gray squirrel, like the chickaree, nests in the tops of trees and is active all year, but it is associated primarily with the hardwood forests of Oregon white oak, tanoak, canyon live oak, bigleaf maple, golden chinkapin, and California-laurel. In addition to truffles, it feeds heavily on acorns and other nuts.

•The Townsend chipmunk, although a forest dweller, lives primarily in belowground burrows, where it may sleep throughout part of the winter. These chipmunks are primarily terrestrial in their daily activities. They, too, feed on truffles but include the seeds of grasses and various berries in their diet as well.

•The Beechey ground squirrel prefers meadows and grassy areas along the edge of the forest, where it lives in belowground burrows, eats grasses and forbs, and hibernates during winter. Although it does climb into trees, it does so only to sun itself and as a lookout point.

Squirrels are represented in western Oregon by six genera.

Chipmunks
Tamias

The generic name *Tamias* is derived from the Greek word *tamias* ("storer," "distributor").

The length of the head and body of chipmunks ranges from $3^1/4$ to $6^3/8$ inches and their weight from almost 1 ounce to $4^1/2$ ounces. These chipmunks characteristically have nine longitudinal stripes on their backs, four dark and five light, originating on the head in the region of the nose and eyes and continuing along the back to the region of the tail. Chipmunks have internal cheek pouches in which they carry food.

Chipmunks are easy to identify as chipmunks when a person sees them or has in hand a scientific study skin and skull, but they are difficult to identify in the regurgitated pellets of owls or the droppings of carnivores because chipmunks' molars are similar in appearance to those of the smaller Oregon squirrels. Chipmunks' incisors differ, however, from those of all other squirrels in western Oregon, having fine, but definite, CHICKAREES longitudinal striations running the full length of the outer pigmented surfaces of both the upper and lower incisors. Where the Townsend chipmunk and yellow-pine chipmunk occur together, they can easily be determined to genus; where they occur singly, they can readily be identified to species, whether eaten by owls or carnivores.

Chipmunks live primarily on the ground but are good climbers and spend some time in shrubs and trees. They seek refuge in their burrows, in hollow fallen trees and rock crevices, and occasionally in bushes and trees. Food is stored primarily in their underground burrows. They retire into their burrows during winter and become torpid, but arouse from time to time and feed on their caches of stored food. In this respect, they are unlike ground squirrels, which hibernate and survive off their accumulated stores of body fat.

Chipmunks occupy a variety of habitats from northern spruce to fir, redwood, and pine forests, and from humid coastal forests to shrub-covered mountains and sagebrush plains. They eat a correspondingly wide variety of foods, including seeds, fruits, bulbs, fungi, insects, and some birds' eggs.

Females produce a single litter annually of four to eight young.

All chipmunks of the genus *Tamias* inhabit North America from the central Yukon and southern Mackenzie drainage basins of Canada south to the Baja Peninsula and the Mexican states of Durango and Sonora.

There are two species of *Tamias* in western Oregon.

Townsend chipmunk
Tamias townsendi

THE SPECIES WAS named in honor of John Kirk Townsend, who was in charge of the hospital at Fort Vancouver during the winter of 1835-36 and collected the first specimen near the lower mouth of the Willamette River, Multnomah County, Oregon, in 1834.

DISTRIBUTION AND DESCRIPTION

total length 221-275 mm
tail 91-131 mm
weight 51-109 g

THE TOWNSEND CHIPMUNK occurs wherever there is suitable habitat throughout zones 1 through 5. The largest chipmunk in Oregon, it ranges in total length from $8^3/4$ to 11 inches, and its tail ranges from $3^1/4$ to 5 inches in length. It weighs from almost 2 to almost 4 ounces.

The Townsend chipmunk has a moderately long, lax pelage and rather diffuse coloration. Summer pelage is lighter, brighter, and more contrasting than is the winter pelage.

There are three black stripes on the back. The middle one extends from between the ears almost to the base of the tail; the outer two extend from about the shoulders to the rump. There is an additional, short, dark brown strip along each side of the back, extending from behind the shoulders almost to the rump. Between the dark stripes are four lighter stripes varying from light brown, to yellowish brown, to whitish. The stripes on the sides of the head are lighter than those on the back.

The sides, below the short, brown stripes, vary from brown to slightly yellowish brown. The long, rather bushy tail is blackish above, with many white-tipped hairs; the underside is bright reddish brown with a black margin and a frosted edge of white-tipped hairs.

Some authorities now separate the Townsend chipmunk into three species, *T. townsendi*, *T. senex*, and *T. siskiyou*.

THE TOWNSEND CHIPMUNK is primarily a denizen of wooded areas, but may inhabit brushy areas where forest or woodland has been converted to an earlier developmental stage. Active from dawn until dusk, these chipmunks, unlike the other, smaller chipmunks, are shy, wary little squirrels that are normally heard rather than seen.

HABITAT AND BEHAVIOR

Their calls are quiet, birdlike, and often muted by the rank vegetation in which they tend to live; thus most people do not even realize they have heard a mammal. The alarm call can be simulated by forming a rigid O-shape with the lips, and then, placing the little finger inside the mouth so that the back of the hand is toward the face, sliding the finger out of the mouth, all the while pressing firmly against the taut cheek. The sound produced is high and crisp and may be written "po," with the sharpest accent on the beginning of the sound. The calls usually emanate from shadowy undergrowth in which the diffuse coloration of the squirrels blends with the varying light of the surroundings. Townsend chipmunks tend to freeze or to crouch when they hear a sound or see a movement.

Although they live primarily in burrows, Townsend chipmunks are expert climbers. They often forage, hide, or sun themselves in bushes and trees. When startled or chased, they

The preeminent naturalist Vernon Bailey writes with apparent affection about these little squirrels in his book **The Mammals and Life Zones of Oregon**:

"Their bright colors are often seen flashing through the leaves and bushes and their shrill chipper of alarm or curiosity is heard along the trails. Sometimes a slow, soft chuck, chuck, chuck is heard from a distance—as one sits on a stump, log, or low tree branch—a chipmunk calling quietly to his friends far and near. Their voices have many degrees of pitch, time, and quality that may mean much to members of their own clan, but little to uninitiated ears."

usually dash quietly for cover, but are just as apt to scurry up a tree, where they are difficult to locate. These silent, graceful, softly colored denizens of forest and woodland are generally not seen by the average person.

In areas blanketed with deep snow, such as the high elevations of the Coast Ranges and the Cascade Mountains, these squirrels normally accumulate body fat and remain in their winter burrows until released by the spring sun. Coastal Oregon, however, seldom has much snow, so these chipmunks are active aboveground throughout much of the winter, remaining in their nests only during the worst winter storms and not, as far as I could determine, accumulating body fat.

In summer and autumn, Townsend chipmunks eat a variety of fruits, especially berries. In late autumn, depending on their habitat, they eat mainly acorns of tanoak, seeds of bigleaf maple, California laurel, thistles, grasses, Douglas fir, hemlock, spruce, and occasionally western red cedar. Their diet during autumn, winter (along the coast), and spring is augmented with the fruiting bodies of subterranean mycorrhizal fungi, which they detect by odor and dig out of the soil. Along the coast, their diet is also augmented in winter with the semidried evergreen huckleberries that remain attached to the bushes throughout most of the winter despite frequent and heavy rains. In addition to vegetable foods, the chipmunks also consume insects, primarily beetles.

The testes of Townsend chipmunks begin to enlarge toward the end of January at elevations that are free of snow. By March, nearly all males are in breeding condition; by July, their testes have decreased in size and appear sexually inactive. Of course, the onset of the breeding season is later at high elevations. Breeding begins in March and apparently lasts into May. Females produce a single litter per year. Families may include from two to six young but usually consist of four to six offspring. Young are born in May and June but are not seen out of their nests until July or August. Even then, they are nursed until mid or late July or early August.

The newborn are naked, blind, and toothless; they weigh about one-tenth of an ounce and are between 2 and $2^1/_2$ inches long. Their loose skin is so translucent that milk can be seen in the stomachs of recently fed babies. When the young are 20 to 22 days old, the lower incisors erupt; the upper incisors erupt in 28 or 29 days. The canal leading to the inner ear opens on the 24th or 25th day, and the eyes normally open between the 27th and 29th day. The pelage is bright and fuzzy with markings more distinct that those of the adults. The young can eat solid food at about 39 days of age but apparently do not mature sexually until their second summer. One wild Townsend chipmunk is known to have lived for seven years.

The main predators of Townsend chipmunks are long-tailed weasels, mink, and probably marten; however, spotted skunks are also quite capable of capturing them. Bobcats catch a surprising number of these small squirrels, and domestic cats take a considerable toll. Townsend chipmunks are occasionally captured by owls that hunt during the late afternoon and early evening, such as the Great Horned Owl.

Yellow-pine chipmunk
Tamias amoenus

THE SPECIFIC NAME *amoenus* is the Latin word for "pleasing" or "lovely." The first specimen of this bright little chipmunk was captured at Fort Klamath, Klamath County, Oregon, by J.H. Merrill in 1887.

DISTRIBUTION AND DESCRIPTION

THE YELLOW-PINE CHIPMUNK occurs in zones 3 through 5. This little squirrel ranges from $7^3/4$ to $8^3/4$ inches long, and its tail ranges from $3^3/8$ to 4 inches in length. It is a small, richly colored chipmunk with a slender tail, pointed face and ears, and nine stripes on its back. In summer, the coloration is bright. There are five black, two gray, and two white stripes on the back and three dark and two white stripes on the sides of the head. The sides, shoulders, and lower surface of the tail are a rich reddish or orangish brown. The top and margins of the tail are black, overlaid with tannish. The backs of the ears are mainly black, and the underside is whitish. In winter, the pelage is slightly duller and more grayish.

total length 195-220 mm
tail 85-100 mm

HABITAT AND BEHAVIOR

THE YELLOW-PINE is the second most widely distributed chipmunk in western North America. In central and eastern Oregon, its habitat ranges from the subalpine forests east of the crest of the High Cascade Mountains, throughout the mixed coniferous forests, ponderosa pine forests, and western juniper woodlands. To many people from western Oregon, yellow-pine chipmunks, as the name suggests, are often associated with the ponderosa pine (often referred to as "yellow" pine) forests on the eastern flank of the High Cascades. In zone 3, however, they are associated with oak woodlands as well as mixed coniferous/deciduous forests. In zone 4, they are found around rocky outcrops and lava fields at high elevations. In zone 5, they are associated with lodgepole pine, going up into the high-elevation habitat of white-barked pine at timberline on such peaks as Washington, Jefferson, and Hood.

Although expert climbers, often seen running about in trees, these delightful little squirrels are generally seen running over fallen trees and rocks or climbing in bushes. Their homes are usually burrows, hollows in fallen trees, clefts and crevices in rocky outcrops, rockslides, or lava fields. Yellow-pine chipmunks are adventuresome, often exploring far from their home bases, though all the while keeping close to cover. They often get bolder, however, when they sense the chance of getting food around someone's picnic or camp table.

Yellow-pine chipmunks have been reported to be omnivorous. A wide variety of food items are given, mostly from observations of their foraging behavior: seeds (generally given as the most important item), fruits, bulbs or tubers, insects, birds' eggs, berries, flowers, green foliage, fungi, roots, small animal life, and so on. But analyses of their stomach contents and fecal pellets reveal that, in addition to these foods, the fruiting bodies of belowground mycorrhizal fungi, called truffles, are very important in their diets.

Truffles often fruit in or under decomposing, fallen trees that retain moisture long into the summer drought period under a closed forest canopy. This moisture retention may extend the truffles' fruiting season, benefiting both the fungus and the yellow-pine chipmunk, in whose diet it is an important component. Truffles could be an important source of nutrients and moisture during July when succulent vegetation in the chipmunk's drier habitats has desiccated and crops of grass seed and berries are not yet abundant. A yellow-pine chipmunk probably needs only a single truffle per meal, but it is possible for a foraging animal to find up to nine in one "dig" because they frequently fruit gregariously and thus are probably energy efficient in terms of the time required for foraging by the yellow-pine chipmunk.

Yellow-pine chipmunks hibernate during winter. But as the deep snows begin to melt, these little squirrels tunnel up through the snow and begin to explore their world, and the males are soon searching for a mate.

The young, usually four or five, are born in May, June, or July. The variation in dates may mean two litters of young in a single season or irregular breeding of females of different ages. When the young appear aboveground, they are about half grown and can pretty well take care of themselves.

The main enemies of yellow-pine chipmunks are most likely weasels, marten, bobcats, hawks, and owls.

Ground Squirrels
Spermophilus

THE GENERIC NAME *Spermophilus* is derived from the Greek words *sperma* ("seed") and *phileo* ("to love").

Ground squirrels vary from small (about 3 ounces) to large (over 2 pounds). They have four basic markings. The most common is yellowish gray to gray that is darker above, with a few to many spots. The second is similar to the first, but the spots are faint or absent. The third marking is two dark, longitudinal lines separated by a light stripe along each side of the back, extending from the shoulders to the hips. The fourth consists of light, longitudinal stripes beginning at or on the head and extending to the base of the tail, separated by longitudinal rows of spots. Pelages of ground squirrels vary from coarse and thin to fine, soft, and thick.

Ground squirrels are well adapted to digging and primarily live in underground burrows. Their burrows may be located in meadows and prairies; around fallen trees, stumps, or live trees; or among piles of rocks, rock outcroppings, or cliffs.

They eat a wide variety of foods, such as seeds, nuts, fruits, roots, bulbs, stems and leaves of herbaceous plants, small mammals, carrion, birds, and birds' eggs. Some foods are transported in cheek pouches located inside each cheek. These squirrels eat incessantly throughout the spring and summer, accumulating great stores of body fat. In late summer, squirrels in the northern area retire to their burrows and do not re-emerge until the next spring. Ground squirrels in regions south of Oregon, are more or less active throughout the year, but they remain in their burrows during inclement weather or when the supply of green food disappears.

Females have a single litter per year. Gestation periods range from 23 to 30 days, and litters tend to be large—up to fifteen. Babies are born naked, blind, and toothless. Although not strictly gregarious, ground squirrels tend to live in loose colonies.

Ground squirrels occupy North America from Ohio to the Pacific Ocean and from northern Alaska and Canada to west-central Mexico. They also inhabit eastern Europe and Asia south to Turkestan and western Mongolia.

There are three species of ground squirrels in western Oregon.

Beechey ground squirrel
Spermophilus beecheyi

THE SPECIES WAS named in honor of Captain Frederick W. Beechey, Rear Admiral of the British Navy, who sailed the British naval ship Blossom along the west coast of Mexico and California in 1828. The first scientific specimen was collected by David Douglas in 1825 along the banks of the Columbia River in Oregon.

DISTRIBUTION AND DESCRIPTION

THE BEECHEY GROUND squirrel occurs in zones 1, 2, and 3, although it may enter zone 4 along the eastern edge of the Willamette and Rogue valleys and the western foothills of the Western Cascades. These ground squirrels range in total length from

total length 370-500
mm
tail 145-227 mm
weight 280-738 g

about 15 to 20 inches and weigh from 10 ounces to 1 pound, 10 ounces. They have a heavy body with a coarse, short pelage.

In summer, the upper parts are dark brownish gray, mottled and scalloped with small whitish spots and wavy, black crosslines. There is a dark brown or black V-shaped area on the back, starting at the level of the ears and becoming broader until it ends at the middle of the back. The sides of the back bordering the V are light gray. The underside is tan. The dark gray tail is moderately bushy and has three concealed black and gray or black and tan bands along each side and around the tip. The squirrel's markings may be inconspicuous in summer when it lives in a dusty habitat. The winter pelage is grayer than that of summer, and the markings are less conspicuous. The claws are dark gray, long, slightly curved, and well adapted for digging.

HABITAT AND BEHAVIOR

BEECHEY GROUND SQUIRRELS (also called "gray diggers") live primarily in open areas, especially open grassy areas, such as woodland savannah. They were once common along the vegetated fencerows that separated agricultural fields. Today, with bigger fields and fewer fencerows, there is far less habitat for these squirrels.

Beechey ground squirrels begin their daily activities at dawn and retire to their underground burrows at sunset. Although they are good climbers, they seek their burrows when danger

---

I became acquainted with the Beechey ground squirrel when I was about eight years old. There was an old, red-brick pumphouse on the golf course near the ditch, where my friend, Billy Savage, and I played. A large, male Beechey ground squirrel—we called him a gray digger—lived in and under the pumphouse, and that was enough to entice us cautiously inside despite the pump's dreadful noise. After all, this wasn't just any gray digger, it was "The" gray digger.

To us, it was "The" gray digger, because in the summer of my eighth year I had managed to corner him in the pumphouse and had tried to catch him by grabbing his tail, only to learn that the skin of a gray digger's tail is very loosely attached. As he whirled to bite me, the skin of the end of his tail came off in my hand. I felt sick with self-reproach for what I had done, but the gray digger—though not without pain—sustained no apparent long-term, debilitating damage. As the bare portion of his tail dried, he chewed it off, leaving him with a stumpy appendage by which we could always recognize him as "The" gray digger, for whom we had enormous respect.

---

threatens, even if they have to leave the safety of a tree to do so. Some construct their burrows under trees and shrubs; others dig them deep into open hillsides where they are hidden by tall grasses.

Like other ground squirrels, these eat as much as they can. Beginning in spring, they stuff themselves with succulent greens, and in late summer and early autumn are still gorging on ripe berries, mature seeds, apples, and nuts, so that by the end of September they have copious amounts of body fat to nourish them through hibernation.

Most breeding seems to occur within the month of March. In Oregon there is an interesting correlation between latitude and the average litter size. In northwestern Oregon the average size of a litter is five, in southwestern Oregon it is six, and in southern California it is eight. One explanation that makes sense to me is that the warmer temperatures and longer growing season in the southern portions of the squirrels' geographical distribution allows them to spend more of the year outside their burrows, thereby increasing predation. Larger average size of litters may be a response to this greater pressure from predation.

Photo 43. Track of Beechey ground squirrel in mud; note the marks of the long digging claws. (Photograph by Chris Maser)

After a gestation period of 25 to 30 days, the young are born hairless, wrinkled, and red. Their eyes are closed and their coordination is poor. At birth they weigh about half an ounce. At two weeks, they are covered with fuzz, their whiskers are beginning to appear, and they weigh almost an ounce. Their eyes are not quite open at four weeks; they can raise their heads off the ground but not their bodies. They weigh a little over an ounce and can shuffle around. At the end of five weeks, some babies have their eyes open and weigh a little over 1 1/2 ounces. By the end of six weeks, the young squirrels weigh almost two ounces, are quite active, and can support themselves on their feet and legs in an upright position. At eight weeks, they weigh over 2 ounces, are about 8 inches long, and are active, leaving the nest frequently.

They begin leaving their burrows in early June (when about eight weeks old) but undoubtedly nurse for a time while becoming adjusted to solid food. The young can be distinguished from their parents when they emerge by their smaller size and lack of wariness. By autumn, however, one can no long tell them apart.

Beechey ground squirrels usually retire to winter quarters towards the end of October, as the last of the apples fall reluctantly to the ground. Occasionally one appears suddenly during a warm spell in December or January (photo 43), only to vanish again with the onset of inclement weather, not to reappear until sometime in mid-February.

Belding ground squirrel
Spermophilus beldingi

DISTRIBUTION AND DESCRIPTION

total length 275 mm
weight 283 g

THIS SQUIRREL IS NAMED in honor of naturalist Lyman Belding. The first scientific specimen was collected on December 28, 1888, at Donner, Placer County, California.

THE BELDING GROUND squirrel (photo 44) occurs in zone 5 in only one place that I know of—the meadow below Three Creek Lake on the north flank of the mountain called Broken Top just south and east of Sisters in Deschutes County. These squirrels are about 11 inches long and weigh about 10 ounces, although a large male with a good accumulation of autumnal fat can weight about a pound.

Belding ground squirrels have upper parts that are generally somewhat yellowish gray, becoming brownish gray on the nose and down the middle of the back. The tail is brownish gray above and distinctly reddish brown below, with a black tip and margins.

HABITAT AND BEHAVIOR

THESE MEDIUM-SIZED GROUND squirrels are abundant east of the High Cascade Mountains, where they primarily live in meadows and other moist grassy areas. Belding ground squirrels live on and under the surface of the ground. They depend on their burrows for protection, nesting, rearing young, and hibernation.

Photo 44. Adult Belding ground squirrel. (Oregon Department of Fish and Wildlife photograph.)

As you look out over the meadow below Three Creek Lake, you may at one moment see all the squirrels standing up, looking like so many stakes or bowling pins in a meadow, and the next they are nowhere to be seen because a few short, sharp, high-pitched whistles have sent them scurrying into their burrows, which are many and always close at hand. After a time, heads begin poking out of the burrows, one at a time, checking carefully to see if the perceived danger has passed before the inhabitants come out again to feed.

They make well-used runways, almost little highways, through the vegetation of the meadow as they run from place to place, always mindful of danger. Fortunately for farmers on the east side of the mountains, these squirrels (who use their fields as their homes) are inactive for seven or eight months of the year—from about mid July until about mid March. When they disappear into their burrows for the year depends on when the summer sun becomes hot enough to dry the vegetation, and when they appear again depends on when the sun is warm enough to start the green vegetation of spring growing.

The beginning of estivation (which grades into hibernation) varies with season, altitude, and age of the animals. At the meadow below Three Creek Lake, for example, which is around 6,000 feet in elevation, the squirrels come out about a month later than they do at lower elevations and stay active about a month longer. Old males, with copious amounts of body fat, which they accumulate earlier than the other squirrels, disappear first. The last to go below ground are the young of the year, because they must put on enough body fat to see them through their dormant periods; this may keep them above

Photo 45. Baby Belding ground squirrel. (Photograph by Chris Maser)

ground until mid-August. Even then, it may be touch and go for those born late in the year.

The time of breeding varies with the time of emergence from hibernation, but begins soon after the squirrels awake. Although six or seven seems to be the usual number of young in a litter, there may be as many as twelve or even fifteen (photo 45). A single litter is probably the norm because there would seem to be little time for a second.

The young usually begin appearing in the meadow sometime in early June. At first, they stick close to their burrows, but with time and experience, they become brave enough to explore their meadow surroundings. By August, when those born late are scarcely fully grown, it is time to retire for the year; whether they survive the winter, I do not know. In this case, the old notion of the survival of the fittest may in fact be the survival of the fattest.

Belding ground squirrels begin eating the moment they emerge in spring and do not stop until they go below in late summer. They seem to eat anything that is succulent and or green, such as clover, grasses, forbs, roots, bulbs, along with some bright meadow flowers and berries. With the advance of summer and ripening vegetation, seeds become more important in the diet. As the end of the summer glut approaches, these squirrels appear to waddle down their trails as they prepare for their long sleep. The only food they store is in the form of body fat.

Their enemies in the meadow are hawks, long-tailed weasels, red fox, coyotes, and bobcats. In other places within their geographical distribution, farmers wage a seemingly constant war against these squirrels, which, because of their often large numbers and incessant digging and eating, can do considerable damage to crops, such as wheat and alfalfa. Many Belding ground squirrels are at times killed by passing automobiles because those still living then put themselves in jeopardy by going into the highway to eat their dead relatives.

Mantled ground squirrel

Spermophilus lateralis

THE SPECIFIC NAME *lateralis* is derived from the Latin word *lateris* ("side" or "flank") and the Latin suffix *alis*, which is added to noun stems to form adjectives meaning "pertaining to." Here the reference is to the bold, clear stripes on the squirrel's sides. The first specimen of the mantled ground squirrel was collected at Fort Klamath, Klamath County, Oregon, by Samuel Parker in 1888.

Mammals of the Pacific Northwest

Mantled ground squirrels occur in zones 3 through 5 (photo 46). They range in length from 10 to 11¹/₄ inches and weigh from almost 5 to 8¹/₂ ounces.

DISTRIBUTION AND DESCRIPTION

total length 250-282.5 mm
weight 134-240.5 g

Mantled ground squirrels are larger and heavier than chipmunks. Their ears are prominent, their tail is moderately bushy, and their feet are large with naked soles and palms. Their internal cheek pouches are of ample size, as can be attested by anyone who has ever seen one of these brightly colored ground squirrels stuff them full of food.

Each side of the back sports three stripes, two black and one white or tannish, which, unlike those of the chipmunks, are confined to the side. The rest of the upper parts, as well as the underside of the rather bushy tail, is bright orangish brown. The top of the head is reddish brown, and the underparts, as well as the tops of the feet, are a somewhat pinkish-yellowish gray or dull orangish brown. The upper surface of the tail is washed with a pale orangish brown. The young are as bright as the adults in summer.

Winter pelages have the same black and whitish stripes but without the bright mantle around the shoulders and with a general coloration duller and grayer than that of summer. The underside of the tail, however, retains its bright color.

Photo 46. Adult mantled ground squirrel. (Photograph by Chris Maser)

THE MANTLED GROUND squirrel in western Oregon has much the same distribution as that of the yellow-pine chipmunk, with whom it shares the habitat. In zone 3, they are associated with oak woodlands and mixed coniferous/deciduous forests. In zone 4, they are found around rocky outcrops and lava fields at high elevations. In zone 5, they are associated with lodgepole pine, going up into the high-elevation habitat of white-barked pine at timberline on such peaks as Washington, Jefferson, and Hood.

HABITAT AND BEHAVIOR

Photo 47. Mantled ground squirrel with a mouth full of nesting material. (Photograph by Chris Maser and Rita St. Louis.)

In habits, as well as physical structure, mantled ground squirrels appear intermediate between arboreal squirrels and true ground squirrels. Although not good climbers, they will seek refuge in a tree if no other escape is possible. The fact that they live in underground burrows deters them not at all from climbing on, running over, or sitting on fallen trees, tree stumps, boulders, fence posts, and the like, from where they can obtain an unobstructed view by sitting motionless yet alert. These lookouts appear to be an important part of their daily life.

Not as quick and "nervous" as the yellow-pine chipmunk, mantled ground squirrels nevertheless are quick to seek their burrows if they detect danger. The usually single shrill whistle of a mantled ground squirrel is seldom heard, and when it is, it is difficult to locate.

Mantled ground squirrels normally dig their burrows under fallen trees, stumps, or rocks or find suitable shelter in lava fields and rockslides. Having excavated their nest chamber, they collect dry plant material with which to line the cavity and create a cozy nest (photo 47).

In autumn, usually by mid September or early October, depending on elevation and weather, the now obese squirrels enter their dens and settle in for their winter's hibernation. They normally reappear some time in May.

These voracious, potbellied squirrels seem to be omnivorous in the extreme, and around campgrounds quickly become fond of human junk food, such as Cheetos and potato chips, to say nothing of peanuts. They seem to literally eat almost anything they can find. I have seen them stuff their cheek pouches so full of peanuts that one would inadvertently pop out and be immediately stuffed in again, lest a colleague abscond with it.

In spring, much green vegetation is eaten as soon at it appears, and roots and old seeds are excavated. Later, as berries ripen, they are eagerly sought, eaten with abandon, and avidly stored. As berries wane, seeds of grasses and flowering plants, as well as nuts, and in some places acorns, chinquapins, and hazelnuts are sought. The squirrels also eat meat when it is available.

They eat copious amounts of truffles in spring and autumn, the live spores of which they then spread around the forest in their feces. They also plant pine trees. I have often watched a mantled ground squirrel collect pine seeds in autumn and bury them, forgetting almost instantly where they are secreted. Going back to some of the same locations in the late spring, I have seen small clumps of germinating pine seedlings, which may explain why—at least in the old days prior to contemporary extensive logging and the cumulative effects of fire suppression—so many tight clumps of three, four, five, and sometimes even six old ponderosa pines were seen growing out of the same place.

Mantled ground squirrels begin mating as soon as they emerge from winter quarters, and their young are born in late June or early July. Litters range from four to eight, but four to six is the usual number. The young normally begin to appear aboveground when a quarter or a third grown, late in July or early August, depending on altitude. It seems that only one litter is born per year, which makes sense because the young have scant time to grow up and store enough body fat to see them through their first winter's hibernation. As it is, many of those born late in the season, neither fully grown nor very fat when winter shuts them into their dens, will emerge in spring—if they survived the long sleep—still sexually immature.

Among their predators, we can list long-tailed weasels, marten, bobcats, and black bear, which occasionally dig out nests, as well as hawks, and early hunting owls.

Marmots
Marmota

THE GENERIC NAME *Marmota* is derived from the French word *marmotte*, literally, a "mountain mouse."

Marmots range in the length of their heads and bodies from 12 to 24 inches, and their tails from 4 to 10 inches in length. They weigh from about $6^1/_2$ to $16^1/_2$ pounds.

The color of the fur varies considerably among species. The upper parts range from bright brownish yellow, dark brown, or reddish brown, to a mixture of hairs that are part black and white. The under parts of some species differ little in color from the upper parts, but there is considerable variation in others, from whitish to dark brown, yellowish gray, or reddish orange. Unlike the yellow-bellied marmot of western Oregon, which retains its color all summer, the Olympic marmot, of the Olympic Peninsula in the state of Washington, changes color, from brown in spring to yellow by autumn; this is apparently due to bleaching by the sun.

In addition, the texture of the fur ranges from thin to long and thick, with varying degrees of coarseness. Climate probably influences the thickness of the pelage. There is apparently only one molt per year, which occurs in early summer.

Marmots are active during the day and retire into their burrows at twilight. Their burrows are dug either in well-drained soil or among large rocks and frequently have several entrances. Some species live in colonies but others are more solitary.

Marmots are downright obese by the end of summer in preparation for the long fast during their hibernation, which lasts from September until the end of March, even longer in the northern part of their geographical distribution, where it can last up to eight months.

Although they are most active on the ground, marmots occasionally climb into shrubs and trees. They also sit upright on their haunches to watch the surrounding area for danger. If danger is spotted, a sharp whistle is given as an alarm.

Their diet consists of herbaceous vegetation and occasionally insects.

They mate in early spring, soon after emerging from hibernation. The gestation period ranges from 35 to 42 days, after which two to nine young are born from April to July in a grass-lined underground den. Their eyes open within about a month, soon after which they venture forth to feed. Some are fully grown in two years and sexually mature in three. The life span is from thirteen to fifteen years.

There are about sixteen species of marmots with a combined geographical distribution that includes the mountains of Alaska, most of Canada, most of the United States except the extreme southern parts, western Europe, and most of Asia except the extreme south. Marmots usually occur only at high elevation in the southern parts of their geographical distribution, but elsewhere they frequent lowlands as well.

There is one species of marmot in western Oregon.

Yellow-bellied marmot
Marmota flaviventris

THE SPECIFIC NAME *flaviventris* is derived from the Latin words *flavus* ("yellow") and *venter* ("belly"), referring to the species' yellowish undersides. The first specimen of the yellow-bellied marmot was collected by David Douglas, probably on Mt. Hood (as determined by A. Brazier Howell in 1915), although neither date nor locality are given. The species was described scientifically in 1841.

THE YELLOW-BELLIED MARMOT irregularly inhabits zone 5, but comes just over the crest of the High Cascade Mountains from the east to take up occupancy wherever there are extensive masses of rock, particularly around the high peaks, such as near timberline on Mt. Washington, where I found their sign in July 1997 (photo 48).

DISTRIBUTION AND DESCRIPTION

total length 650-700 mm
weight 2.3-3.6 kg

These marmots are the largest ground squirrels in Oregon. They range in length from about 26 to 28 inches and weigh anywhere from 5 to 8 or perhaps more pounds. Because they are both heavy and short legged, they have the appearance of being squat. Their tails are short and bushy, and their ears are set low on their heads. The long underfur is concealed by coarse outer guard hairs.

The rich color of the pelage is about the same at all ages and seasons. The upper parts are dark brown, coarsely grizzled over the back and sides with somewhat pinkish white subterminal sections of the stout outer guard hairs. The tail is reddish brown fading to yellowish. The side of the neck and hips are bright pinkish yellow, whereas the legs, feet, and underparts are light brown to light orangish brown. The nose and chin are whitish, and there sometimes is a whitish or grizzled band across the face in front of the eyes.

YELLOW-BELLIED MARMOTS, with their heavy, squat bodies and sedentary ways, are slow and physically rather defenseless. So they need solid protective areas as retreats in times of danger, which they find among the crevices, small caves, and ledges of cliffs and the deep labyrinths of great lava fields and rockslides, called talus. These features are irregularly distributed in the High Cascade Mountains because of the nonuniform nature of the various geological episodes that created them. The marmots' occurrence in zone 5 is correspondingly erratic and unpredictable.

HABITAT AND BEHAVIOR

Large boulders or ledges on which to lie, or massive fields of sharp lava, or the immense slopes of huge broken and angular rocks tossed one upon another at the base of a high cliff—these are the favorite haunts of a family of marmots. From here, three or four venture out along the edge of safety to forage in the succulent growth of spring while one or two lie atop some high point and act as lookouts, giving their sharp, piercing whistles at the first hint of danger.

Although such whistles need be given only once or twice to send family members scampering for cover as fast as their short legs and flopping tails will allow, sometimes the whole area seems to come alive with whistles. Some marmots bolt headlong into the safety of their rocky homes; others pause in the entrance to see if danger is imminent. All the while the sentinels keep up their warning calls until either the danger has passed or all the marmots, including the lookouts, have disappeared into the safety of their rocky fortresses.

When possible, marmots dig burrows under or within the protection of their rock strongholds, but this may not always be possible. Nevertheless, there undoubtedly are dry areas even deep with a talus in which a nest of

Photo 48. Droppings of the yellow-bellied marmot. (Photograph by Chris Maser)

One particular marmot many years ago gave me a healthy respect for their ability to dig. I was sixteen years old and working as a ranch hand at an elevation of 7,000 feet in northwestern Colorado. The ranch on which I worked had a hay field bordered by a rather high cliff on which several "whistle pigs," as we called them, had their residences—that is, all but one. That particular marmot had dug its tunnel in the edge of the hay field. Seeing this, the rancher told me to get rid of the marmot; he didn't care how.

With those marching orders, I watched the marmot for about a week, doing my best to learn its habits. Discovering that it seemed to feel fairly safe moving about in the tall grasses of the hay field, I waited until it was as far removed from its burrow as I thought it would get, before I slipped quietly up to its burrow and plugged the entrance with a large rock, which I thought would dissuade the marmot from living there.

The next step of my plan was to chase the marmot back to the cliff, where I thought it belonged. Being a little intimidated by such a stout animal, I armed myself with my trusty shovel, the one with which I dug post holes. As I approached the marmot, it headed straight for its well-plugged burrow, but came up short at the rock sealing the entrance. From here, I expected it to run for the cliff, with me after it for good measure, and join its kin, but it had another idea. It ran a few yards from its now-useless burrow and began digging a new one even deeper in the hay field than before.

I stood momentarily aghast as it rapidly dug itself almost out of sight in the moist soil. Recovering, I leapt into action and began digging after it as fast as I could, thinking I could out-dig it and so persuade it to emigrate to the cliff. But no matter how fast I dug, I never got any closer to that marmot than the tip of its tail. The marmot was still digging when I, totally exhausted, gave up.

cured vegetation can be made and kept safe from moisture dripping through the rocks as rain falls and snow melts.

Marmots begin eating from the moment they emerge in the spring until they retire in autumn, either when the vegetation dries out or the first frosts appear. They seem to eat anything that is green and tender, beginning with grasses and clovers and moving on to stonecrops. As the seasons change, their diets consist more of flowers and seeds, either green or ripe, but whatever they eat, it's the best available nutritionally.

Marmots seem to eat continuously. They have enormous stomachs that seem to be always filled to maximum capacity, especially as the year matures and the time of hibernation draws near, usually in late July, August, or September, depending on the weather, the age of the animal, and the available supply of food. Since they are true hibernators and do not actively store food for winter snacks, they must store it in accumulated body fat, which seems to be their primary mission during their aboveground activities—a mission they take seriously. A marmot just entering hibernation is absolutely rotund! Once they are in hibernation, the High Cascade Mountains will not see the marmots again until the following February, March, or even April.

March seems to be the breeding season and May the time of birth. Litters range from four to eight, but the usual size is probably four to six. By June, half-grown youngsters are normally out of their burrows, and by late August, though not yet fully grown, the young of the year are large enough to take care of themselves and to accumulate enough body fat to see them through their first winter.

The main enemies of the yellow-bellied marmot, especially the young, seem to be Golden Eagles, bobcats, and coyotes, although pumas probably take one from time to time.

Gray Squirrels and Fox Squirrels
Sciurus

THE GENERIC NAME *Sciurus* has the same derivation as that of the familial name.

These squirrels are fairly large, ranging in head and body length from 8 to $12^3/4$ inches, with tails from 8 to $12^3/8$ inches long. In weight, they range from 7 ounces to over 2 pounds. Colors of the various species differ greatly. The backs and sides are usually gray, grayish brown, blackish brown, or various shades of red; the undersides range from white, through tan and yellow, to orange. Individuals or whole populations, however, may be much darker, even black.

They inhabit deciduous, coniferous, and both humid and arid tropical forests. Their periods of greatest activity are early morning and late afternoon when they often descend to the ground and forage for food, which consists of nuts, seeds, fruits, buds, shoots of young trees, fungi, insects, birds' eggs, and occasionally young birds. Some nuts are buried instead of eaten; in this way squirrels often plant nut-bearing trees.

Nests are constructed with twigs, leaves, or mosses on limbs or in forks of branches or hollows in trees when available. Although individual squirrels have favorite nest-trees, they also maintain several other nests to escape from enemies. These squirrels can be noisy, especially during territorial disputes.

Some species have two litters per year. After a gestation period of 38 to 44 days, a litter of one to ten is born. Females usually do not breed until their second summer.

Members of this genus occur in most of Europe and Asia, south of the northern limits of tree growth, Japan, and the New World, from southern Canada south, through the United States, Mexico, and Central America, into South America as far as northern Argentina.

There are three species of squirrels in this genus in western Oregon: one indigenous, two introduced.

Western gray squirrel
Sciurus griseus

THE SPECIFIC NAME *griseus* is the Middle Latin word for "gray." The western gray squirrel was scientifically named in 1818 and is based on the descriptions of a squirrel seen by Lewis and Clark at The Dalles, Wasco County, Oregon.

DISTRIBUTION AND DESCRIPTION

total length 445-593 mm
tail 240-310 mm
weight 315-965 g

WESTERN GRAY SQUIRRELS occur in the extreme southern part of zone 1, but move eastward in the vicinity of Port Orford, in Curry County, until they enter zone 2, where they occur primarily along the edges. These beautiful squirrels occur generally in zone 3, and may be found in zones 4 and 5 along the Oregon-California border.

The western gray squirrel, also called "silver-gray," is the second largest squirrel, after the marmot, in western Oregon, ranging in total length from $17^3/4$ to 25 inches. It also has the longest, bushiest tail of any Oregon squirrel, ranging from $9^5/8$ to $12^3/8$ inches. In weight, it ranges from 11 ounces to 2 pounds.

The back is clear, bright gray that appears frosted because of many white-tipped hairs. The backs of the ears are a light reddish brown. The tail hairs are long and have five bands of color on each hair, including those on the tip, but they are usually clear on the underside. The top of the tail appears black with white frosting. The squirrel's underside, except for the tail, is clear white.

WESTERN GRAY SQUIRRELS seem particularly fond of the tanoak and Oregon white oak habitats, although they also take readily to the drier mixed forests in zone 3, which grade into pure ponderosa forest as the squirrels venture eastward through zones 4 and 5 along the Oregon-California border to the eastern flank of the High Cascade Mountains. Although in zone 2 they seem to prefer the Oregon white oak habitat, they do get into some drier areas of Douglas fir on the more southerly exposures along the edges of the Willamette Valley. In recent years, they have also begun to invade towns, such as my home town of Corvallis, from which they were absent when I was a boy.

Western gray squirrels are most active in early morning and again in late afternoon. These gorgeous squirrels are amazingly stealthy in their movements and very clever at vanishing when pursued. Once in a tree, they stretch out on a limb with their plumy tails over their backs and remain motionless. Except for an occasional, hoarse warning bark—"chuff, chuff, chuff"—most often heard in August, September, and October, they are silent.

These squirrels build their large nests on limbs of oak, fir, or pine trees, where they may be situated 60 feet above the ground. The outer portions of the nests are composed of sticks and twigs with leaves attached, or occasionally of mosses and lichens. The inner sleeping quarters are lined with fine lichens, mosses, or the shredded bark of dead bigleaf maple, California laurel, or other broad-leaved trees. During the wet winters, however, they probably use available hollows in trees rather than outside nests. Although active throughout the year, they retire to their nests during the worst winter storms.

Western gray squirrels are adroit climbers, often leaping from one treetop to another with amazing swiftness and dexterity. Walter Dalquest, a naturalist in the state of Washington, wrote in 1948 of a gray squirrel's leaping out of the top of a tall fir tree: "the squirrel leaped far out into the air. Its legs were stretched out stiffly, the tail was extended and the body slightly arched. It struck the ground with an audible thud and bounced fully 18 inches. At the height of its bounce, the squirrel's legs began moving rapidly, and it struck the ground the second time at a full run."

When bounding over the ground, these squirrels are grace and flowing beauty in motion. Their long, plumy tails seem to float out behind them as they cover the ground in long, easy, undulating bounds, which cause people, including me, to stop and admire them.

In Oregon, western gray squirrels eat the seeds of sugar pine, ponderosa pine, Jeffrey pine, and lodgepole pine, and the acorns of Oregon white oak, tanoak, Sadler oak, canyon live oak, and California black oak, as well as hazel nuts and the large nuts of California laurel and golden chinkapin. Truffles are eaten whenever available, as well as green vegetation, such as false dandelions and the tender shoots of coniferous trees. Berries and insects are also included in their diets.

The testes of western gray squirrels begin to enlarge in December and are fully enlarged by mid January. There is one litter per year, which consists of two to five young born from March into June. Babies are born naked and blind in nests as high as 60 feet above the ground. June is the earliest that I have seen young squirrels out of their birth nests. Adults are silent and unobtrusive while the young are being reared, presumably to avoid attracting the attention of potential predators.

Although the bobcat is the only predator in Oregon that I have found to capture western gray squirrels, there are undoubtedly others. People probably take the greatest toll through hunting and control measures to protect nut crops. As the squirrels increasingly enter towns, I see them killed more often by automobiles.

Eastern gray squirrel
Sciurus carolinensis

DISTRIBUTION AND DESCRIPTION

total length 385-500 mm
tail 165-240 mm
weight 396-453 g

HABITAT AND BEHAVIOR

THIS SQUIRREL IS NAMED after the state of Carolina, where the first specimen was caught, with the Latin suffix *ensis*, meaning "belonging to." It was described scientifically in 1788.

THE EASTERN GRAY SQUIRREL has been introduced from the eastern United States into some cities within the Willamette Valley, zone 3. These squirrels range in length from 15 to 20 inches with a tail that varies from $6^5/8$ to about $9^5/8$ inches long. Weights range from 14 ounces to 1 pound.

The upper parts are gray with a reddish wash in summer. The underparts are whitish. The tail is long, somewhat flattened, and bushy with long white-tipped hairs. These squirrels are occasionally all black, a condition called melanism.

THE EASTERN GRAY SQUIRREL was extremely abundant in some of its indigenous eastern habitats a century ago, and reports of one man killing a hundred individuals in a half-day's hunt are not unusual. Although the hordes of squirrels from those days are gone, the species still thrives in considerable numbers wherever suitable habitat exists. In the Willamette Valley of Oregon, however, these squirrels are denizens of the cities into which they have been introduced.

The squirrels are most active in the hours of early morning and late afternoon, when they forage for food, a trait well-known to eastern hunters, who sit motionless in a favorite place and wait for the squirrels to become active. Should a squirrel chance to see a hunter and feel disturbed by the sight, it gives a rasping whicker or a harsh squall, the most characteristic of its various notes.

Eastern gray squirrels are essentially tree dwellers, venturing to the ground only to forage for and bury food. Active throughout the year, they rely on stores of food secreted during summer and autumn, on food eaten by humans, such as peanuts, as well as food put out for birds, which they raid from bird feeders in people's back yards. And some people specifically feed the squirrels. Nuts, such as acorns, are the mainstay of the squirrel's diet. In spring swelling buds are included in their diet, and in summer and autumn, various fruits, such as apples, and berries are added.

Eastern gray squirrels mate during midwinter, and the young are born after a gestation period of about forty days. One to four, but usually two to three, babies are born in large, warm leaf nests or in a sheltered cavity in a tree or even someone's house. Youngsters make their debut into the world outside their natal nests in late April or early May. These squirrels are known to live for more than fourteen years.

Automobiles, domestic dogs and cats, and Great Horned Owls are perhaps the main enemies of eastern gray squirrels in and around the cities of Oregon.

THE SPECIFIC NAME *niger* is the Latin word for "black." The first scientific specimen of the eastern fox squirrel is thought to have come from South Carolina and was described in 1758.

Eastern fox squirrel
Sciurus niger

DISTRIBUTION AND DESCRIPTION

THE FOX SQUIRREL has been introduced from the Eastern United States into some cities within the Willamette Valley, zone 3. The species ranges in length from 19 to $22^{1}/_{2}$ inches with a tail that varies from 9 to $10^{5}/_{8}$ inches long. Individuals weigh about 2 pounds.

The fox squirrel is considerably bigger and heavier than the eastern gray squirrel with which it may share city acres. Its fur is coarse and its tail flattened and well furred with long white-tipped hairs. Although its color pattern is exceeding variable, the top of its head in all phases of color is almost invariably black, while the nose, lips, and ears are usually light yellowish white, a combination that almost always distinguishes the fox squirrel from the eastern gray squirrel. Even the melanistic or

total length 475-562.5 mm
tail 225-265 mm
weight 743-908 g

black phase of the fox squirrel still has the light yellowish white nose, lips, and ears.

Habitat and behavior

THE FOX SQUIRREL is seldom found in the deep woods chosen by the eastern gray squirrel, with whom it shares much of its geographical distribution in the east, preferring the savannah-like stands of oak trees. In the Willamette Valley of Oregon, however, these squirrels are denizens of the cities into which they have been introduced.

The summer home of the fox squirrel is a loosely built shelter of leaves high in a tree, though conspicuous from a long distance away. In autumn, they retire to a cavity in some tree, or, in lieu of a cavity, construct a substantial nest in which limbs, smaller branches, and a goodly quantity of leaves and bark are firmly molded together into a bulky but well-constructed abode that will keep them dry and warm during the harshest storms. These nests may provide a safe haven against wind, rain, sleet, and snow for many years, as well as being a nursery for the young.

The fox squirrel's main diet is nuts, which they often bury to recover in times of need. But this squirrel does not rely on nuts alone; it also eats the ripening seeds of maples and the fruits of tulip trees in summer. Berries and fruits are eaten in season as are the bursting buds of spring, and a variety of fungi.

Prenuptial chases may occur as early as late December, and mating may occur throughout January. A second peak of breeding may take place in May and June. Gestation is about 45 days. One to six, usually two to four, young are born, usually in late February and early March and again in June and July, although pregnant females have been found in every month of the year. Females born in spring tend to give birth in spring, whereas those born in autumn tend to give birth in autumn, and two-year-old females are capable of producing two litters per year.

As with other tree-dwelling squirrels, development of the young is slow. The babies are born naked and blind. Their eyes open at the end of the fourth week, and they remain in the nest for seven to eight weeks before venturing outside and eleven to twelve weeks before descending to the ground and beginning to fend for themselves. Fox squirrels are known to live over seven years.

Automobiles, domestic dogs and cats, and Great Horned Owls are perhaps the main enemies of fox squirrels in and around the cities of Oregon.

THE GENERIC NAME *Tamiasciurus* is derived from the Greek words *tamias* ("storer," "distributor") combined with *skia* ("shadow") and *oura* ("tail"). The name alludes to the shade cast when a squirrel holds its bushy tail over its back and to the fact that this genus is noted for its storage of excessive quantities of winter food.

Red squirrels and chickarees, often called pine or timber squirrels, are relatively small for North American tree-dwelling squirrels. In length of head and body, they range from about $6^1/2$ to 9 inches, and their tails are from $3^1/2$ to 6 inches long. Their weight ranges from 5 to 11 ounces. Dorsally, they vary from slightly yellowish brown or orangish brown, to rich brown, to reddish brown, to an almost reddish green-brown. Ventrally, they are white, through yellowish or orange, with a longitudinal blackish line on each side separating the dorsal and ventral coloration.

These attractive little squirrels are noisy, alert, and active. Chickarees (called Douglas squirrels by some) are less vociferous than red squirrels and can be overlooked. Each individual

Red Squirrels and Chickarees *Tamiasciurus*

Photo 49. Burrow of red squirrel in its kitchen midden, where the squirrels spends the worst winter weather in a safe, warm, belowground nest. (Photograph by Chris Maser)

Photo 50. The kitchen midden of a red squirrel in the middle of which are stored the cones of lodgepole pine. (Photograph by Chris Maser)

maintains several nests of four basic types: a loosely constructed nest in a tree for use during the summer; a hollow in a tree used during the winter; a weather-tight nest in the dense foliage of a tree—an alternate winter nest; and an underground nest, usually in the main midden pile, which is used during severe winter storms or prolonged periods of exceedingly cold weather (photo 49). (A midden is the refuse pile of accumulated, uneaten scales and cores of the cones on which the squirrel has been dining, photo 50.)

Red squirrels and chickarees eat a variety of foods, such as seeds from cones, truffles, nuts, fruits, the tender shoots of coniferous trees, and occasionally birds' eggs and young birds. These squirrels are notorious cone cutters. The cones, cut while still green, are stored for winter food—a necessity, since the squirrels are active throughout the year.

Except for mating, these squirrels are solitary in their habits. There may be two litters a year, consisting of from one to eight young per litter, but usually from four to six. Gestation periods vary from 36 to 40 days. These squirrels are fairly long lived; one lived nine years in captivity.

There is a single species in western Oregon.

Chickaree
Tamiasciurus douglasi

THIS SPECIES WAS NAMED in honor of the Scottish botanist David Douglas, who collected the first specimen near the mouth of the Columbia River in the early 1800s. It was described scientifically in 1838.

DISTRIBUTION AND DESCRIPTION

total length 273-355 mm
tail 100-150 mm
weight 150-300 g

THE CHICKAREE OR Douglas squirrel (photo 51) frequents habitats in all five zones, but is least likely to be found in the bottom of the Willamette Valley, where its habitat is all but nonexistent. It is a relatively small squirrel that ranges from 10 to 14 inches in length and weighs between 5 and 10 ounces.

The eyes are encircled with short, orange hairs, and the small ears have short tufts of blackish hair at the tips. In summer, the back varies from slightly reddish brown to slightly grayish brown with many orange- and black-tipped hairs. There is a short blackish stripe on each side, extending from the forelegs to the hips. The underside and the tops of the feet are light to dark orange; however, there occasionally are white patches on the throat and chest, and near the forelegs. The bushy tail is wide and somewhat flat. In winter, the squirrel's coat is slightly grayer then in summer, and the black stripes on the sides are less apparent. The orange of the underside is obscured by many gray-, brown-, or black-tipped hairs. The tops of the feet are dark gray. The short, sharply curved claws vary from brown to dark gray.

CHICKAREES ARE FOREST dwellers, especially in coniferous forests, although they also live in mixed forests and occasionally white oak woodlands. They were much more common in white oak woodlands when I was a boy, but then white oak woodlands were much more common.

Chickarees arise at dawn to begin their day and usually retire with the setting sun. They are one of the three main sentinels in the forest, together with Steller Jays and Belted Kingfishers (the latter mainly along the streams and rivers). Little escapes their attention. Chickarees spend much time climbing up and down trees, searching the ground for truffles and other foods, minding the affairs of others, or just sitting quietly on a branch next to the trunk of a tree with their tails over their backs.

When building a summer nest of twigs, a chickaree cuts many live twigs (primarily Douglas fir) and carries them to the selected site, and in a short time constructs a loose nest. Inside the nest, it uses soft, dry mosses, lichens, or shredded inner bark from bigleaf maple, red alder, or western redcedar for its sleeping quarters. Summer nests may also be merely large balls of mosses, lichens, or shredded bark sometimes interwoven with grass into which the squirrel burrows a hole and makes its sleeping

Photo 51. Adult chickaree or Douglas squirrel. (Oregon Department of Fish and Wildlife Photograph by Ron Rohweder.)

Chipmunks & Squirrels • Sciuridae

Photo 52. Sitka spruce cones eaten by a chickaree along the Oregon coast. White cones are freshly eaten, dark cones were eaten some time ago. Note discarded cone scales. The collection of materials constitutes a kitchen midden. (Photograph by Chris Maser)

quarters. Chickarees often take over and remodel an abandoned nest, such as that of a bird, flying squirrel, or red tree vole.

Winter nests are often located in hollows in trees, frequently abandoned woodpecker nest cavities. If winter nests are constructed on limbs, they are well within the crown of the tree and are bulkier and much thicker than summer nests.

Chickarees have several calls, ranging from a low "chirrr" or "burrr" to an explosive "bauf, bauf, bauf." Except during spring and summer when the young are being reared, chickarees are relatively vociferous, and noisy territorial disputes and occasional temper tantrums are not infrequent, especially in the autumn. Every now and then a chickaree becomes so irritated by the trespass of an intruder that it bounces up and down with such vigor on the top of its nest that the whole nest is knocked out of the tree. After a "cooling off period," the chickaree usually retrieves its nest to the last twig, piece of tissue paper, or shoe lace, packs it all back to the original site, and rebuilds it.

The diet of chickarees is usually associated with cones such as Douglas fir, Sitka spruce (photo 52), Engelmann spruce, lodgepole pine, grand fir, and so on, and they are indeed an important part of the diet. When seed-bearing cones of the Douglas fir near maturity in early autumn, the chickarees cut them off the branches, extract the ripening seeds, and eat them. A cone is held with the forefeet, and individual scales are cut off the central core of the cone. Good seeds are eaten and defective ones are discarded.

A chickaree normally eats in one or two selected places near its nest or food storage area. The discarded scales of the cones and other food debris accumulate on the ground as a midden heap (see photo 50, page 155). Chickarees do not normally eat on the ground where they cannot see what is happening around them.

As the majority of the cones ripen, the chickarees ascend to the tops of the trees in the early morning, cut the cones off, and let them fall to the ground. I have found it patently unwise

Photo 53. Cones of Douglas fir eaten by a chickaree. Top cone was stored in a stream and was soft; note that the scales have been pulled off. Bottom cone was stored on land and was hard; note that the scales have been chewed off. (Photograph by Chris Maser)

to take a nap under a tree in which a chickaree is cutting cones, especially those of ponderosa pine! The cones are then collected one at a time and carried to a storage area, such as an underground burrow, hollow stump, or hollow fallen tree. Cones are normally cached in a moist place that prevents them from drying out and shedding their seeds. They can even be deposited in small streams or springs; cones stored under water remain fresh for a year or more, while those stored on land become moldy and spoil much faster. An examination of a chickaree's midden in winter will usually show how the cones were stored. Those stored under water often have the scales pulled off, whereas those stored on land are hard and the scales have to be chewed off (photo 53).

A chickaree usually stores more food than it consumes during a winter. There is, however, survival value in this excessive harvesting of cones; should there be a failure in the next year's crop, the chickaree can rummage through its unused stores of cones and often find enough good ones to augment an inadequate harvest.

In addition to cones, chickarees eat a variety of foods. During early spring they frequently cut the newly active terminal shoots of Douglas-fir. They eat the developing inner bark and needles but discard the old, mature needles. They also gnaw the thin bark of burls on the limbs of lodgepole pine (photo 54). Some chickarees also eat the mature pollen cones in great quantities, turning their feces yellow because of the high pollen content. During

Photo 54. The tender bark has been eaten off of the burl on a lodgepole pine by a chickaree in spring. (Photograph by Chris Maser)

Photo 55. Mushroom stored in a the lower twigs of alpine fir in the High Cascade Mountains by a chickaree in autumn, to be eaten at a later date. (Photograph by Chris Maser)

summer, they eat some green vegetation and various ripening fruits and berries. They may also eat sap that oozes from the holes made by sapsuckers.

Their main food during spring, summer, and autumn is truffles that are detected by odor and dug out of the forest floor in a manner similar to that of the flying squirrel and other forest rodents, although chickarees also eat and store mushrooms, especially in autumn (photo 55). When a chickaree, flying squirrel, or other forest rodent eats a truffle and defecates near the boundary of its home range, and the spores germinate and forms a fruiting body—a truffle—that is subsequently eaten by some other rodent that in turn deposits feces somewhere else, the fungus's genetic material is moved throughout the forest to combine and recombine with other colonies of the fungus.

The majority of male chickarees are sexually active from March through May. Females usually have litters of four to six that are born from May through June. Chickarees are born naked and blind and stay in their tree-top nursery nests until they are about one-half to two-thirds the size of their mothers. When first out of their nests, siblings stay close together and are tended by their mother. Families are still together by the end of August, although they are more independent of the mother and sibling associations are less evident. Families tend to remain relatively close through December.

Chickarees fall prey to owls, hawks, marten, long-tailed weasels, and coyotes, to name a few. They are very important in the winter diet of bobcats.

Flying Squirrels *Glaucomys*

THE GENERIC NAME *Glaucomys* is derived from the Greek words *glaukos* ("silvery," "gray") and *mys* ("a mouse").

North American flying squirrels have fine, soft, thick pelages that are gray with varying amounts of tannish or brownish wash. Being strictly nocturnal, they have large eyes that are sensitive to light.

Flying squirrels cannot fly; they can only glide downward by climbing to an elevated point and launching themselves. As they leap into space, they extend their legs outward from the body, spreading the large, loose folds of skin along the sides of the body so that a monoplane is formed, allowing the squirrel to glide gently and quietly with good control. Steering is accomplished by raising and lowering the forelegs. The tail, flattened horizontally, is used as a stabilizer to keep them on course.

Before a squirrel starts its glide, it carefully examines the chosen landing site by leaning to one side and then to the other; this possibly acts as a method of triangulation in measuring the distance. As it reaches its landing point, normally the trunk of a tree, it changes course to an upward direction by raising the tail. At the same time, the forelegs and hind legs are extended forward, which not only allows the gliding membrane to act as a parachute to slow the glide but also allows the legs to absorb the shock of landing.

The instant the squirrel lands, it races around the trunk of the tree, thereby eluding any predator, such as a Spotted Owl, that may be following it. Then, to make another glide, it dashes to a higher position with astonishing swiftness and agility and again launches itself into space. From a height of 60 feet, a squirrel can glide about 160 feet at a rate of almost 6 feet per second.

Flying squirrels inhabit hollows in trees and abandoned woodpecker nest cavities (photo 56). They also establish residences in buildings and artificial birdhouses. Nests, constructed on the limbs of trees, are composed of shredded bark, mosses, dry leaves, and other soft materials.

These squirrels have a varied diet that includes nuts, seeds, acorns, lichens, fungi, fruits, insects, and occasionally meat.

There may be one to two litters per year, depending on the species. The gestation period is about 40 days, and litters range from one to six young. Babies, usually born in April or May, are naked, pink, and blind at birth. Their eyes open at about 25

Photo 56. A northern flying squirrel that took over the abandon nest cavity of a common flicker. (Oregon Department of Fish and Wildlife Photograph by Walter Van Dyke.)

days of age. They nurse for 60 to 70 days and are neither vigorous nor confident in their actions before that. Flying squirrels must be well developed before they can successfully cope with the hazards of an arboreal and gliding style of life. Their development is correspondingly slow compared with many ground-dwelling mammals of similar size.

The two species of North American flying squirrels are primarily confined to forested and wooded areas from central Alaska and northern Canada south throughout the eastern United States; south in the Rocky Mountains through much of Idaho, and Wyoming into Utah; and south through most of Washington and Oregon into southern California. They also occur in small, isolated areas in Mexico and Guatemala.

There is one species of flying squirrel in the Pacific Northwest.

Northern flying squirrel
Glaucomys sabrinus

THE SPECIFIC NAME *sabrinus* is a Latin word meaning "a river nymph," and probably alludes to the fact that these squirrels are often abundant along the edges of forest streams and rivers. The first scientific specimen was caught at the mouth of the Severn River in Ontario, Canada, and was described scientifically in 1788.

DISTRIBUTION AND DESCRIPTION

total length 226-430 mm
tail 105-180 mm
weight 47-186 g

THE NORTHERN FLYING squirrel (photos 57 and 58) inhabits zones 1 through 5, but is least likely to be found in the floor of the Willamette Valley proper. It ranges from 9 to 17 inches long and weighs from a little over $1^1/2$ to $6^1/2$ ounces. It has long, very fine, soft hairs. Because all the hairs are about the same length, the fur is not separated into guard hairs and underfur; hence the coat appears sleek. One of the distinctive features of this squirrel is the loose fold of skin that stretches from the wrist of the foreleg to the ankle of the hind leg. (See photo 1, page 8; note the fold of skin attached to the squirrel's wrist extending downward to its ankle.)

The hairs on the back are bicolored; the shafts are dark gray, and the tips are dark reddish brown, giving the back a predominantly dark reddish brown appearance. The cheeks are light grayish brown, and dark gray hairs encircle the large eyes. The top edge of the gliding membrane is dark gray, and the hairs along its margin are tipped with light tan, giving the appearance of an almost whitish stripe. The underside is tannish. The tail is relatively wide and horizontally flat; the hair of the tail is dense and of the same texture as that of the body. The top of the tail is dull; it is brownish gray along the basal one-third, becoming dark gray toward the tip. The hairs on the underside are gray, tipped with light to dark tan, and give a tan

Photo 57 (left). Adult northern flying squirrel. (USDA Forest Service Photograph by Jim W. Grace)

Photo 58 (below) Baby northern flying squirrel. Note gliding membrane, which is the clear, dark line extending from the forefoot, down the side to the hind foot. (USDA Forest Service Photograph by Jim W. Grace)

appearance. The underside of the tail is dark tan with a dark gray margin.

STRICTLY A FOREST DWELLER, the northern flying squirrel is most often associated with coniferous forests in western Oregon, especially the remnants of ancient forests, the squirrel's preferred habitat. These bright-eyed nocturnal squirrels feed mainly on truffles during spring, summer, and autumn. They descend each night to the forest floor and dig out the truffles,

HABITAT AND BEHAVIOR

Flying squirrels often become very tame. My friend and graduate mentor, Ken Gordon, had one that used to sleep in the pocket of his sports coat while he taught class. It was a different story at his home, however.

Ken lived alone … well, not quite alone. Besides the flying squirrel, which had chewed holes in all his drapes (but only because he folded them over the curtain rods so the squirrel could nest in them), there was his spotted skunk, which slept in the kitchen by day and roamed somewhat freely at night, and the Townsend big-eared bat that roosted in the walkway between Ken's garage and the back door to his kitchen. I loved to watch his flying squirrel zoom around the living room. One time, however, it landed on my knee, defecated, and departed, upon which Ken asked: "Chris, do you think the squirrel is trying to tell you something?"

---

which they detect by odor. Truffles are abundant in and around large, fallen trees; these decomposing trees, protected from drying by the dense tops of the young trees, act as reservoirs that hold water all year and thus prolong the fruiting season of the truffles well into the summer. Flying squirrels are therefore most abundant in those areas that have large numbers of slowly decomposing ancient trees. In winter, the squirrels eat the lichens also used to build their nests; they may also be able to glean truffles later into the year in the young forest because the dense canopy protects the floor from snow.

Because these squirrels get most of their food from below the surface of the forest floor, they are vulnerable to predation by pumas, bobcats, coyotes, and marten while on the ground. This makes it imperative that flying squirrels are familiar with their home areas. One way the squirrels mark their home areas is by cheek-rubbing. They have a scent gland in the corner of each side of their mouth and as they move about their home areas at night, they find an appropriate place and, twisting their heads, drag one side of their face across the object. They generally mark only with one side, but on some occasions they mark with both. Such scent marking is done primarily at sites of grooming, feeding, and resting and probably acts to keep a particular squirrel oriented to and reassured that it is in its home area.

Breeding begins in March and April, and most youngsters are born in May and June, after a gestation period of 37 days. Some mothers have from one to six young, but most have from three to five. Born in snug, soft, warm nests, babies weigh about two-tenths of an ounce at birth (photos 59 and 60; see also photo 6, page 13 and photo 58, page 163). Their eyes open

---

While on an expedition for Oregon red tree voles with my friend and mentor, Murray Johnson, I was high in a Douglas fir tree about to search a nest for the voles (which I will discuss later), only to discover it was full of flying squirrels. When I yelled this fact down to Murray, he said: "Catch one! I need a live one to trade with a zoo in Japan for a Japanese flying squirrel to add to our exhibit in the Point Defiance Zoo" in Tacoma, Washington.

"Okay," I said, and grabbed one as it shot out of the nest just above me. Before I could secure a better grip on the squirrel, however, it bit my finger so hard that its teeth met in the middle and simultaneously urinated in my face. I hung onto a limb more than 50 feet above the ground with one hand, a biting squirrel with the other—unwilling to let go of either, and my eyes well flushed with squirrel urine. The Japanese zoo got the squirrel!

---

164   **Mammals of the Pacific Northwest**

when they are about 32 days old. By this time they are fully clothed in exquisitely soft fur, and their locomotion and coordination are well developed. They begin leaving their nest for short periods when they are about forty days old.

Flying squirrels are associated with large amounts of rotting wood on the floor of the forest, especially large fallen trees, because that is where truffles fruit most abundantly. As a flying squirrel searches out the belowground truffles, it digs in the forest soil. If it uncovers the uninoculated root tip of a Douglas fir and happens to drop fecal pellets into the hole it has dug, the fecal pellets can come into contact with the root tip. The spores of the mycorrhizal fungus that the squirrel ate yesterday or the day before will inoculate the fir's root tip when they germinate and will form the mycorrhizae necessary for the uptake of nutrients and water by the tree. The tree in turn will feed the fungus sugars produced in its green needles. If, on the other hand, the root tip that the flying squirrel digs up has already been inoculated with the same species of fungus, the nonreproductive portions of the fungi fuse and exchange genetic material.

Many a flying squirrel ends up as a meal for Spotted Owls. In addition, I have found their remains in the droppings of coyotes, bobcats, marten, and puma, all undoubtedly captured as they foraged on the ground.

Photo 59 (left). Twig nest of a northern flying squirrel 50 feet up in a young Douglas fir in the Coast Range. The sleeping chamber is made of moss and the shredded inner bark of a dead bigleaf maple. (Photograph by Chris Maser)

Photo 60 (right). Summer nest of a northern flying squirrel in the Coast Range. The nest, made out of grasses and moss, was 30 feet up in a young Douglas fir tree. (Photograph by Chris Maser)

Pocket Mice and Kangaroo Rats Hetero-myidae

THE FAMILIAL NAME Heteromyidae is derived from the Greek words *heteros* ("different") and *mys* ("mouse") plus the Latin suffix *idae* (which designates it as a family).

Heteromyid rodents exhibit considerable variation in their external appearance; some genera are mouse-like and scurry about, whereas others are highly modified to travel by jumping, with long and powerful hind legs, and reduced forelimbs. The tail in members of this family is long and well haired. They also have external, fur-lined cheek pouches in which food is carried, and which can be turned inside out for cleaning and then drawn back into place by a special muscle. The pelage varies from soft and velvety to harsh and spiny. The length of the head and body ranges from $2^1/4$ to $7^1/4$ inches and that of the tail from $1^3/4$ to $8^5/8$ inches.

The family Heteromyidae includes five genera and approximately seventy species, which inhabit prairies, arid plains, deserts, dry open forests, brushy or grassy areas, or humid tropical forests. Members of this family shelter in self-constructed burrows under bushes, trees, and fallen trees or logs. The entrance may be marked by a mound of soil or may simply be a small round hole. The burrow system, with its chambers for sleeping, storing food, and other purposes, often has several openings. After returning at dawn from their nocturnal foraging, many species plug the entrance to the burrow with moist soil, thereby creating a safer, more favorable environment inside. Heteromyid rodents generally remain in their burrows during inclement weather.

The diet consists mainly of seeds and vegetation but may also include insects and other invertebrates. Many members of the family store food. Members of some species derive all their water from the solid foods they eat and from the conversion of some of the food into water within the body. In addition, their excretory systems may be modified to extract the maximum liquid from the food eaten.

Offspring are born in a nest in the burrow. Litters range from one to eight young, and one to several litters per year are born to these rodents throughout much of their geographical distribution, though usually only one litter per year is produced in the northern parts of the range. The life span in the wild may be only a few months because of predators, which include coyotes, weasels, badgers, skunks, owls, hawks, and rattlesnakes.

Heteromyid rodents occur in southwestern Canada and the western United States, southward through Mexico and Central America to Ecuador, Colombia, and Venezuela. There is a single genus in western Oregon.

THE GENERIC NAME *Dipodomys* is derived from the Greek prefix *di*, meaning "two" or "double," and the Greek words *podos* ("a foot"), and *mys* ("a mouse"). The name "two-footed mouse" refers to the large hind feet of members of this genus and their habit of hopping upright like a kangaroo.

The head and body range in length from 4 to 8 inches and the tail from 4 to $8^5/8$ inches. The tail is usually longer than the head and body. Kangaroo rats range in weight from less than $1^1/2$ to 5 ounces.

The upper parts are pale yellow to dark brown, whereas the underside is clear white. The tail is usually dark above and below, with white sides. The end of the tail has a tuft of longer hairs. In most species, there is a white band that crosses the region of the thigh and joins the base of the tail. There are also distinct facial markings. The hind legs are very long, and the fifth toe of the large hind foot is diminutive or absent. A prominent, oil-secreting gland is present on the back between the shoulders.

The 22 species comprising this genus live in arid and semiarid brushy or grassy country. Well-drained, easily dug soil is usually preferred. They normally select relatively open ground, which permits an unobstructed view of their surroundings and facilitates their bipedal method of travel. They travel by hopping on their hind legs, using their forelegs only when traveling slowly and for short distances (photo 61). They apparently clear vegetation by cutting it down so their movement is not impeded.

Kangaroo rats are generally silent. They are active only at night, but bright moonlight keeps them in their burrows, where they also remain during periods of inclement weather, including heavy fog. Bathing in dust is apparently necessary to their health. Individuals denied the ability to dust themselves in captivity develop sores on the body as their fur becomes matted from the oily secretions of the gland on the back.

Photo 61. Tracks of a kangaroo rat. Note the impressions of the large hind feet and those of the much smaller front feet. (Photograph by Chris Maser)

They feed on seeds, some fruits, green vegetation, and occasionally on insects. Due to the scarcity of foods caused chiefly by their dry environment, most members of this genus store food in their burrows, and sometimes raid the stores of seeds cached by ants sharing their habitat. All food and nest materials must be transported in the fur-lined cheek pouches, which are stuffed full using the small forefeet. On reaching its destination, a kangaroo rat uses its forefeet to simultaneously shove forward against the back

Photo 62. The small, dry, hard, moisture-saving droppings of a kangaroo rat. (Photograph by Chris Maser)

and outside of the cheek pouches, emptying their contents.

Kangaroo rats seldom drink water; they use water resulting from the chemical breakdown of their food and conserve body moisture by coming out of their burrows at night, when humidity is highest. In addition, their kidneys are at least four times as efficient as those of a human and thus need much less water to remove nitrogenous wastes. Finally, their droppings are very small, dry, and hard, another moisture-saving measure (photo 62).

Although members of this genus living in areas of favorable climate may breed throughout the year, mating activity is low during the winter. Gestation is approximately 29 to 33 days. Three litters may be produced within a year, consisting of one to five young, but usually two to four. Youngsters remain in the nest for about six weeks.

Kangaroo rats occur west of the Missouri River from southwestern Canada to Baja California and south-central Mexico. There is one species in western Oregon.

Heermann kangaroo rat
Dipodomys heermanni

NAMED AFTER DR. Adolphus L. Heermann, a surgeon and naturalist on one of the Pacific Railroad Surveys (1853-54), this kangaroo rat is thought to have been first captured along the Calaveras River in Calaveras County, California. Its scientific description dates back to 1853.

DISTRIBUTION AND DESCRIPTION

total length 269-340 mm
tail 160-217 mm

HEERMANN KANGAROO RATS occur in the south-central portion of zone 3. They range in length from $10^{3}/_{4}$ to $13^{5}/_{8}$ inches with tails that vary from $6^{3}/_{8}$ to $8^{3}/_{4}$ inches.

Heermann kangaroo rats are the largest of the three species found in Oregon. They are compact in form with a large head. Their tails are very long and slender. The hind legs and feet are long and the four toes are armed with sharp, nearly straight nails. The hands are small and the thumb short. The cheek pouches are ample and fur-lined, and there is a noticeable groove down the middle of the upper incisors.

The upper parts are dark yellowish, heavily overlaid with black on the top of the head and back. The face is marked with black and the lower parts, sides and tip of the tail, stripe across the thighs, and the ear and brow spots are white. The soles of the hind feet and to top and bottom of the tail are blackish.

Mammals of the Pacific Northwest

LIKE OTHER KANGAROO RATS, the Heermann kangaroo rat inhabits mellow to sandy soil in open areas and areas of chaparral, although they can at times be found in clay or gravel soils. They are active at night, and the tell-tale tracks of their large hind feet can be seen in the early morning light, before the wind erases them for the day. They typically travel only on their long hind limbs in graceful leaps, with the tail serving as a balancing organ in locomotion and as a prop when standing. Should a kangaroo rat lose its tail, the counterbalance to the forward thrust of its powerful hind legs, it would turn a somersault every time it propelled itself forward by hopping.

These interesting rodents collect food at night by stuffing it into their cheek pouches and carrying it back to the burrow systems, where it is stored. They eat such things as the seeds of manzanita, buckbrush, rabbitbrush, lupines, grasses, and other plants. During the long, hot summer and autumn, they must depend on the moisture in succulent green vegetation for their supply of fresh water.

Heermann kangaroo rats tend to be solitary, and may fight if forced into close proximity, except during the breeding season, which extends from April into September. Although females come into heat several times in a breeding season, they form a gelatinous "copulation plug" at the entrance of their vaginas when they are not receptive. A receptive female may have as many as three litters of two to five youngsters each. Females born in the first litter may have two litters of their own before winter. Thus, a single female can be responsible for up to 25 young in one season.

Generally speaking, an animal that is so prolific also has a short life span due to severe pressures from predation, and such is the case with kangaroo rats. Their predators include foxes, coyotes, bobcats, weasels, owls, and snakes.

Pocket Gophers Geomyidae

THE FAMILIAL NAME Geomyidae is derived from the Greek words *geios* ("of the earth") and *mys* ("mouse") combined with the Latin suffix *idae* (which designates it as a family).

Pocket gophers are robust rodent. They have small eyes and ears, short legs, and two external, fur-lined cheek pouches (photos 63 and 64), extending from the lower portion of the face back to the shoulders. They can be turned inside out to be cleaned and are pulled back into place by a special muscle. They apparently are used only for transporting food (photo 65).

Pocket gophers have strong forelimbs; their forefeet have five toes, each terminating in a powerful claw adapted for digging. Their naked or sparsely haired tails are generally short and sensitive to touch. Gophers have very loose, flexible skin that is thickest in the region of the head and throat; this may be an advantage when they fight since pocket gophers are normally pugnacious.

They have stout bodies and almost no visible neck. They are adapted to their subterranean lifestyle by small eyes that are cleansed of soil by a thick fluid from the tear glands, and by lips that can be closed behind their protruding, curved front teeth, allowing them to gnaw through hard earth without swallowing any of it. The softness and laxness of their pelage allows them to move both forward and backward in their tunnels. In fact, they can run backward almost as fast as they can run forward.

Pocket gophers exhibit significant differences in size, both among and within species. The variations are influenced, at least to some extent, by sex (males are usually larger than females) and by the type of soil they frequent. The head and body range from $3^1/2$ to 12 inches in length. Their pelages, which lack underfur, vary from black to almost white.

Gophers dig two types of tunnels: shallow tunnels for gathering food, such as roots and tubers, and deep tunnels for shelter, as well as chambers for nesting, storing food, and deposition of fecal pellets. The burrow systems are usually marked by a series of earth mounds on the surface of the ground. Unlike moles, which expel excess earth through vertical shafts into conical or volcano-shaped mounds (see photo 10 on page 44), gophers expel excess earth through inclined lateral shafts, making fan-shaped mounds (photo 66). Digging is accomplished primarily with the strong foreclaws, but the large front teeth are used to loosen soil and rocks, as well as to cut roots. Loose earth is held between the chest and forelegs and pushed to the surface. The exits of their burrows are plugged with soil (referred to as a "button"), thus creating an effective air-conditioning system and some protection against unwanted visitors (photo 67).

Photo 63. The external, fur-lined cheek pouch of a Townsend pocket gopher. (Photograph by Chris Maser)

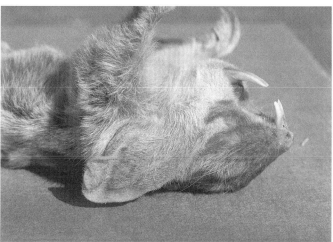

Photo 64. An everted cheek pouch showing that it extends to the area of the shoulder. (Photograph by Chris Maser)

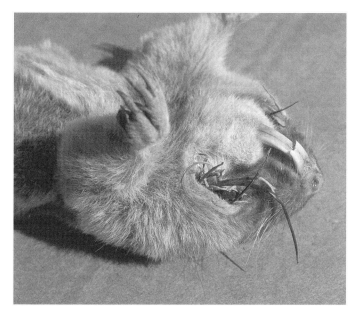

Photo 65. Food stuffed into the cheek pouches for transportation to the burrow. (Photograph by Chris Maser)

Photo 66. The fan-shaped mound of a Mazama pocket gopher. Note the rocks in the excavated soil. Pocket gophers inhabiting rocky soils tend to be smaller than those inhabiting deep, rich valley-bottom soils, like the Camas pocket gopher. The smaller size allows smaller tunnels and thus less expenditure of energy digging in rocky terrain. (Photograph by Chris Maser)

Photo 67. An earthen plug or "button" by which a Mazama pocket gopher has closed its burrow. (Photograph by Chris Maser)

Gophers do not hibernate, but in the colder parts of their range, which does not seem to include western Oregon, they may become less active during winter. Although gophers occasionally travel through unlined tunnels in the snow, they normally line their tunnels by pushing earth into them and pressing it against the snow, where it freezes in place (photos 68). When the snow melts, these ropelike strands of earth (known as gopher cores) settle to the ground, and the different direction and relative levels of the gopher's winter burrows can be seen.

Pocket gophers lead solitary lives except during the breeding period. Each pregnant female annually has one to several litters, consisting of two to eleven young.

Pocket gophers occur from about 54 degrees north latitude in western Canada south to Panama. There is a single genus of pocket gophers in western Oregon.

Photo 68. (a) Soil lining of a burrow through deep snow of made by a northern pocket gopher, which has not collapsed as the snow melted. (b) A trench burrow under deep snow made by a Mazama pocket gopher. (c) Collapsed solid cores of soil at different levels, which are the soil linings of burrows in deep snow that caved in on themselves; made by a northern pocket gopher. (d) Sometimes a gopher bumps into an obstacle while tunneling in snow. (Photographs by Chris Maser)

Western Pocket Gophers
Thomomys

THE GENERIC NAME *Thomomys* is derived from the Greek words *thomos* ("a heap") and *mys* ("mouse"); the name alludes to the mounds of earth deposited around the openings of the gopher's burrows.

In length of head and body, western pocket gophers range from $6^1/_2$ to 12 inches; their tails range from $1^1/_2$ to almost 4 inches in length; and they weigh from about 2 ounces up to a pound. Males are considerably larger than females. These gophers have robust bodies, small eyes and ears, short legs, and long front claws. The slightly tapered tail is highly vascular and well supplied with nerves, making it sensitive to touch. Unlike most members of other genera within this family, the incisors of the genus *Thomomys* lack longitudinal grooves on their outer surfaces. The pelage is short, soft, and smooth; it varies from black through gray and brown to almost white. The underside is only slightly lighter than the back.

Western pocket gophers spend most of their lives underground. These burrowing mammals inhabit many types of soil in deserts, prairies, open forest, grasslands, and mountain meadows. The length and design of a burrow system does not seem to follow any particular pattern. The earth thrown out of a burrow forms a fan-shaped mound. The burrow entrance may be carefully plugged from within. These gophers do not hibernate, even in the coldest areas, as evidence of their winter activities—gopher cores—will attest when the snow melts. Their burrows have separate chambers for nests, storage of food, and elimination of bodily wastes (photo 69).

As far as is known, western pocket gophers are strict vegetarians. Their diet consists of roots, bulbs, and tubers, as well as the aboveground portions of plants. The gophers may forage above ground on and off from early evening to dawn, and they may be occasionally seen foraging abroad on overcast days; more often than not, however, if they feed in the daytime, they bite off the roots of a plant and pull it down into the burrow.

Photo 69. Nest chamber and toilet of a pocket gopher in winter under the cover of deep snow in the high mountains. (Photograph by Chris Maser)

Plants may be consumed immediately, or they may be cut into convenient lengths and pushed with the forefeet into the fur-lined cheek pouches to be transported to an area for storage or eating. Because these gophers apparently obtain sufficient moisture from their food, they need not drink.

Northern species seem to have a more limited breeding season than do southern species. Litters, ranging from three to ten young, are born and raised in underground nests. At birth, they are blind, almost naked, and helpless, and weigh from one-tenth to two-tenths of an ounce. After they are weaned, they commence their solitary lives.

Western pocket gophers inhabit most of the western half of North America from southwestern Canada, through the western United States, into the Baja Peninsula and the mainland of Mexico. There are three species of pocket gophers in western Oregon.

THE SPECIFIC NAME *bulbivorus* is derived from the Latin words *bulb* ("a bulb") and *voro* ("to devour"). The first specimen of this gopher was taken from the banks of the Columbia River, probably near where Portland now stands. It was described scientifically in 1829.

Camas pocket gopher
Thomomys bulbivorus

THE CAMAS POCKET GOPHER occurs in zone 3 from the Columbia River south to the vicinity of Eugene in Lane County. These gophers reach up to 12 inches in length and weigh a little over a pound. The dorsal pelage is slightly grayish brown and their ears and noses are blackish. Their undersides are grayish except for an irregular-shaped patch of white on the throat. Their pelages in winter are longer and softer than in summer. Otherwise, they fit the description given under the family.

DISTRIBUTION AND DESCRIPTION

total length 300 mm
weight 545 g

EVEN THOUGH A CAMAS pocket gopher is big, it has relatively weak claws for digging in the dense clay soil of the Willamette Valley, but by using its protruding incisors, it is able to loosen the hard-baked earth through which it tunnels. The earth plugs or buttons with which a gopher closes the entrance to its burrow are usually loose and only a few inches in length, but they can sometimes be a foot long and solidly compacted. In addition, a burrow that appears to be open, may in fact be plugged some distance underground.

While foraging on the ground during the evening, these gophers are alert and stay close to their burrows. They quickly cut off vegetation, cram as much as possible into their cheek pouches, and disappear belowground, reappearing in a short time to continue gathering food. In late summer, they

HABITAT AND BEHAVIOR

sometimes remain outside for longer periods, sitting on their haunches and, with their forefeet, deftly bending down one grass stalk after another, cutting off the soft, green heads and stuffing them into the cheek pouches. I have never seen one of these gophers take time to eat while exposed on the surface of the ground.

The breeding season seems to extend from April into early June. Young are born in April and May, naked, toothless, and without cheek pouches. Growth is rapid, however, and they begin to develop hair by two weeks of age. They can crawl and eat solid food by three weeks, and develop cheek pouches by four weeks. Their eyes open during their fifth week, and they are weaned at six weeks of age. They probably begin to disperse shortly thereafter.

Owls, especially the Great Horned Owl, kill some of these large gophers as they forage aboveground, and long-tailed weasels and large gopher snakes hunt them in their burrows. But people—who consider them pests because of the damage they cause—are their greatest enemy.

Mazama pocket gopher
Thomomys mazama

DISTRIBUTION AND DESCRIPTION

total length 183-239 mm
tail 55-81 mm
weight 52-96 g

THIS SPECIES WAS NAMED after Mt. Mazama, which erupted and formed Crater Lake in Klamath County, Oregon. The first specimen of this species was secured near Crater Lake on September 3, 1896, by Edward A. Preble.

MAZAMA POCKET GOPHERS, sometimes called western pocket gophers, used to occur in the northern half of zone 1 and at a small spot near the southern end of the zone. I remember the black subspecies Thomomys mazama niger being common in clearcuts on the flanks of Marys Peak, Benton County, Oregon, as well as in the meadow on top in the 1950s, '60s, and early '70s, but I have not seen any sign of them for over a decade now. If they are still there, they certainly are not as abundant as they once were. Furthermore, I did not find Mazama pocket gophers during the three-year ecological survey I conducted of the entire Oregon coast (1970-73).

Adult Mazama pocket gophers range in length from about 7 to $9^1/2$ inches and weigh from almost 2 to a little over 3 ounces. The three subspecies occuring in zone 1 differ in appearance, so I will describe their coloration separately.

Thomomys mazama hesperus was first collected at Tillamook, Tillamook County, Oregon, on November 9, 1894. The back varies from reddish brown to dark reddish brown, to brown, to black. The reddish brown and brown individuals have large patches of black hair around and behind their small ears. The underside varies from light grayish brown to tan, to reddish brown, to brown, to black. The cheek pouches and the tops of the feet are whitish. The tail varies from dark gray to brown at the base, becoming white at the tip. Some, however, may be either partially or wholly black.

Thomomys mazama niger was first collected at Seaton (now known as Mapleton) at the head of tidewater on the Siuslaw River, Lane County, Oregon, on October 6, 1894. The back is a uniform, glossy black with purple and green iridescence. The underside is duller and more gray, with irregular white spots. The feet and the tip of the tail are white. (This is the subspecies that I used to find abundantly on Marys Peak.)

Thomomys mazama helleri was first collected at Gold Beach at the mouth of the Rogue River, Curry County, Oregon, in 1901, and is known only from here. The back is a dark, rich reddish brown; the sides and underside are tannish. There are patches of black hair surrounding the ears. The nose and face are blackish. Rarely is there a trace of white on the lips, but the tip of the tail is usually white.

A fourth subspecies, *Thomomys mazama mazama*, occurring pretty much throughout zones 4 and 5, is the one on which I have based my life history description of the species. It has a back that varies from reddish brown to dark reddish brown. The nose and cheeks are somewhat darker and less reddish, and there are black patches around the ears. The underside is somewhat yellowish. The feet and tail are a soiled whitish to gray.

THE MAZAMA POCKET GOPHER has a wide elevational distribution, but not a contiguous population. Members of this species occur from almost sea level in zone 1 to above timberline in zone 5, where they are primarily inhabitants of grassy areas, such as meadows, but also move into areas of forest that have either been clearcut or burned.

Breeding occurs in March, and the gestation period is 28 days. By mid July, the rapidly growing youngsters leave their place of birth and establish their own burrow systems. They are about two-thirds grown and resemble their parents.

Two kinds of tunnels are dug: shallow ones for gathering food, such as roots and tubers, and deep ones for shelter. The deep tunnels include chambers for nesting, food storage, and toilets. The burrow systems are marked by a series of earthen mounds on the surface of the ground; the gophers expel the excess earth through inclined lateral shafts that result in fan-shaped mounds.

HABITAT AND BEHAVIOR

Photo 70. Mazama pocket gopher pushing soil out of its burrow system as it enlarges it. (Photograph by Chris Maser)

The gophers' deep burrows bring subsurface soil to the surface and are often bare of plants because there are not enough spores of the necessary mycorrhizal fungi in this deep soil to inoculate the roots of plants and allow them to become established. However, this bare soil is vital to the survival of the Cascade tiger beetle, which requires open areas to fly into and out of and to find its food, ants. But Mazama pocket gophers that live along the edge of a mountain meadow and a forest may eat truffles in the forest, the spores of which they deposit in fecal chambers thus inoculating the soil with mycorrhizal fungi.

These gophers also alter the soil in other ways. Their mounds can cover 5 to 10 percent of the surface in some areas, and their burrows—6 inches to a foot below ground—are sometimes so numerous that deer and elk break into them. The tunnels are constantly extended and gradually fill up as they are abandoned, and the old nests, partially filled pantries, and toilets are buried well below the surface, where the buried vegetation and feces become deep fertilization. The creation of the mounds constantly buries surface vegetation deeper and deeper, with a similar effect. The soil thus becomes mellow, porous, and penetrated with the burrows of gophers; so an increasing part of the snowmelt and rainfall is held in the ground instead of running over the surface where it is likely to cause soil erosion.

With time, seedlings from the forest become established in the meadow, gradually claiming it. As the forest encroaches, the gophers decline in number and eventually disappear altogether. (I have watched this happening over the last three decades at Lost Prairie along the South Santiam Highway, just a few miles east of Tombstone Pass; it began in the early 1960s.)

Mazama pocket gophers, like other gophers, dig primarily with their strong foreclaws, but their large front teeth are used to loosen soil and rocks, as well as to cut roots. They hold the loose earth between their chest and forelegs and push it to the surface (Photo 70). They plug the exits of their burrows with soil, thus creating an effective air-conditioning system and some protection against unwanted visitors.

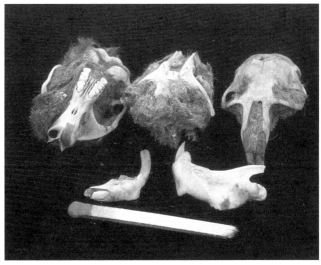

Although the gophers are active aboveground primarily from the evening, throughout the night, into the early morning, they are active at any time on warm, overcast days. Underground activity seems to be almost continuous and is often heralded by muffled gnawing or scratching. The sound ceases, and the stem of a lupine or other favored food plant begins to wiggle as a small, brown nose appears. The hole is quickly enlarged to allow the gopher's head to emerge beside the plant. A stem is cut off and drawn belowground. A good meal is gathered within a minute and the hole securely plugged with soil.

While gathering food in the evening, a gopher is alert and stays close to its burrow. It cuts vegetation quickly, crams as much as possible into its external, fur-lined cheek pouches, and disappears belowground. It reappears in a few minutes and gathers more food, which it takes into its burrow for storage in its food chamber. See the account of the Camas pocket gopher (page 000) for more information on feeding behavior, because they are much the same.

Owls (photo 71), weasels, coyotes, fox, and marten prey on Mazama pocket gophers.

Photo 71. Skulls of three pocket gophers recovered from the regurgitated pellets of a great horned owl. (Photograph by Gerald Strickler and Chris Maser)

THIS SPECIES WAS NAMED for Paolo E. Botta, one of the first naturalists in California—from 1827 to 1828. The first specimen was collected along the "coast of California," and was scientifically described in 1836.

BOTTA POCKET GOPHERS occur in zone 3. They are 7⁵/₈ to 10⁷/₈ inches with tails that vary from 2¹/₄ to 3⁷/₈ inches in length. Their weights range from 2¹/₂ to almost 9 ounces.

The back is dark brown with relatively few black-tipped hairs. There are small patches of dark gray to black hairs around and behind the tiny ears. The underside is a mixture of light brown, light gray, and tan hairs, with irregular patches of white hair.

Botta pocket gopher
Thomomys bottae

DISTRIBUTION AND DESCRIPTION

total length 190-273 mm
tail 55-97 mm
weight 71-250 g

The tail is light gray to tan. The incisor teeth of this pocket gopher in Oregon may be tipped with white.

THESE POCKET GOPHERS burrow extensively in relatively hard clayey and volcanic soils, which may account for the white tips on their incisors. Since gophers often use their strong, protruding, and ever-growing front teeth to loosen up compacted soil while excavating their elaborate burrow systems, the tips of the teeth wear away, which removes the yellow enamel surface, and indicates the regularity with which some gophers use them as tools in digging.

Because these gophers dig in clayey soil, their mounds are often composed of hard lumps. Their burrows can also be found in soft soils, but they are rarely found in forested areas. Beyond this, the habits of the Botta pocket gopher closely resemble those of both the Camas and Mazama pocket gophers.

These gophers eat a variety of plants, both the underground and aboveground portions, much as other gophers do. As indigenous plants become scarce in an area due to human activity, the gophers are quite willing to substitute introduced and cultivated plants.

The usual number of young per litter seems to be from four to eight, which, like other baby gophers, are born in a warm, soft underground chamber.

THE FAMILIAL NAME Castoridae is derived from *kastor*, the Greek word for the beaver, combined with the Latin suffix *idae* (which designates it as a family).

Since there is only one living genus within this family and a single living species within the genus in North America, refer to the species description.

There are two species of beaver. North American beaver occupy most of North America, except parts of northern Alaska and northern Canada, central Nevada, parts of California, western Utah, Florida, and most of Mexico. European beaver inhabit Norway, Germany, and France, east through much of northern Asia to the Lena River in Siberia.

THE GENERIC NAME *Castor* has the same derivation as the familial name.

The geographical distribution of the genus is the same as that given for the family. There is one species of beaver in western Oregon.

THE NORTH AMERICAN beaver was named for Canada, combined with the Latin suffix *ensis* ("belonging to"). The first scientific specimen came from Hudson Bay, Canada, and its scientific description was published in 1820.

BEAVER INHABIT ZONES 1 through 5. They are among the largest rodents in the world and the largest living rodent in North America. Beaver are compact, thick-set animals with small eyes and ears. They have short legs and a large, broad, scaly, paddle-like tail. Adults range from 2 feet 6 inches to 4 feet in length; their tails are 8$^1/_2$ to 18 inches long and 4 to 5 inches wide. Adult beaver often weigh more than 60 pounds.

Beaver are well adapted to aquatic life. They are excellent swimmers and divers that use oxygen economically and can remain submerged more than fifteen minutes. Their small eyes are protected by nictitating membranes, inner eyelids that help to keep a beaver's eyes clean. A beaver's nostrils and ears are valvular and can be closed under water. They have large webbed hind feet (photo 72), and the claws of the second and third toes are split, presumably as an aid to grooming (photo 73).

Beaver have unusually dense coats that consist of soft, short, gray underfur overlaid with coarse shiny guard hairs that vary from rich glossy brown, to yellowish brown, to reddish brown. The guard hairs on the underside are not as long or as close together as they are on the top. The underparts, therefore, are lighter in color. The feet and the tail are black.

True Beavers Castoridae

True beaver
Castor

North American beaver
Castor canadensis

DISTRIBUTION AND DESCRIPTION

total length 950-1150 mm
tail 340-460 mm
weight 27.2 kg

BEAVER CAN BE FOUND in any suitable body of water, from roadside ditches to lakes, streams, and rivers, and occasionally even estuaries, where they live in colonies usually consisting of two parents and the young of the year, as well as those born the previous year. Perhaps more than any other animal, the beaver has had an important effect on North American history. The quest for its valuable pelt stimulated exploration; the fur trade was born; battles were fought; fortunes were made and lost; a unique way of life evolved—the solitary trapper and Mountain Man. Overexploitation rapidly exterminated beaver from much of their geographical distribution and severely reduced their populations. Today, through the efforts of federal and state agencies, the beaver has increased in numbers and is once again found throughout most of its former range, including the rivers and their tributaries in western Oregon. While working on the Oregon coast, I had the opportunity to study beaver in such rivers as the Coquille in Coos County and in the small lakes along the sand dunes in Lane County.

These large rodents have the ability to cut down large trees and to construct complex dams and abodes in streams and lakes. Although beaver dams may be large—over 6 feet high and occasionally more than 595 yards long (photo 74)—those in small tributary streams are usually inconspicuous. Beaver dams impound water that may form ponds covering many acres (photo 75). In mountainous regions, these ponds gradually fill in with sediment and vegetation and ultimately form meadows.

Beaver dams alter a stream channel not only by impounding water but also by giving the channel gradient a stair-step profile, which decreases the velocity of the current, expands the area of flooded soils, and increases the retention of sediments and organic matter. The wetted soils surrounding the pond increase the beaver's supply of food while the pond itself offers protection from predators.

The dams are closely inspected by the beaver and are often maintained for years. Dams are built with sticks (normally after the animal has eaten the bark) pushed into an appropriate soft spot in the bottom of a stream or into a foundation of mud carried to the damsite by the beaver. (Some streams have rocky bottoms, however, and cannot be dammed.) To the foundation of sticks (butt ends facing upstream), the beaver add mud, vegetation, shrubs, and more sticks (photo 76). As the dam grows, the sticks become crisscrossed and the dam becomes securely anchored (photo 76); such a dam is often extremely difficult to remove by hand, though severe winter floods often wash out a dam, even a securely anchored one.

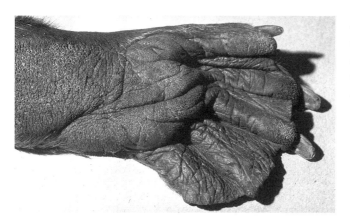

The wetted area stores nitrogen and other important elements, such as potassium, calcium, and, iron, which become available as nutrients for uptake by plants when dams either break or are abandoned by beaver, allowing the surrounding soil to be exposed to air. When a dam breaks during a winter storm, however, the elements are flushed downstream along with the accumulated silt, which can have both positive and negative effects, depending on one's point of view.

Beaver also influence the community dynamics of riparian vegetation, instream bottom-dwelling organisms, fish, wildlife, the structural diversity of streams, and the nutritional value of certain species of trees. Some ponds gradually fill in, forming wet meadows, even with a resident beaver at work, because the calm water of a pond allows suspended sediments to settle out—a major factor in building up soil that eventually becomes a meadow. A dam and pond have the additional benefit of capturing eroded soil close to the source of erosion rather than allowing it to wash downstream, causing turbidity in the water.

Beaver create structural diversity not only directly by felling trees and building dams but also secondarily by flooding areas

Although beaver dams have in the past been removed (often with dynamite) because they were thought to impede the passage of salmon, they routinely wash out during high flows when coho salmon in coastal streams travel to the river's headwaters to spawn. The beaver repair their dams when flows decrease, and in the process create excellent habitat in which young coho thrive.

Photo 74. Although beaver dams may be large, those in small tributary streams are usually inconspicuous. (Photograph by Chris Maser)

Photo 75. Beaver dams impound water that may form ponds covering many acres. (Photograph by Chris Maser)

Photo 76. A beaver dam under construction; note the new mud plastered along the top of the dam and the freshly peeled wood from the beaver's having stripped its bark for food. (Photographs by Chris Maser)

Photo 77. A free-standing lodge in open water away from the bank. (Photograph by Chris Maser)

Photo 78. Scent mound of a beaver, often called "beaver mud pie." (Photograph by Chris Maser)

that kill trees, which in turn creates standing dead trees or snags that are important to numerous insects, birds, and mammals. And beaver even affect the chemical composition of some trees.

Quaking aspen, for example, which is a favored food along the eastern flank of the High Cascade Mountains, appears to have chemical defenses that reduce grazing by beaver, which avoid aspen shoots regenerated from trees they themselves have cut. In fact, a chemical isolated from the roots of cut trees acts as a deterrent against future cutting by beaver when applied to aspen shoots that would otherwise be acceptable as food. This dynamic could explain how beaver can continue to populate an area without killing off all their food.

A beaver dam can also have a profound effect on the riparian habitat surrounding a pond, by dramatically altering both how water flows and is stored in a stream's immediate drainage way. A beaver dam and the pond it creates cause more water to be held in the subsurface soils surrounding the dam and pond

Photo 79. Mountain hemlock that beavers have had trouble cutting down even on winter snows of various depths. (Photographs by Chris Maser)

than would exist without them. This contributes to the general diversity and richness of species within the floodplain and is particularly important to species adapted to or requiring wetlands. In addition, the subsurface flow of water from the beaver pond around the dam and back into the stream below the dam cools the stream in summer and warms it in winter.

It works like this. In summer, as water from a beaver pond enters the subsurface flow, it is generally about as warm as the water in the pond. It becomes cooler as it moves slowly through the subsurface soil and ultimately returns to the stream below the dam after two or three months. The reverse is true in winter.

Beaver construct three types of abodes: a standing lodge in open water away from the bank (photo 77), a bank-lodge with a burrow extending into the bank, or a simple burrow dug into the bank without a lodge of any type. The bank-lodge is most frequently used in the streams of western Oregon above the influence of the rise and fall of the oceanic tide, whereas a simple burrow in the bank is the norm in those areas influenced by tidewater.

Beaver have paired anal scent glands known as "castors" or "beaver pods" that occur in both sexes but are slightly larger in the male. The content of the castors is a musk with a pleasant, rather sweet odor called "castoreum." It is deposited on "scent-mounds," also called "beaver mud pies" or simply "beaver pies," (photo 78) to mark territorial limits. These scent-mounds are made with piles of mud, occasionally mud and vegetation, rarely vegetation without mud, placed on the shore along the water's edge. Scent-mounds vary in size from barely perceptible to piles of mud 2 feet high and in number from two to seven per colony. Although most are located along the edge of a colony, some may be located near the dwelling, especially if the abode has been recently constructed.

Young and old beaver, both males and females, regularly and frequently visit the scent-mounds to deposit mud and castoreum. This behavior eliminates the necessity of actively defending a territory by removing antagonism between adjacent colonies while retaining the integrity of the resident colony. Strange castoreum on a scent-mound causes a resident beaver to immediately deposit its own, the emission of which produces a sound that can be heard about 15 yards away, and may also elicit aggressive behavior in the resident beaver as indicated by loud hissing.

Beaver are social and somewhat placid animals; their calmness can be attributed to their familial and colonial characteristics. For example, other than untrained young, which might vocalize while outside the lodge, vocalization occurs chiefly within the lodge and is related to those behaviors in which sound or communication is of little survival value. Sound production in beaver does not seem to have the same usefulness as it does in animals less secretive that roam at will. Vocalization is associated more with familial and colonial behavior than with territoriality.

Beaver do, however, produce one sound with which many people are familiar—slapping their tails on the surface of the water just prior to quickly diving beneath it. This is usually a warning given to another beaver by a startled or frightened individual. The normal dive of an undisturbed beaver is quiet.

Beaver are vegetarians and eat a wide variety of plants. In western Oregon, the bark of red alder, cottonwood, willow, and quaking aspen (a favorite food where available) are their main diet, as well as salmonberry, salal, deer fern, swordfern, sedges, and during the spring and early summer, skunkcabbage. They also eat small quantities of bark from Douglas fir and western hemlock. They sometimes have trouble felling a tree (photo 79). Since beavers' main food is the bark of trees, their fecal pellets are composed of coarse fibers. These pellets are deposited under water, where they break down and add organic material to the aquatic system (photo 80).

Photo 80. Droppings deposited by a beaver in the water by its dam; note the disintegrating droppings on the bottom of the pond, some of which are marked with arrows. (Photograph by Chris Maser)

This is a story about a beaver that miscalculated while cutting one of its favorite trees, a quaking aspen, a species that grows almost exclusively on the east side of the Cascade Mountains in zone 5. Some years ago, just over the Cascades a short distance west of Sisters, Oregon, I found three quaking aspens growing next to one another in a row along a stream. The middle aspen had been cut off by a beaver but, instead of falling, remained suspended in the air, supported by its neighbors.

Imagine how confused the beaver must have been! But this tree had grown in such a way that its branches interlaced with its neighbors', that, despite the beaver's best efforts, supported it in the air, at least long enough for me to enjoy a good chuckle.

Beaver do not hibernate but store food under the water for use during winter. They construct canal systems inland from their ponds, on which they float food, such as small, trimmed trees, from the cutting sites to the pond. The canals are usually filled with water from about 15 to 25 inches deep, with soft, muddy bottoms. Underground chambers are frequently constructed into the sides of a canal, often with a large hole penetrating the roof to the surface of the ground above. Beaver store food for the winter in these chambers, as well as on the bottom of a pond where it remains fresh and is available when the pond freezes over. In western Oregon they seldom travel more than 200 yards from the water to secure food.

Perhaps the most visible signs of beavers' feeding activity are the stumps of trees they fall to obtain the bark. The trees are cut by the beavers' sharp incisor teeth. If these front teeth are not constantly worn down, they grow so fast that the upper and lower teeth grow beyond each other, and the beaver, unable to open its mouth wide enough to eat, starves to death.

Both male and female beaver probably reach sexual maturity in their second summer, although males may be capable of reproduction late in their first summer. The breeding season is normally from January through March; breeding activity peaks in early February. There is a single litter per year, which ranges from one to nine offspring, but the usual litter consists of two to four young called "kits." Kits are born well haired and with their eyes open. At birth they weigh from 8 to 24 ounces. They nurse for about six weeks and remain part of the family group until sexual maturity is attained. Beaver may approach twenty years of age.

Unlike most animals, beaver may actually be helped by the logging of an ancient forest since removal of the conifers stimulates a sudden increase in red alder and willows, both of which are used by beaver.

Beaver are so large that they have few enemies. Pumas occasionally kill a beaver, and coyotes and Great Horned Owls taken a youngster now and then. But for the most part, humans are the enemy; people still trap them for their pelts and to remove them as pests.

Rats and Mice Muridae

THE FAMILY NAME Muridae is derived from the Latin word *murinus* ("mouselike") and the Latin suffix *idae* (which designates it as a family).

Muridae is a relatively large family because taxonomists have found that members of three previous families were closely enough related to group them within a single family but still distinct enough in characteristics to warrant keeping them as distinct subfamilies. I will therefore describe the characteristics of each subfamily instead of the family as a whole.

Indigenous Rats and Mice Sigmodontinae

THE SUBFAMILIAL NAME Sigmodontinae is derived from the Greek words sigma, which is equivalent to the English S, hence meaning "curved like the letter sigma," and *odontos* ("tooth"), combined with the New Latin suffix *inae* (which designates subfamily).

Members of this subfamily vary greatly in size, shape, and habitat requirements; most are terrestrial—they scamper or jump over or burrow under the surface of the ground. The scampering forms often are good climbers; some are cliff-dwellers; others are semi-arboreal. Some members of this subfamily are even semiaquatic.

These large-eyed mammals are primarily nocturnal and slender in proportions. They have pointed noses and large, sensitive ears that seldom are concealed in the body hair. Their legs are usually long and slender, their tails long and generally well haired. Pelages are generally light in color, appearing rich and lax.

Members of this subfamily eat mainly seeds and vegetables, but many are omnivorous.

The gestation period in members of most genera is 20 to 33 days, and the number of young per litter ranges from one to eighteen. Young generally weigh less than half an ounce at birth. Females of some genera begin to breed at about six to seven weeks of age. In warm parts of their geographical distribution, members of this subfamily apparently breed throughout the year, provided temperatures are not excessively high; in cold regions, however, the breeding season is restricted. The length of daylight seems to be the major factor affecting the timing and the duration of the breeding season. In the wild, most individuals are victims of predation and probably live less than two years, frequently less than one.

Members of this subfamily, although primarily southern in distribution, occur nearly worldwide but are absent from certain islands, such as Ireland and Iceland, a few arctic islands, Antarctica, and the Austro-Malayan area. There are three genera in western Oregon.

THE GENERIC NAME *Reithrodontomys* is derived from Latin *re* ("back again") and the Greek words *thouros* ("rushing"), *odontos* ("tooth"), and *mys* ("mouse"). The name refers to the distinctive patterns in the enamel on the crowns of the teeth.

Habitats for members of this genus, which includes sixteen species, vary from salt marshes to tropical forests, but they are usually associated with stands of short grass. The elevational range is from below sea level in places to above timberline on some mountains in Central America.

The length of the head and body is from 2 to $5^3/4$ inches, and the length of the tail is from $2^5/8$ to $3^3/4$ inches. Adults usually weigh from about half to almost one ounce.

The pelage is fairly soft in most species, though it can be dense and somewhat woolly or thin and coarse. The coloration of the upper parts ranges from pale yellowish to orangish gray, through yellowish to orangish white, and from pinkish brown through browns to black. The sides are lighter and usually more yellowish to orangish than the upper parts. The under parts are white, grayish, or one of these colors tinged with yellowish to orangish white or pinkish brown. The tail is dark above and light below or unicolor; it is slender, scaly, and thinly haired. The pelage of juveniles is more or less gray, but the adult pelage is brighter.

The most noticeable evidence of harvest mice is the presence of their globular nests of grasses, which are about 6 or 7 inches in diameter and are usually situated aboveground, suspended in grasses, low shrubs, or small trees. Some winter nests are located in burrows or crevices. Active throughout the year, these nocturnal mice use the runways of other rodents on the ground and are nimble climbers. They make a high-pitched bugling sound.

Food consists mainly of seeds.

Harvest mice breed throughout the year, except where winters are severe. Studies suggest that breeding stops during cold weather. Gestation periods range from 21 to 24 days, and litters very from one to seven youngsters, but the average size of a litter is four.

There is one species in western Oregon.

THE SPECIFIC NAME *megalotis* is derived from the Greek words *megas* ("great") and *otos* ("ear") and refers to the conspicuous ears of members of this species. The first specimen, on which the scientific description is based, was caught in the northeastern corner of Sonora near San Luis Spring, Grant County, New Mexico, by C.B.R. Kennerly, in 1855. The species was described scientifically in 1857.

THE WESTERN HARVEST mouse occurs in the southern portion of zone 3. It ranges in length from 5 to 6 inches; the tail ranges from $2^3/8$ to 3 inches in length. It weighs between a quarter and a half an ounce.

The harvest mouse in Oregon resembles the common house mouse, from which it can be distinguished by the presence of distinctive grooves down the length of the upper incisors. The tail is well haired and does not noticeably taper; the ears are large and relatively hairy. House mice, on the other hand, have scantily haired, scaly tails with a noticeably taper toward the tip and have smaller, naked ears.

Harvest mice might also be confused with the Pacific jumping mice of the genus Zapus (to be discussed later), but can be distinguished from them because jumping mice, which also have grooved upper incisor teeth, have tails that are about 150 percent of the length of the head and body and pure white bellies.

The upper parts of the harvest mouse are a rich brown with some yellowish pink, darkening with grayish along the median line of the back, clearing to orangish with an admixture of yellowish pink along the sides and chest. The chin, belly, feet, and lower half of the tail are light gray or whitish.

HABITAT AND
BEHAVIOR

HARVEST MICE ARE usually found in meadows or grassy uplands or along vegetated fencerows bordering fields, where they commonly use the existing runways and burrows of other animals, such as meadow voles. They tend to gather during the dry season in moist areas or even wet marshes but spread out again with the onset of winter rains.

These little mice are nocturnal, and, unlike many other species of small mammals, are more active on moonlit nights than on dark ones. They build their bird-like nests of woven balls of grass on the ground or supported above the ground in dense clumps of grass or herbaceous vegetation. The nests are frequently lined with fine grasses or down and are entered through a hole in one side.

Their food consists mainly of seeds, which are gleaned from the ground or cut from the stems of grasses by bending the stem to the ground. Green shoots of vegetation and some insects are also eaten. The western harvest mouse can drink 100 percent seawater and survive.

Harvest mice breed throughout the year; females give birth to from one to six young, usually four to six and probably have more than one litter per year. Babies weigh about 3/100 of an ounce at birth. They leave the nest when about three weeks old and attain the size of adults in about five weeks. Some females breed when seventeen weeks old.

These little mice have many enemies, including owls, weasels, domestic cats, snakes, and even scorpions, which have been found eating still-warm babies.

THE GENERIC NAME *Peromyscus* is derived from the Greek words *peron* ("something pointed") and *myskos* ("little mouse").

Deer Mice and White-footed Mice
Peromyscus

Deer mice and white-footed mice are found in almost every habitat throughout their geographical distribution. Since these little mice are the most readily trapped of all North American mammals, they appear to be the most abundant, but are not necessarily so in every habitat, despite appearances.

Members of this genus vary in length of the head and body from $3^1/4$ to $6^3/4$ inches and in length of the tail from $1^5/8$ to $8^1/4$ inches. Adults weigh from half an ounce to $1^3/4$ ounces. They vary in color from nearly white, to gray, to yellowish or orangish brown, to dark brown, to black. Those that inhabit forested or wooded areas are generally darker with larger ears and longer tails than those species that live in open or arid country.

They emit faint to shrill squeaks and buzzes. When excited, members of many species produce a drumming sound by rapidly thumping their front feet. These agile, nocturnal mice spend their days sleeping in clean, soft nests made of grasses, mosses, birds' feathers, shredded bark, thistle down, string, or whatever is available. A nest may be in a burrow, in a hollow tree, log, or stump; under bark, boards, or stones; in rock crevices, nests of other small mammals, or human abodes. When a nest becomes soiled, it is abandoned and a new nest is constructed.

Their diet consists of anything edible!

In many areas, these mice breed throughout the year. The gestation period ranges from 21 to 27 days but may be longer. Litters vary from one to eleven young; average litters consist of four. The youngsters disperse when three to six weeks old. Although most of these mice probably live less than one or two years in the wild, in captivity they may live as long as five and a half years.

Peromyscus is a genus of the New World and occurs from southeastern Alaska, northern Canada, and Labrador south to extreme northern Columbia, South America. There are two species in western Oregon.

Deer mouse
Peromyscus maniculatus

DISTRIBUTION AND DESCRIPTION

total length 149-228 mm
tail 70-126 mm
weight 12-32.5 g

HABITAT AND BEHAVIOR

THE SPECIFIC NAME *maniculatus* is a New Latin word meaning "small handed." The first specimen of this ubiquitous little mouse came from the Moravian settlements in Labrador, Canada, and its scientific description was published in 1845.

DEER MICE OCCUR in all zones (photo 81). They vary in length from 6 to 9 inches, with tails that range from $2^3/4$ to 5 inches in length. Adults weigh from about half to just over one ounce.

The deer mouse is a slender animal with large eyes, big ears, and a long tail. In summer, the back varies from brown to dark brown, but is darkest along the midline from the top of the head to the base of the tail. Some individuals are reddish brown, others grayish brown. The top of the tail is light brown to dark brown. The tops of the feet, the underside of the body, and the underside of the tail are clear white. There is a sharp line of demarcation between the dark upper parts and the white under parts. In winter, the long, soft pelage is slightly brighter in color. An immature mouse is gray above and white below.

DEER MICE OCCUPY every conceivable terrestrial habitat in the wild, even at times where I would not have thought one would live. Their domain includes the beach; numerous subterranean burrow systems; piles of brush; in, under, and on top of rocks, stumps, and fallen trees; and up in the crowns of forest trees themselves. On several occasions, after I disturbed these mice in their daytime nests as high as 80 feet above the ground in Douglas fir trees, they raced along the branches and up and down the trunks. They are also good swimmers, though they readily cross streams on fallen trees that act like bridges.

Almost strictly nocturnal, they become active as soon as it is dark. Even though they are extremely graceful and agile, they can be noisy enough to suggest a much larger animal as they scamper through dry vegetation or rummage among cooking utensils left on a camp table.

Deer mice appear to be social and have home ranges that shift and overlap loosely. The average size of a male's home range is about $4^3/4$ acres and that of a female is about $3^1/2$ acres. They are unrestricted by humans and most human activities in their home ranges and roam more or less freely over most of their habitat (photo 82).

Deer mice build nests from a variety of materials, such as grasses, mosses, fibers or roots, mattress stuffing, wool of domestic sheep, and even thistledown. Vernon Bailey, writing about deer mice in western Oregon, even found their nests "under or in the great fleeces of moss that drape the trees in the coastal forests." And I have found nests of these mice under

Photo 81. Deer mice nearing adulthood. (Photograph by and courtesy of Murray L. Johnson.)

boards and large slabs of bark, in hollow trees, abandoned woodpecker cavities, seats of little-used or abandoned automobiles, cupboard drawers, and mattresses in abandoned buildings, and in the abandoned nests of tree squirrels, woodrats, Oregon red tree voles, and birds.

These beautiful mice are active throughout the year. During cold, wet weather, activity is curtailed aboveground, although they may be active belowground. At high elevation, they are active under the insulating cover of snow.

Deer mice are omnivorous and eat a wide variety of foods, which they find by odor rather than sight. In western Oregon, they eat seeds of such trees as Douglas fir, lodgepole pine, Sitka spruce, bigleaf maple, tanoak, California laurel, and black walnut. They also eat such fruits as salmonberry, skunkcabbage, salal, thimbleberry, blackberries, huckleberries, and so on, as well as mycorrhiza-forming fungi, serving the same type of ecological function as I described in the introduction. They also eat insects, such as beetles, and other invertebrates, and they get calcium from gnawing on the shed antlers of deer and elk and the bones of dead animals (photo 83).

Although deer mice breed throughout the year in the Willamette Valley, my data indicate a lull in breeding activities during the most inclement winter weather along the coast.

Photo 82. If one reads carefully the tracks left at night by visiting animals, one is apt to find tracks of deer mice. The normal hopping gait of a deer mouse in a moderate hurry; having left the small pit it dug, visible in the lower left of the picture, the mouse proceeded from the pit toward the top of the picture. The elongated tracks in front represent prints of the hind feet, whereas the front feet are the smaller, diagonally place tracks to the rear of the direction in which the mouse was moving. (Photograph by Chris Maser)

Photo 83. The leg bone of a deer gnawed by these little mice. (Photograph by Chris Maser)

Breeding is undoubtedly seasonal in the higher mountains. Litters range from one to nine offspring but are usually three to five. Most females have more than one litter per year.

Deer mice are born helpless, naked, and blind. They weigh about 7/100 of an ounce at birth, but they grow rapidly and in a few weeks are independent of their parents. The gray pelages

of the youngsters make them distinguishable from those of the adults, which are predominantly brown. In late summer and early autumn, the young are more often captured in traps than are the adults because they are dispersing to establish their own home ranges. In the wild, deer mice are probably short lived; however, one male is known to have survived in the wild for 32 months.

Deer mice are important prey for snakes, owls, weasels, skunks, mink, marten, bobcats, domestic cats, coyotes, foxes, and ringtails (a relative of the raccoon). Humans take a toll on these mice, using poisons and traps because they occasionally invade human habitations, where they may damage foodstuffs and electrical wiring.

Take care in handling deer mice; they have been shown in recent years to carry the dangerous hantavirus.

In the summer of 1958, at the age of 18, I got a job as a counselor in a YMCA forest camp for boys in the Cascade Mountains. I slept on a low, wooden pallet in a three-sided shelter made of logs and cedar shakes. The shelter was surrounded by ancient western hemlock and western redcedar trees through whose high crowns the wind blew in soft, swooshing sighs as darkness crept into the forest.

One night, shortly after I arrived in camp, a plump, female deer mouse scrabbled about on my sleeping bag. Over the course of an hour or so, she rummaged here and there, nibbling on my bar of soap, and generally keeping me awake with her hustling and bustling. I did not think too much about it, because deer mice were everywhere scurrying hither and yon, which is exactly what deer mice are supposed to do!

But the next night she was back, and the night after that, and the following night. By the fourth night, I decided I wanted to get a good look at this wee mouse, so I left a candle burning. At first nothing happened. Then, just as I decided the gently-flickering light was keeping the mouse away, she suddenly appeared out of the shadows and scampered onto my chest, where she "screeched" to a halt. With pointed nose twitching, large, dark eyes glistening, and big, sensitive ears straining forward, she inspected me, all the while having everything in reverse for an instant getaway.

Neither of us moved; I held my breath until I could not hold it any longer, and began breathing as quietly and slowly as possible. But instead of dashing away, she relaxed also. Sitting on my chest about 6 inches from my face, she began washing her face and ears. She was most fastidious in her grooming, which ended only when she had cleaned her body right down to the very tip of her long, slender tail.

Finished with her grooming, she ventured closer and closer to my face until she almost touched me. That, however, was quite enough bravery for one night, so she scurried away.

She appeared regularly, and performed her nightly toilet in the dancing light of my candle while sitting on my sleeping bag only inches from my face. Her visits made each night a special time I looked forward to as the shadows of evening began stealing the light of day from the forest.

She became quite trusting and seemed to enjoy the little snacks I left for her on the ledge alongside my bed. Not knowing at first what she would like, I left an assortment of shelled peanuts, rolled oats, raisins, and pieces of apple. Although over the course of a night she either ate whatever I left or packed it away to her pantry, her favorite food appeared to be raisins, which she consumed on the spot.

Camp lasted about three months, and it was toward the end of the summer that I first saw the short-tailed weasel near my sleeping shelter. I was greatly saddened, though not surprised, when my little friend suddenly vanished.

Thinking back, I am still deeply touched that so small a mouse would trust me enough to perform her nightly toilet on top of me as I lay in my sleeping bag during those soft, quiet summer nights some forty years ago. I had participated in new kind of relationship, one in which a wild animal befriended me of its own accord, in its own habitat. That indeed was an honor!

Piñon mouse
Peromyscus truei

THIS MOUSE WAS named in honor of the early naturalist F.W. True. The individual on which the scientific description is based was caught at Fort Wingate, McKinley County, New Mexico, on September 14, 1885, and was described scientifically in the same year.

DISTRIBUTION AND DESCRIPTION

total length 170-231 mm
tail 76-123 mm

THE PIÑON MOUSE occurs in the southern part of zone 3. These mice range in length from $6^3/4$ to $9^1/4$ inches and their tails, which are about 90 percent of the length of the head and body and long-haired at the tip, vary from 3 to 5 inches.

These mice are beautiful with their large, nearly naked ears. The soft and lax fur is a dark reddish brown on top, considerably darker along the midline of the back. The feet and under parts of the body are white. The tail is bicolored, a brownish black stripe on top, which is about one-third of its circumference, and white below. Immature mice are dark gray on their upper parts.

HABITAT AND BEHAVIOR

THESE BIG-EARED MICE generally live in chaparral, open mixed forest, and oak woodlands, where they inhabit piles of rocks, cliffs, and fallen trees, but like other members of the genus, they wander freely as they explore for food. During seasons of excessive dryness, they may even occupy the burrows of other small mammals.

Piñon mice are strictly nocturnal and are rarely seen except when captured or driven from their nests. They occasionally enter houses, where they make a racket at night, but are seldom common enough to be noticed.

Like other mice in the genus *Peromyscus*, they live mainly on seeds, nuts, berries, and some insects, as well as mycorrhiza-forming fungi (see introduction).

Litters ranging from two to six young, usually two to four, are born after a gestation period of 25 to 40 days. The babies are born helpless, naked, and blind, but their eyes open between fifteen and twenty days of age. Owls, weasels, spotted skunks, foxes, domestic cats, and snakes, such as gopher snakes and the Pacific rattlesnake, are among their main enemies.

THE GENERIC NAME *Neotoma* is derived from the Greek words *neos* ("new" or "recent") and *tomos* ("a cut"). The reference of this name is obscure.

The length of the head and body in woodrats ranges from 6 to 9$^1/_4$ inches, and their tails vary from 3 to 9$^5/_8$ inches. They weigh from 5$^1/_2$ ounces to almost 1 pound. Their fur can vary from long and soft to somewhat short and harsh. Coloration of the pelage is delicate, blending from pale tannish gray to dark gray or reddish tan to reddish brown above and almost pure white below. Adults have a prominent gland in the middle of their bellies that usually causes the hair to appear soiled; this is most noticeable in the males.

Woodrats occur from seacoasts to deserts or humid jungles up to and above timberline in the high mountains. They build their nests on the ground around fallen trees, stumps, and in hollow trees, in piles of rocks, talus, or the nooks and crannies of cliffs; some species construct nests in shrubs, trees, or cacti. The outer nests or lodges vary from a pile of sticks and assorted materials to a compact, conical structure that may reach 8 or more feet in height and 8 feet in diameter at the base.

A woodrat picks up materials while foraging; if, on the way to its nest, it finds a more desirable item, it will be taken instead. Around a human camp, woodrats often take shiny objects and leave old bones, dried pieces of dung, or whatever they had been carrying, earning them the names "traderat" and "packrat." Two enterprising scientists took advantage of this compulsion to collect or trade and studied the movements of woodrats by using shiny balls of aluminum foil with numbers coded inside.

When woodrats are excited or alarmed, they drum their feet on the substrate. Although some people think this behavior is simply a nervous reaction, I think it is a form of communication or warning.

Their food consists of a wide variety of plant materials as well as some invertebrates. They apparently glean enough moisture from their food that they seldom need to drink.

One to four young are born in a well-lined, warm, soft nest after a gestation period of 30 to 37 days. In the northern portions of their distribution, woodrats apparently have a single litter per year, but they seem to breed throughout the year farther south.

Woodrats of the genus Neotoma occur from southeastern Alaska and British Columbia throughout most of the lower 48 United States, south as far as Nicaragua and Guatemala. There are two species in western Oregon.

Woodrats
Neotoma

Dusky-footed woodrat
Neotoma fuscipes

THE SPECIFIC NAME *fuscipes* is derived from the Latin words *fuscus* ("brown, dark, dusky") and *pes* ("foot"). The first specimen of this woodrat was collected at Petaluma, Sonoma County, California, by E. Samuels in 1856, and was described scientifically in 1858.

THE DUSKY-FOOTED WOODRAT occurs in the southern portion of zone 1 to the vicinity of Coos Bay, Coos County, and throughout most of zones 2 and 3. It is basically an invader from southern chaparral type of vegetation from California and seems to be restricted along the immediate coast to this southern type of vegetation, which terminates in the Bandon-Coos Bay area of Coos County.

These woodrats range in length from 16 to 19 inches, and their tails vary from 8 to $12^3/4$ inches. They weigh from just over 8 to almost $9^1/2$ ounces.

The dusky-footed woodrat's round, tapering tail is almost as long as the head and body and is covered with short, blackish hairs. The ears are large and thinly haired. The pelage is moderately long and soft, reddish to yellowish brown on the back, which is darkened with numerous black-tipped hairs; it is darkest along the middle, sometimes appearing grayish. The sides are lighter than the back with fewer black-tipped hairs, and the underside is whitish. The middle of the belly is washed with light tan in most individuals; this is the area of the large skin gland. The tops of the feet are grayish brown, but the toes and claws are white.

HABITAT AND BEHAVIOR

THESE LARGE, DARK, round-tailed woodrats are primarily inhabitants of chaparral and open woodlands over much of their geographical distribution, but they also live in vegetated fencerows and along the edges of dense forest, where they build their stick houses in the trees, even near the tops of the trees. Although they are chiefly nocturnal, one may occasionally see a dusky-foot during daylight hours.

In open habitats, these woodrats construct large, conical lodges of sticks and other materials on the ground at the base of trees, in dense brush or blackberry brambles, on and over logs, or in trees, where I have found them as high as 50 feet above the ground. A woodrat's lodge is normally the most conspicuous evidence of its presence. Along the Oregon coast, however, the lodges are usually small, well hidden, and difficult to detect. Inland, the lodges are more visible, probably because the country is more open than the coast with its rank vegetation.

The nest or nests are situated within the outer lodge, which may measure 8 feet in height and 8 feet in diameter at the base (photo 84). The nest or nests are situated within the lodge. The lodge is composed of broken sticks or sticks cut by the rat itself, pieces of bark, or almost anything the rat can manage to carry, such as bones, cans, glass, old shoes, wire, sheep's wool, bits of dried cow hide.

A lodge may have from one to five chambers or compartments, depending partly on its size, which in the Willamette Valley at least grows over time as succeeding generations of young rats add to it. The various chambers are used as nursery, living room, storage room, and toilet. Except for the toilet, all chambers are kept clean. They are connected with one another and with the outside by runways.

Chambers are made of dry grasses, mosses, shredded inner bark of dead trees, dry leaves or fronds of ferns, sheep's wool, or feathers. There used to be many abandoned buildings in the Coast Range of Oregon; if a woodrat moved into one,construction of the lodge was meager at best and might include such things as wallpaper, lace curtains, or the lining of an old, discarded hat.

There are various exits and runways from the lodges to the outside. These often are hollow branches or hollow fallen trees when available and incorporated into the lodge, or branches that interlace with nearby trees when a lodge is situated in trees. Such routes offer quick retreats from enemies, since these woodrats are expert climbers. Some lodges, built on the ground, have tunnels beneath them going down into the soil, but they rarely exceed 2 feet in length. When a nest is disturbed, the woodrat may abandon the nest; if not, it will repair it, sometimes changing the position of the living room.

In addition to the woodrats themselves, other animals may inhabit the lodges. I have found deer mice, house mice, Oregon

While I was living at Bandon in the early 1970s as I studied the mammals of the Oregon coast, I received a call from a lady who said that blossoms of her prized roses were mysteriously disappearing and asked if I would please look into the thefts. The blossoms had been neatly cut off at a tell-tale angle, which showed the tooth marks of a woodrat, leaving the bushes almost naked of flowers. After an hour's search, I found the well-concealed lodge of a dusky-foot within 50 feet of the edge of the lady's lawn and the rose bushes. The proof, however, was the 25 to 30 freshly cut roses in the dusky-foot's storage chamber.

Photo 84. The outer lodge of a dusky-footed woodrat. (Photograph by Chris Maser)

red tree voles, spotted skunks, striped skunks, and Pacific treefrogs in the nests of these rats. In addition, an incredible host of invertebrates create a world of their own in the chambers housing the old, decomposing fecal pellets. At times, the woodrats apparently are evicted from their own abodes, but this is not always the case.

When cornered by a potential predator, these woodrats can be savage fighters, but they appear to be more or less tolerant of one another when forced to be together. Dusky-footed woodrats are colonial in that several individuals usually occupy an area; the colonies, however, wax and wane over the years. Where the habitats of the dusky-footed and bushy-tailed woodrats overlap, the colonial nature of the former apparently allows them to outcompete the more solitary bushy-tails for nesting sites.

Dusky-footed woodrats eat a wide variety of plants, which undoubtedly is an important factor in allowing this species to successfully occupy the range of habitats that it does. They eat the needles, leaves, and inner bark of Douglas fir and domestic apple trees; the leaves of western redcedar, Port-Orford-cedar, rhododendron, wax myrtle, salal, blackberries, and thimbleberries, the leaves and fruits of California laurel, acorns of tanoak, fronds of bracken fern and deer fern, and the flowers of domestic roses.

Female dusky-footed woodrats apparently mate with a single male in any given year. After mating, males become hermits, living apart in small nests, which they usually construct in trees. Females normally bear their young from February to May, but one pregnant female has been found in July and another in December. Although these rats usually produce one litter per year, I found one female who had produced two litters within the same year. Litters range from one to four young, usually two or three.

Babies are born naked, blind, and helpless. Their eyes open at seventeen days, and they are weaned in about three weeks. Before they are weaned, babies remain firmly attached to their mother's teats most of the time. Their deciduous incisors curve sideways as they grow out from the jaws, forming a diamond-shaped opening when they are brought together, a natural locking mechanism around the mother's teats. This has definite survival value: a mother can carry her babies wherever she goes, greatly reducing the chance of predation on the defenseless youngsters. It is not uncommon, for example, to see a mother, with babies firmly attached to her teats (photo 85), escape from a disturbed nest by moving from one tree to another along interlaced branches.

Dusky-footed woodrats have a number of enemies, such as hawks, owls, coyotes, long-tailed weasels, spotted skunks, and domestic cats and dogs, as well as people. The two main predators, however, appear to be Northern Spotted Owls and bobcats. I once found a pair of bobcats in the vicinity of Port Orford, along the Oregon Coast, that were subsisting to a large extent on an ample supply of these woodrats.

Photo 85. Baby dusky-foots firmly attached to their mother's teats. (Photograph by Chris Maser)

THE SPECIFIC NAME *cinerea* is the Latin word for "ash-colored." The first specimen was captured near Great Falls, Cascade County, Montana. Its scientific description was published in 1815.

Bushy-tailed woodrat
Neotoma cinerea

BUSHY-TAILED WOODRATS occur in all zones. They range in length from 13³/4 to 18³/4 inches, and their tails vary from 6⁵/8 to 9³/8 inches. They weigh from just over 6 ounces to almost 1 pound.

The bushy-tailed woodrat is large and has a long, bushy tail, large ears with little hair, long whiskers, and a soft, almost woolly pelage. There is little seasonal variation in the color of the pelage. The back varies from a dark grayish brown to a somewhat reddish brown; the pelage is darkest along the middle of the back from the head to the base of the tail. The sides are lighter and more brownish. The underside varies from gray to light gray with areas of clear white. There is a prominent gland in the skin near the lower middle of the abdomen; it has shorter hairs and is stained yellowish. The tail has two types of hairs— short, woolly hairs, which are light gray, and long, straight hairs, which are dark gray to blackish. On the underside of the tail, the hairs vary from light brown, to light gray, to white. The tops of the feet and toes, including the claws, are white.

DISTRIBUTION AND DESCRIPTION

total length 345-472 mm
tail 165-236 mm
weight 156-444 g

THESE LARGE WOODRATS with the decidedly bushy tails live among rocks, cliffs, and fallen trees, and in hollow trees and buildings when such are available, both in the open and in the deep forest. My experience with both the dusky-footed and the bushy-tailed woodrats indicates that, while the dusky-foot is

HABITAT AND BEHAVIOR

Photo 86. Nest of a bushy-tail in the attic of an abandoned building. (Photograph by Chris Maser)

In years past, I used to trap bushy-tailed woodrats in aluminum live traps in an old cabin in the Coast Range of Oregon. Lying in my sleeping bag inside the cabin, I could hear the woodrats rummaging about—until one got caught in a live trap. The instant the woodrat discovered that it was confined, it would invariably drum its feet, and all would be instantly still. I never caught another woodrat during the same night.

more prone to live in relatively open country, the bushy-tails are more denizens of the coniferous forest, although both species do occur together in some places.

Bushy-tailed woodrats are nocturnal in their foraging habits. Over most of their distribution in Oregon, they are directly associated with cliffs, rocky outcroppings, natural talus, and human-made talus along the banks of roads and railroad tracks. In western Oregon, where suitable rocky habitats are scarce, they live in hollow standing and fallen trees and build outdoor nests on the branches of trees or in buildings (photo 86). These woodrats are notorious along the coast because they frequently invade buildings, including houses occupied by people.

Bushy-tail woodrats living in rocky habitats characteristically leave their urine in such a way as to create white, calcareous deposits (photos 87 and 88) and produce tarlike feces; those living in forests devoid of rocky areas do not. All bushy-tails, however, have typical droppings, both hard and soft.

Unlike dusky-footed woodrats, bushy-tails do not build compact, conical lodges. Instead, they normally make an open cuplike nest, 6 to 8 inches in diameter, in a hollow tree, building, or some other sheltered place.

In the Cascade Head Experimental Forests (between Otis, Lincoln County, and Neskowin, Tillamook County, along the Oregon coast) I found bushy-tails living in the hollow trunks of western hemlock trees. At times, these nests had small collections of sticks at their entrances. Although most entrances were level with the ground, cavities in the trunks of trees had entrances 10 to 15 feet above the ground.

Mammals of the Pacific Northwest

Photo 87. Calcareous deposit created by generations of bushy- tails urinating in the same place. (Photograph by Chris Maser)

Photo 88. Tarlike feces deposited by bushy-tails, often on a vertical rock or even while clinging to the underside of an overhanging rock face within the residential crevice of a cliff. I have never figured out the significance of these droppings or why they differ from the normal droppings. (Photograph by Chris Maser)

I also found a few outdoor nests in live Sitka spruce and Douglas fir trees. Nests in trees—some as low as 15 feet above the ground, others as high as 50 feet—are compact and made of dry sticks and freshly cut twigs, often taken from the tree in which the nest itself is built. One nest, in a Sitka spruce tree, was made from the twigs of Sitka spruce, western hemlock, red alder, and elderberry; another in a Douglas fir tree, from Douglas fir sticks. Such nests vary from 15 inches to 3 feet in diameter and $2^1/2$ feet in height and may have from one to six connecting chambers. Living quarters are made of such materials as moss and the shredded inner bark of dead trees. The ground under a nest is often littered with cut twigs and the rat's dung.

By day, bushy-tailed woodrats are quiet and usually in their nests. But at night these expert climbers can and often do make an amazing racket as they rummage about in the darkness seeking food and collectibles. After all, they, like their cousins the dusky-foots, are not called "packrats" for nothing.

Bushy-tailed woodrats are primarily vegetarians, eating a wide variety of plants, but they probably also consume meat when they find it. In western Oregon, they eat such plants as the needles and bark of Douglas fir, Sitka spruce, and western hemlock, the leaves of red alder, and the green portions of Pacific bleeding heart, angled bittercress, red elderberry, Pacific waterleaf, trailing blackberry, and Himalaya blackberry. In addition to green vegetation, bushy-tails also feast on truffles when they are available.

Bushy-tails in the lower elevations of western Oregon begin breeding in January or February. The young are born from March (perhaps even earlier) through June or July. There usually is one litter per year, sometimes two. Litters range from one to six young, but two to four offspring is probably the usual size.

A female giving birth sits in a crouched position with the top of her head against the substrate. With the onset of labor, her whole body trembles spasmodically until a youngster appears, after which the trembling stops for two to three minutes while the newly born baby creeps under its mother's body and attaches itself to a nipple. The mother's body then begins to tremble again and continues until the next youngster is born, when it too attaches itself to a nipple. The process is identical for the birth of each baby, which, for a litter of four, takes a total of about fifteen minutes. After the last baby is attached to a nipple, the mother rises and pulls them into the nest. The only sign of the birthing process is a small, wet spot.

Baby bushy-tails are born naked, blind, and helpless. Their eyes open at thirteen to sixteen days of age. Male babies gain an average of about one-tenth of an ounce per day for the first eleven days, females a little less. Youngsters are apparently

---

Some years ago, my friend and mentor Murray Johnson, who was then curator of mammals at the Puget Sound Museum of Natural History, had a gentle, captive bushy-tail at his home in Tacoma, Washington. On one occasion, I put my hand into the cage to scratch the sleeping bushy-tail's ear. On being touched, the bushy-tail opened its eyes, took my finger gently in its front teeth and carried the finger to the edge of the cage. With its forefeet, it pushed the finger out. After expelling the unwanted finger, the bushy-tail went back to its nest and resumed its sleep.

Shortly after going back to sleep, the rat defecated, but instead of soiling its cuplike nest, it grasped each fecal pellet with its front teeth as the pellet was expelled and, with a flip of its head, tossed them away from the nest. By all appearances, this procedure did not interrupt the rat's sleep.

---

weaned in about a month. Before they are weaned, however, they remain firmly attached to their mother's teats, just as young dusky-foots do. See the previous account on page 202 for a description of how this locking mechanism works.

The two main predators of bushy-tailed woodrats in Oregon probably are Northern Spotted Owls and bobcats. Other predators include large owls, hawks, coyotes, marten, long-tailed weasels, spotted skunks, and domestic dogs and cats. Humans must also be included because they kill many in and around buildings with traps and poisons.

When space permits, they keep a loose collection of treasures surrounding the nest. For example, Walter P. Taylor, a noted student of mammals, once found the following items secured by one bushy-tailed woodrat in a dormitory near Snoqualmie Pass in the Cascade Mountains of Washington: chewed rags, grass, and leaves, chewed paper, the thumb of a glove, pieces of string, thongs, an apple core, onion peel, bacon rind, raisins, ten bars of chocolate, figs, oakum, puff balls, a dime, a newspaper clipping on prevention of forest fires, the lid of a coffee can 4 inches in diameter, paraffin from a jelly jar, bread crusts, bones, meat scraps, the rind of a cantaloupe, a scone, nineteen pieces of candles, four potatoes, dried apricots, several cakes of soap, lemons, mushrooms, beans, peanuts, a banana, and fifteen lumps of sugar.

None of the nests that I have found in western Oregon had so rich a treasure trove. Most nests in buildings were, to the human eye, merely piles of junk, but those on rafters and between double walls were simple cuplike structures without the usual pile of sticks, bones, and other sundry materials.

Voles
Arvicolinae

THE SUBFAMILIAL NAME Arvicolinae is derived from the Latin words *arvum* ("a field") and *colo* ("to inhabit") plus the New Latin suffix *inae* (which designates a subfamily).

Although many members of this subfamily are commonly called meadow mice or field mice, they are more properly termed voles. The word "vole" is from a noun, "vole-mouse," of Scandinavian origin. Vole is also a French noun used in playing cards. "To go the vole" means to hazard all for great gains, an apt description of the life of many members of this subfamily.

Voles are primarily terrestrial in habits, occurring from the Arctic to the desert, from sea level to above timberline in the high mountains. Most members of this family burrow considerably beneath the surface of the ground. Others are mainly arboreal, and some are aquatic.

In western Oregon, for example, two voles are aquatic but live in widely separated geographical areas: the water vole lives high in the mountains and the muskrat in the valleys and foothills.

The Oregon red tree vole lives in coniferous trees, while the white-footed vole, which has much the same geographical distribution, lives on or below the ground but feeds much of the time in deciduous trees.

The California red-backed vole is a denizen of the forest, where its diet is almost solely truffles (the fruiting bodies of hypogeous fungi). It dies within a year after the forest is logged by clearcutting because the fungi stop fruiting and the voles starve. The Oregon vole lives in the same forest as the red-backed vole, but in low numbers. It, too, lives primarily on truffles in the forest, but also eats some green vegetation. When the forest is clearcut, however, the Oregon vole explodes in numbers because it switches its diet to grasses and forbs—something the specialized red-backed vole cannot do.

Like the water vole, the heather vole lives high in the mountains, but is strictly terrestrial and lives primarily along the forest-meadow interface.

In the one place that I know of, the montane vole occurs in the High Cascades in zone 5, where it lives in a meadow and is strictly terrestrial. The water vole, which lives in the same meadow, is primarily aquatic and its life centers on the stream running through the meadow. The heather vole, which also lives in the same area, is found primarily along the edge of the meadow, where it meets the forest.

The California vole is confined to the Umpqua and Rogue River valleys some miles south of Eugene, whereas the gray-tailed vole is a denizen of the Willamette Valley from the vicinity of Eugene northward. The Townsend vole, whose geographical distribution overlaps those of the other two, lives in habitat that is wetter than that of either the California vole or the gray-tailed vole.

The long-tailed vole is confined mainly to the coast and Coast Range, west of the Willamette, Umpqua, and Rogue valleys. It lives along small streams but is not aquatic, whereas the Townsend vole in the area lives in the meadows, and the Oregon vole in the area lives in the forest and clearcuts.

In comparing the habitats of the voles of western Oregon, one can begin to get an idea of how they partition the habitat in such a way that it drastically reduces their competition for food, shelter, water, and space, which is not to say that some competition does not exist where they chance to overlap on the same acre. But even then, competition is limited, from my experience at least.

Voles, primarily those in the northern part of their distribution, have stocky bodies. They have blunt noses and small, round ears that are usually concealed in the body fur. These small-eyed mammals are often active throughout the 24-hour cycle. Their legs are normally partially concealed within the contours of their bodies, making them appear short. Their tails are usually short, with little hair. Pelages are normally dark in color, often appearing coarse in texture.

Voles are primarily vegetarians, but a few are known to eat meat, such as carrion, when it is available.

The gestation period may be as short as 21 days or as long as 30 days. Voles do not hibernate, and in mild climates, litters may occur throughout the year. Litters range from one to eleven offspring, usually four to eight. Babies usually weigh less than half an ounce at birth. Females may begin breeding when only 3 weeks old and usually breed within 24 hours after giving birth.

In the wild, most voles probably live less than two years because of predators. Some members of this subfamily, such as lemmings, are famous for their cyclical fluctuations in numbers; a population may build for three or four years and be seemingly everywhere and then suddenly crash and appear to be all but nonexistent.

Voles occur in Europe, Asia, and North America, extending south only into northern Central America and North Africa. There are five genera of voles in western Oregon.

Red-backed Voles
Clethrionomys

THE GENERIC NAME *Clethrionomys* is derived from the Greek words *kleithrion* ("a bar for closing") and *mys* ("mouse"). The reference of this name is obscure.

Red-backed voles are attractive little mammals. They have thick pelages that are long and soft in winter but shorter and harsher in summer. The general coloration of the back is a pronounced reddish wash over dark gray hairs; the wash is less pronounced on the grayish sides. Their undersides vary from dark gray to almost white. They have slender, relatively long tails. Their ears are moderately conspicuous through the body hair. The length of the head and body ranges from about 3 to $4^1/2$ inches; the tail from 1 to almost $2^1/2$ inches. These voles weigh from half an ounce to about $1^1/2$ ounces.

Red-backed voles are active throughout the 24-hour cycle and throughout the year. They are good climbers, living around the stumps, fallen trees, rocky outcroppings, cliffs, and rock slides of forested areas. They inhabit tundra, bogs, moist or dry forests, and woodlands. Some species burrow considerably.

Breeding usually begins in late winter and continues until late autumn. After a gestation period of 17 to 21 days, one to eight, usually four to six, babies are born in a nest of shredded vegetation.

Red-backed voles occur throughout Canada, Alaska, and most of the northern United States, extending south along the Pacific coast, and the highlands of the Rocky Mountains and the Appalachians. They also inhabit Europe and Asia, including Japan. There is a single species in western Oregon.

California red-backed vole
Clethrionomys californicus

THIS VOLE WAS NAMED after the state of California, combined with the Latin suffix *cus*, denoting possession. The first specimen was collected at Eureka, Humboldt County, California, by T.S. Palmer in 1890.

DISTRIBUTION AND DESCRIPTION

total length 120-187 mm
tail 34-74 mm
weight 16-32 g

CALIFORNIA RED-BACKED VOLES are delicate, attractive little mammals that occupy zones 1 and 3 to 5 (photo 89). They range in total length from almost 5 to $7^1/2$ inches, and their tails range from just over 1 to just under 3 inches. They weigh from half an ounce to just over 1 ounce.

These voles are of slender proportions. They have weak teeth, varying from light yellowish to whitish. Their ears protrude moderately beyond the body hair. They have a relatively long, thick, soft pelage. The upper parts are usually somber with an often poorly defined median dorsal stripe, which is obscured by intermixed black hairs. The sides are light to dark gray washed with light brown to light yellowish brown; the venter is dark gray washed with light brown to light yellowish brown.

They have a slender, long tail that may be sharply or indistinctly bicolored, light brown or blackish above and whitish below. Juveniles are darker, duller, and do not have the brown wash.

Habitat and behavior

CALIFORNIA RED-BACKED VOLES are strictly inhabitants of coniferous forests, especially those with substantial numbers of large fallen trees. Within the coastal mountains, two primary factors affect the presence or absence of red-backed voles. The first is the amount of light that reaches the forest floor, which, in turn, controls the quantity and variety of herbaceous plants and shrubs that survive. Red-backed voles apparently prefer dense forest with little or no ground vegetation. The second factor is the abundance of rotting, punky fallen trees. The voles increase in number as the number of large rotting fallen trees increases.

The species occurs from the Oregon side of the Columbia River south into northern California, and from the coastal forest to the eastern flank of the High Cascade Mountains. There are two subspecies: *Clethrionomys californicus californicus* lives west of the Cascade Mountains and *C.c. mazama* lives in the mountains. There are a number of differences between the subspecies. Those that live west of the Cascade Mountains are extremely secretive, whereas those in the Cascade Mountains are much less so.

The former subspecies was long considered uncommon because of the difficulty of capturing them. In fact, prior to 1941, they had been reported from only four localities within Oregon. I have found these voles to be common when I selected the right microhabitat (restricted habitat) within the forest and took my unbaited traps to the voles. While selecting the right microhabitat is important, so are the voles' habitats.

Photo 89. Adult California red-backed vole, eating a truffle. (USDA Forest Service Photograph by Douglas C. Ure.)

Most of its life is spent under the forest floor, where it lives in a relatively stable regime of temperature and humidity and is close to its subterranean source of food—truffles, which is one of the secrets in trapping them. Being almost strictly fungus feeders, they are uniquely unattracted to trap bait, so I had to learn to capture them without using it.

The aboveground activity of red-backed voles along the coast seems to be influenced by the weather. I had been trapping repeatedly four miles southeast of Bandon, in Coos County, beginning July 1970 but had neither found sign of nor captured a single red-backed vole. Then, on November 28, 1971, there was a subtle change in the usual winter weather, and the red-backed voles suddenly became active on the surface of the forest floor. A domestic cat killed the first red-back I had seen. Aboveground activity continued until December 21, 1971, when there was again a subtle change in the weather. During this time, a domestic cat killed another red-backed vole, and I trapped nine of them. Although I trapped on and off in this location until June 9, 1972, I found no more evidence of red-backed voles; it is as though they had simply vanished. In contrast, about two hundred California red-backed voles of the latter subspecies were captured in live-traps set on the forest floor in the on the west slope of the Cascade Mountains in zone 4. And no special effort was made to capture them.

Another contrast between the two subspecies is that *C.c.californicus* are normally active during the night, whereas *C.c. mazama* are active throughout the 24-hour cycle. As far as I know, nothing is known about the belowground habits of these voles.

There is also a difference in the diet between the two subspecies of voles. *Clethrionomys c. californicus* has a specialized diet composed primarily of truffles and lichens, whereas *C.c. mazama* eats more vascular plant material. Despite these differences, both subspecies are closely associated with large, fallen trees because the wood, under closed forest canopies, remains wet throughout the year and is a site of prolonged fruiting of the truffles—the vole's specialized food. In fact, California red-backed voles die out of an area within one year after clearcut logging because as the trees die, the belowground fungi stop fruiting, and the voles starve to death.

While we have learned much about the food habits of the California red-backs, little is known about their reproduction. *Clethrionomys c. mazama* has a long breeding season that begins in early April and lasts until late October or early November. In contrast, *C.c. californicus* breed throughout the year. Litters range from one to eight offspring, but the usuall litter size is two to four young.

The main predators of the California red-backed vole are marten, short-tailed weasels, long-tailed weasels, and spotted skunks, as well as bobcats, Great Horned Owls, Saw-whet Owls, and Northern Spotted Owls. Domestic cats also take a significant toll of these small rodents along the Oregon coast.

THE GENERIC NAME *Phenacomys* is derived from the Greek words *phenakos* ("a cheat, impostor") and *mys* ("a mouse"). Only two species comprise the genus *Phenacomys*: the Ungava vole and heather vole.

The head and body are about 4 inches in length, and the tail varies from about 1 to 2 inches in length. These voles range in weight from almost 1 ounce to about $1^1/2$ ounces. Their fairly long, finely textured pelages are grayish to slightly brownish above and whitish to grayish to very light brownish below.

The genus occurs throughout most of Canada and the high mountains of the western United States as far south as New Mexico. There is one species in western Oregon.

Heather Voles
Phenacomys

THE SPECIFIC NAME *intermedius* is derived from the Latin words *inter* ("between") and *medius* ("middle"). The first scientific specimen was captured 20 miles north-northwest of Kamloops, British Columbia, by George M. Dawson on October 2, 1888 and described scientifically in 1889.

Heather vole
Phenacomys intermedius

THE HEATHER VOLE occurs in zones 4 and 5. It ranges in total length from $5^1/4$ to 6 inches; its tail varies from 1 inch to $1^3/4$ inches. Adults weigh from just over 1 to about $1^1/2$ ounces. A heather vole's back is grayish to brownish gray, and its underside is whitish or light brown. There are moderately stiff, orange hairs in the peripheral half of the inside of the ears. The bicolored tail is grayish to brownish gray above, whitish below. The feet are light brown to whitish on top. Juveniles may be duller and darker in all respects.

DISTRIBUTION AND DESCRIPTION

total length 130-150 mm
tail 25-42.5 mm
weight 34-40 g

HEATHER VOLES INHABIT subalpine and alpine areas. Within their wide geographical distribution, they seem to have an equally wide range of habitats. I have caught these voles and found their sign in thickets of heather in alpine meadows above 13,000 feet and in subalpine meadows strewn with boulders and interlaced with swift, cold, rocky streams above 10,000 feet of elevation. I have also caught them along the edge of lodgepole pine forest, where it interfingers with meadows of grasses and forbs, with streams or snowmelt ponds nearby (photo 90).

HABITAT AND BEHAVIOR

Photo 90. Habitat of the heather vole at the interface between the subalpine forest and meadow. (Photograph by Chris Maser)

Although the voles were often trapped in dry sites along fallen trees or large rocks, I also caught them in the entrances of their burrows amidst mountain huckleberry.

Summer nests made of grasses and mosses, and those in which young are born, are usually situated belowground. However, I once found an occupied summer nest under a fairly large piece of plywood that had been discarded by deer hunters the year before. Winter nests, on the other hand, are built on the ground under the cover of snow; they are thick walled and usually made of grasses, but lichens and other materials are also used. The inner nest chambers are warm, soft, and clean since these voles deposit their feces in specific blind-ended tunnels in snow outside the nest. In spring, when the snow melts, the accumulation of fecal pellets becomes apparent.

Newly exposed toilets are light reddish and appear fresh (photo 91). Those that have been exposed for some time, however, are darker and beginning to break down and the surrounding vegetation is growing up around and through them (photo 92).

Runways are characteristically used under cover of snow; as the snow disappears these small voles take up subterranean abodes and there is scant indication of runways in summer when the voles are freed of winter's constraining snows. Nevertheless, I sometimes find their faint runways connecting one burrow entrance with another, and I occasionally find their dropping-studded runways in thickets of huckleberry.

Heather voles seem to be strict vegetarians. They eat such plants as white heather, huckleberry, bear grass, and lousewort. In summer they eat mountain huckleberry. They cut off the terminal twigs in lengths of $1^1/2$ to $3^1/2$ inches and pull them down into their burrows, where they eat the leaves and discard

Photo 91. Recently melted snow reveals the heather vole's winter toilet; the lighter pellets are more recently deposited than are the darker ones along the edge of the pile. (Photograph by Jan Henderson and Chris Maser.)

Photo 92. A heather vole's winter toilet as it appears in early summer with vegetation beginning to grow through it. (Photograph by Chris Maser)

the twigs as refuse. At times, I have found several uneaten twigs in a burrow entrance, which the occupant may have been saving for a later meal.

In winter, under the cover of snow, heather voles gnaw the bark off lodgepole pine. On one occasion, I found where they had climbed up 5 or more feet into the snow and eaten the needles off the terminal shoot and several lateral branches of sapling lodgepole pines. Sometimes they even kill young trees (photos 93 and 94), and all around these trees were their tell-tale toilets.

Pregnant heather voles have been captured from May through August. Litters range from two to eight, with four to six probably being the usual number. Even in the short summer at high altitudes, it is probable that more than one litter per adult female is produced per year.

Weasels, martens, and owls are probably the main predators of heather voles, although some may also fall prey to the red fox, coyote, and bobcat.

Photo 93 (left). A young subalpine fir girdled by heather voles under the cover of snow. (Photograph by Chris Maser)

Photo 94 (right). A young subalpine fir almost completely eaten by heather voles under the cover of snow; clearly, they killed the tree. (Photograph by Chris Maser)

Heather voles are active throughout the 24-hour cycle, so I have on occasion captured them alive during the day. I find these fascinating little creatures of gentle demeanor and a pleasure to handle. I distinctly remember one that I captured alive. It did not struggle, but settled down calmly in my hand.

Having nothing in which to put it while I prepared a small holding cage for its transportation, I put it in the pocket of my shirt, from which it immediately climbed onto my shoulder. From my shoulder, it went up the collar of my shirt and sat quietly between my neck and the collar of my shirt the whole time I arranged the cage. Not wanting to frighten it, I moved about very slowly.

When the cage was prepared, I put my hand slowly to my shoulder until I touched the vole, which climbed unhesitatingly onto my hand and thence into the cage with no attempt to flee. All this happened while I was standing in the middle of its habitat with no way to prevent its escape.

THE GENERIC NAME *Arborimus* is derived from the Latin words *arboris* ("tree") and *mus* ("mouse").

Voles of the genus *Arborimus* have been separated from the genus *Phenacomys*, but not all mammalogists have agreed with the separation. They are considered among the most primitive members of the vole family. The genus consists of three species, one primarily terrestrial and two primarily arboreal.

Members of the genus have long, soft pelages that vary dorsally from rich brown to bright orangish red. They have long, hairy tails, small eyes, and pale, almost-naked ears. These secretive voles are nocturnal, uncommon to rare, and difficult to obtain; thus they are seen by few people.

Although they breed throughout the year, they have small litters, ranging from one to four young, usually two to three offspring per litter. Tree voles, and presumably also white-footed voles, grow and develop slowly compared with members of the other genera of voles.

Voles of this genus have a very restricted geographical distribution, occurring from the Columbia River in Oregon south into Sonoma County, California, and from the Pacific Ocean east into the Cascade Mountains but not, as far as is known, reaching the crest of the High Cascades of zone 5.

There are two species in western Oregon, the white-footed vole and the Oregon red tree vole, both of which are nocturnal and have a remarkably similar distribution. The spatial use of its habitat by the Oregon red tree vole extends from belowground, where I have occasionally seen them disappear into burrows to avoid my capturing them and found one nest, into the coniferous canopy; the spatial use of its habitat by the white-footed vole extends from belowground, where it apparently nests, into the hardwood canopy. Both voles are obligate browsers and have behavioral patterns unlike other voles, which helps demonstrate their close relationship.

White-footed Vole and Tree Voles *Arborimus*

According to my late friend and colleague, Murray L. Johnson, with whom I studied the genus Arborimus *for well over a decade, tree voles and white-footed voles probably evolved in the ancient Klamath Mountains of southwestern Oregon. Murray wrote in 1973: "Geologic history provides data for an attractive hypothesis:* Arborimus *evolved in this region from some primitive microtine during the Pliocene [about 13 million years ago], adapting to a habitat of mixed deciduous and coniferous forest during a time of moderate temperature and high rainfall; consistent climatic conditions along the Pacific Coast have maintained the restricted habitat."*

White-footed voles and Oregon red tree voles may be among the smallest browsing arboreal mammals. The wet climate and moderate extremes in temperature, which contribute to a long growing season and little winterkill of their food plants, along with limited competition for browse by other mammals in the tree canopy, obviously influence the occurrence of these voles along the coast of Oregon and northwestern California. There are two species in western Oregon.

White-footed vole
Arborimus albipes

THE SPECIFIC NAME *albipes* is derived from the Latin words *albus* ("white") and *pes* ("foot"). The first white-footed vole was caught near Arcata, Humboldt County, California, on May 24 1899, and it was not until 1915 that a second specimen was caught, this time two miles west of Vida, Lane County, Oregon.

DISTRIBUTION AND DESCRIPTION

total length 149-182 mm
tail 57-75 mm
weight 17-28.5 g

WHITE-FOOTED VOLES OCCUR in zones 1, the extreme southwestern part of zone 3, and the western edge of zone 2. It crosses from zone 2 into zone 4 in the region of Eugene and Springfield, which lie at the southern terminus of the Willamette Valley proper (zone 2) and north of the hot, dry, southerly influence of the Rogue Valley (zone 3).

White-footed voles range in total length from 6 to 7 inches and weigh from a little over half an ounce to 1 ounce. They have long, soft pelages. Dorsally, they are a rich, warm brown; ventrally, they are gray to gray washed with light brown. These voles have relatively long, distinctly bicolored tails, blackish above and white below. They have very small eyes; their ears, although usually concealed in the pelage, are naked of hair. The tops of the feet are usually white.

HABITAT AND BEHAVIOR

I HAVE CAPTURED 22 of these white-footed voles in western Oregon, all along the coast, where they are associated with the red alders along the sides of small streams, often with an understory of salmonberry, willow, thimbleberry, and vine maple. Forbs in the groundcover include northwest nettle, Oregon oxalis, and American twinflower.

In the 74 years between 1899 and 1973, a total of only 72 of these rare voles were captured, 63 in Oregon and 9 in California, almost all of which were caught very close to water. Although still considered one of the rarest members of the vole family in North America, it is more common than previously thought.

Food-habit studies indicate that it depends primarily on riparian areas and forest edge for its habitat and food, rather than on aquatic or semiaquatic vegetation.

The tree-climbing ability of this vole is substantiated by studying its food habits, which show that it is dependent on

Mammals of the Pacific Northwest

leaves of red alder and red willow during the summer, before they fall from the trees. That it is has thus far, to the best of my knowledge at least, been captured solely on the ground indicates that it certainly is not restricted to the tree canopy. This notion is further corroborated by its food habits.

About 50 percent of the vole's diet is supplied by red alder and red willow. A study found that hardwood trees (red alder, red willow, bitter cherry, and blue elderberry) accounted for more than 57 per cent of the diet. But forbs (such as northwest nettle, western spring beauty, western golden saxifrage, wood groundsel, and American twinflower), accounted for more than 23 percent and shrubs (such as long-leaved Oregon grape, salal, ocean-spray, red huckleberry, and shot huckleberry) for more than 15 percent. The eighteen species that are the most abundant in the voles' diet accounted for 95 percent of its food; the remaining fourteen species comprised less than the remaining 5 per cent of the diet. No seeds, fruits, fungi, or animals parts were found in the diet.

It is therefore concluded that the white-footed vole occupies a habitat extending from its belowground nests into the tree canopy, and that its diet consists entirely of leaves from vascular plants, the leaf blades of which are finely chewed. These habits seem to closely parallel those of the Oregon red tree vole.

White-footed voles breed throughout the year. Although females have three pairs of teats, litters range from two to four, but three offspring seems to be the usual size of a litter.

Owls, weasels, mink, spotted skunks, domestic cats, and such snakes as the rubber boa are probably the main enemies of white-footed voles.

Oregon red tree vole
Arborimus longicaudus

THE SPECIFIC NAME *longicaudus* is derived from the Latin words *longus* ("long") and *cauda* ("tail"). The first specimen of the Oregon tree vole was captured at Marshfield (now Coos Bay), Coos County, Oregon, in 1890 by Aurelius Todd.

DISTRIBUTION AND DESCRIPTION

total length 158-206 mm
tail 60-93.5 mm
weight 25-47 g

THE OREGON RED TREE vole (photo 95) occurs in western Oregon from zone 4 westward to the shores of the Pacific Ocean and from the Columbia River southward to the vicinity of the Smith River just south of the Oregon and California border. It is, however, absent from the floor of the Willamette Valley due to lack of habitat. Its geographical distribution is becoming increasingly fragmented due to clearcutting of Oregon's forests.

Adult tree voles are from 6 to 8 inches long, including their tails, and weigh from about 1 to 1½ ounces. The fur on their backs varies from brownish red along the northern coast, to brighter brownish red along the middle coast, to more orangish red along the southern coast and along the western flank of the Western Cascades. Their undersides are light gray, and their long, hairy tails vary from rich medium brown to black.

OREGON RED TREE VOLES are primarily inhabitants of coniferous forests, although they occasionally live in mixed coniferous-deciduous forest. They inhabit Douglas fir throughout most of their geographical range, but along the Oregon coast they can be found in Sitka spruce and in occasionally in old-growth western hemlock. The voles will only inhabit trees that are at least 25 to 30 years of age because younger trees are not structurally able to fulfill their nesting requirements or give them adequate protection from inclement weather.

Adapted to a life in the tops of a few select species of coniferous trees, primarily Douglas-fir, the Oregon red tree vole is one of the most specialized arboreal (tree-dwelling) mammals in the world. Although Douglas fir needles are the chief food of Oregon red tree voles (photo 96), at times they also eat the needles of Sitka spruce, grand fir, and western hemlock. They also eat the tender bark off the twigs (Photo 97), and some individuals split the twigs open, apparently to obtain their pithy center.

Oregon red tree voles cut twigs at night. Some feeding is done away from the nest, but most of their food is eaten in and/or on top of the nest. Twigs, 1 inch to 9 inches long, are cut and carried to the nest by the vole; most are stored on top of the dwelling, from which some may be blown to the ground

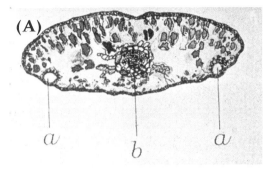

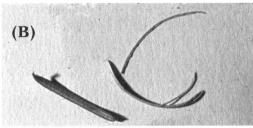

Photo 96. (A) Cross section of a Douglas fir needle, where "a" depicts the resin ducts that are split off and discarded by red tree voles as feeding refuse, and "b" is the vascular bundle, which is eaten. (Photograph by Murray L. Johnson.) (B) Douglas-fir needles: one whole needle, one needle showing a resin duct being split off, and one resin duct split off. (Photograph by Murray L. Johnson.) (C) Twigs of Douglas fir cut by tree voles with a pile of discarded resin ducts in the center. (Photograph by Murray L. Johnson.) (D) Tell-tale discarded resin ducts on the floor of the forest under an old-growth Douglas fir with a tree vole's nest high in its crown. (Photograph by Chris Maser)

by a winter storm. Other, shorter twigs are often partially or completely pulled into the nest and stored in the tunnels.

A tree vole bites the needles off near their bases, one at a time. Holding a needle with one or both forefeet, it rapidly and mechanically bites along the edge of the outer resin duct with its front teeth, flips the needle over, bites off the other resin duct, and then eats the rest of the needle (photos 96 and 97). Young, tender needles are often eaten entirely. Although a vole spends much of the day sleeping in its nest, it periodically arouses and goes to its pantry for a snack.

Tree voles obtain most of their moisture requirements from their food, but they also lick dew and rain off the needles of the

Photo 97. An Oregon red tree vole spliting resin ducts off the needle of a Douglas fir. (Photograph by Kenneth L. Gordon and Chris Maser.)

trees in which they live. Along large rivers, such as the Columbia, the fog that condenses on the needles of the ancient trees is also an important source of water for the voles, and allows them to extend their geographical range farther eastward than would otherwise be possible.

A female tree vole may breed again within 24 hours of giving birth. But the sexual encounter is brief because male and female lead separate lives, each with its own nest, getting together only when the female is receptive.

Although a mother may give birth to one to four youngsters, the usual litter is two to three naked, blind, helpless babies. A nursery nest normally contains a single litter (photo 98), but occasionally two litters of two different ages occupy the nest simultaneously. The youngsters stay in the nest until they are a month or more old, at which time they construct their own

Photo 98. Baby Oregon red tree voles on a bed of discarded resin ducts from Douglas fir needles. (Photograph by Kenneth L. Gordon and Chris Maser.)

nests. There seems to be a survival advantage in slow development; when they leave the nursery nest at a more advanced age than their ground-dwelling cousins the meadow voles (who can reproduce when 25 days old) tree voles have relatively good balance in addition to being more self-sufficient. Extensive wandering outside a nest before they are adequately developed would increase accidental mortality—such as fatal falls—and predation.

Oregon red tree voles disperse slowly through the young forest, partly because of the small size of their litters and slow development, and partly because they tend to move only short horizontal distances in young forests, which results in a colonial or clustered distribution throughout the forest. But in the ancient forest, many hundreds of generations of tree voles can live in one ancient Douglas fir because they disperse vertically up and down within the crown of the tree and horizontally from the trunk to the outermost living boughs of the tree's branches.

The young tree voles construct their nests at a height above ground limited by the height of the living branches. In the ancient forest, nests are usually built at the outer limits of the branches and often over 150 feet above the ground. Nests in the young forest are normally built on a whorl of living branches next to the trunk of the tree as high as 60 to 70 feet above the ground (photo 99).

Old, abandoned nests are also visible in the young forest as dilapidated clumps of debris on the dead branches below the living crowns of the trees. As the trees continue growing toward the sunlight, the lower branches become shaded and die from lack of sunlight. This leaves the voles' nests exposed and without a close supply of fresh food—the living green needles of the tree— so the nests are abandoned as the voles follow their food supply ever upward toward the sun.

For the most part, a nest is constructed from twigs that a vole cuts from the tree in which the nest is situated, but voles may also renovate the abandoned nests of birds, squirrels, and woodrats. The outer nest is made of twigs; the inner nest chambers are lined with the discarded resin ducts from Douglas fir needles.

Photo 99. Occupied nest of an Oregon red tree vole 50 feet up in a young Douglas fir. (Photograph by Chris Maser)

A nest constructed entirely by a tree vole is a more or less haphazard affair. The nest begins as a platform of food twigs on which the vole feeds; as additional twigs are carried to the foundation, food refuse accumulates. The vole's movements, along with deposition of feces and urine, continuously pack the material down. As larger twigs collect, the vole crawls under them to feed. The discarded resin ducts are pushed and pulled around as the vole moves, making a small cavity for itself. It pushes and scratches the resin ducts up and over the sides and top until a completely enclosed chamber is formed. Food twigs are brought to the nest nightly, and although some are stripped of their needles, others are not. As the vole alternately feeds in the nest and on top of it, the nest gradually increases in size and settles until it becomes firmly packed and well anchored. From then on, the growth of the nest is incidental, but the stages of growth are fairly standard. The nest changes from the original small structure to a large roundish structure on one side of the tree. As growth progresses, the nest spreads out, continuing around the trunk until it connects with its beginning. Thereafter, growth of the nest is up and out. Rain, snow, and constant movement of voles inevitably packs the structure. As generation after generation of voles inhabit the nest, a thick layer of fecal material accumulates, further anchoring the nest firmly to the tree.

The various portions of the interior of a nest are connected by a series of tunnels that lead to the outside. Although there is no predictable pattern to the system of tunnels, one particular tunnel exists in every nest; this is an escape tunnel leading from the interior of the nest to an exit at the bottom of the nest next to the trunk of the tree through which the vole can escape undetected. There are usually one or two tunnels leading to the top of the nest and the daily supply of food. These tunnels are normally situated in such a way that a vole can reach food without exposing itself for any length of time.

The system of tunnels within a nest changes constantly. If a nest is large enough to surround the trunk of the tree, however, a circular runway going around the trunk is normally present; at times such a runway is found both inside the nest and on top of it. In old, long-established nests with a thick layer of decomposing, earthlike manure, tunnels and chambers lined with resin ducts penetrate the fecal mass. These tunnels and chambers are dry and relatively permanent. Possibly such a nest can be inhabited constantly, despite changes in the weather.

All inner chambers are lined with resin ducts. The inner chambers are of two types, nest chambers and fecal chambers. Old nest chambers become fecal chambers as new living quarters are established; they become filled with feces and urine that gradually decompose, along with the resin ducts, forming a soil-like layer typifying nests of long use. Oregon red tree voles are clean and do not defecate or urinate in their sleeping quarters or the tunnels that are being used; they do defecate on top of the outer nest, however, while feeding during the night.

On being evicted from their nests by predators, such as raccoons, the voles often move head-first down the trunk of the tree, and if they reach the ground, go into a handy burrow or under any available debris. Or they may go out onto a branch, cross to an adjoining tree, and suddenly stop, crouch, and remain motionless. Such behavior, along with their small size, is protective because their reddish pelage blends into the dimly lighted surroundings so well that they are difficult to see when motionless. In deep twilight, when they begin to be active, a motionless tree vole is almost impossible to see among the branches of the tree because red is one of the first colors to fade or become neutral as darkness approaches, and without the red of their pelage or their motion to attract attention, a crouching tree vole is easily mistaken for a fir cone that has become lodged on a branch.

Some Oregon red tree voles, usually adults, launch themselves into space when confronted by a predator instead of going onto a branch or down the trunk of the tree. Although many have their falls broken by lower branches to which they are adroit at clinging, others merely "free-fall" to the ground and almost invariably land on their feet—uninjured. As they free-fall, they spread their legs out, and use their tail for balance. Some voles free-fall from as high as 60 feet up in the trees, land, and head for the nearest cover. Age, and perhaps a degree of learning, seems necessary before such a feat can be accomplished successfully because young voles seldom land on their feet. They appear to lack the ability to spread their legs and do not seem to have control of their tails; thus, they land on their backs.

Although the red tree vole has a variety of potential enemies, such as raccoons that tear the nest open and get the young, as could long-tailed weasels and marten, the Northern Spotted Owl is their main predator, in addition to Saw-whet Owls, Screech Owls, Long-eared Owls, and occasionally Great Horned Owls.

Meadow Voles
Microtus

THE GENERIC NAME *Microtus* is derived from the Greek words *mikros* ("small") and *otos* ("the ear") and refers to the small, generally concealed ears of members of this genus.

Voles of the genus *Microtus* vary greatly in size. The length of the head and body ranges from about 3 to 7 inches; the length of the tail from three-quarters of an inch to a little over $4^1/2$ inches. Meadow voles weigh up to $4^1/2$ ounces. Their pelages are usually fairly long and lax. The general color of the upper parts is grayish brown, but the darker members of the genus are blackish and the lighter members more reddish or yellowish. Their underparts range from grayish through light brownish to whitish. They have small eyes, and their ears are normally concealed in the body hair.

Meadow voles do not hibernate and can be seen at any time. Most species make well-defined, often elaborate, runways through and under low vegetation. When ground cover is scant, they may dig extensive underground burrow systems. A used runway is a neat path with all the vegetation clipped to the ground; an abandoned runway, however, usually has vegetation growing in the middle and looks unkempt.

Meadow voles do not walk along their runways—they dash. When startled, a vole may emit a high-pitched squeak, gnash its teeth, and either flee or freeze, depending on its location and previous activity. If it flees, it will, however, do so only along the runway.

Some meadow voles are promiscuous breeders, and incest among siblings still in the nest is not uncommon. Females may begin mating when about three weeks old and produce up to thirteen litters per year, usually ranging from four to eight young per litter. The gestation period is about 21 days. In the southern part of their geographical distribution, they breed throughout the year, but breeding in the northern portion is limited to the summer months. The amount of daylight probably limits breeding in winter. Young are weaned at about two weeks and are thought to live for a little more than a year in the wild.

Meadow voles occur throughout most of North America south to Guatemala and throughout most of the northern two-thirds of Europe and Asia.

There are seven species of meadow voles in western Oregon. Because their life histories are similar in many ways, some accounts are briefer than others, where to give more information would simply be redundant.

THIS VOLE WAS NAMED in honor of John Kirk Townsend, primarily a naturalist, who was in charge of the hospital at Fort Vancouver during the winter of 1835-36. The first specimen was obtained on the "Lower Columbia River," near the mouth of the Willamette River, "on or near Wappatoo or Sauvie Island" and was described scientifically in 1839.

TOWNSEND VOLES OCCUR in zones 1 through 4. In the latter, however, they are confined to the western edge of the zone in the low foothills.

DISTRIBUTION AND DESCRIPTION

total length 169-238 mm
tail 48-85 mm
weight 42-103 g

These voles are fairly large, ranging from about $6^1/2$ to $9^1/2$ inches in total length with a tail that varies from 2 to almost $3^1/2$ inches long. They weigh from just under $1^1/2$ ounces to just over $3^1/2$ ounces. Townsend voles have harsh pelages. The back is dark brown to dark reddish brown in summer and dark brown to blackish brown in winter. There are many black-tipped guard hairs mixed in the fur of the back. The underside is grayish to grayish brown. The tail is blackish to black above, a little lighter below; the feet are gray to dark gray above.

THE TOWNSEND VOLE is primarily a creature of moist grassy areas often interspersed with sedges and rushes. Within their habitat, they sometimes live in lawns. Along the edges of such lawns, where adequate protective cover is lacking, they build extensive underground burrows with the entrances close together.

HABITAT AND BEHAVIOR

Although these voles are often abundant, little is really known about their habits. They are active at any time and are highly dependent on their runways and burrows. In areas that are free of winter flooding, the same runways may be used by several generations of voles, eventually being worn into ruts 1

Townsend voles are good swimmers and divers and readily enter the water when pursued. Along the coast, they swim across small streams with ease. In fact, one of the first things I heard about when I arrived on the Oregon coast to study the mammals was the "bog rats." According to reports, they ranged anywhere from 5 inches to well over 3 feet in length and had tails that were either long or very short. One cranberry grower told me that his bog rats had tails of various lengths; I suggested that these differences in length were likely the work of his cat! They were excellent swimmers and divers, I was told, and exceedingly difficult to catch. The bog rats, which inhabited the cranberry bogs along the southern coast, turned out to be Townsend voles. They were often seen swimming and diving when the cranberry bogs were flooded in October, floating the berries to one end of a bog to harvest them.

Photo 100 (above). The nest of a Townsend vole situated above the saturated soil in early spring. (Photograph by Chris Maser)

Photo 101 (right). The tell-tale activity of Townsend voles as they ate their way through the grasses under the cover of snow. (Photograph by Chris Maser)

to 2 or more inches deep. Food refuse and droppings are often found in these older runways. Heaps of well-packed droppings form at the intersections of frequently used runways that can reach 6 to 7 inches in length, 2 to 3 inches in width, and 5 inches in height and resemble a road overpass in shape.

When Townsend voles inhabit an area with a high water table, the entrances to their burrows may be under water much of the year. In such cases, their nests are located within any small hummock that stands slightly above the water table, until the water table gets too high in winter. During the dry part of the year, they usually construct their nests underground, but during wet winters when the water table is high and floods them out of their burrows, they often locate their round, grass nests on the surface of the ground, frequently on high points (photo 100) and preferably under cover of old boards or other structures if available.

Townsend voles are vegetarians, eating the succulent stems and leaves of a wide variety of green plants, such as velvet grass, clover, buttercups, false dandelion, alfalfa, horsetail, and cattail, to name a few. Under cover of the rare snows in the Willamette Valley, these voles literally eat their way through the buried grasses (photo 101). At times they store food by piling it under protective cover; one such cache contained almost $14^1/2$ quarts of mint roots.

Townsend voles apparently breed throughout the year. Litters range from two to ten offspring, but the usual number is four to six. Babies are naked, blind, and helpless at birth. They are born into a warm nest made of dry grasses.

Owls and hawks probably are the main predators of Townsend voles, although bobcats, foxes, coyotes, weasels, mink, skunks, and snakes also take a toll. Domestic cats and people also kill many of them, especially when the voles take up residence in lawns and cranberry bogs. In marshy areas, particularly near the mouths of large rivers along the Oregon coast, Great Blue Herons often hunt these voles, stabbing at them with their long beaks as the voles dash along their runways.

Long-tailed vole
Microtus longicaudus

THE SPECIFIC NAME *longicaudus* is derived from the Latin words *longus* ("long") and *cauda* ("tail"). First caught at Custer, 5,500 feet, in the Black Hills of Custer County, South Dakota, the long-tailed vole was scientifically described in 1888.

There are two subspecies of long-tailed voles in western Oregon—*Microtus longicaudus abditus* and *M.l. angusticeps*. The subspecific name *abditus* is the Latin word for hidden, secret, or removed, whereas the subspecific name *angusticeps* is derived from the Latin *angustus,* meaning narrow or small, and the New Latin *ceps,* meaning head.

Microtus l. angusticeps was first caught at Crescent City, Del Norte County, California, in 1889. The first specimen of *M.l. abditus* was captured at Pleasant Valley, 8 miles south of Tillamook, Tillamook County, Oregon, by A. Brazier Howell on September 8, 1920.

Microtus l. angusticeps occurs in Oregon, as far as I know, along the immediate coast in zone 1 from approximately 5 miles north of Port Orford, Curry County, southward into California. The subspecies *abditus,* on the other hand, occupies zone 1 from about Reedsport, Douglas County, northward to the Columbia River and westward into the Coast Range. In $3^1/2$ years of studying mammals along the Oregon coast, I found no long-tailed voles from Reedsport south to the vicinity of Port Orford.

The two subspecies are quite different—*M.l. abditus* is large, measuring from almost $7^1/2$ to just over $10^1/2$ inches in length; the tail ranges from just over $2^1/2$ to just over $4^1/2$ inches in length. This vole weighs just over 1 ounce to 3 ounces. *Microtus l. angusticeps* is smaller, measuring from 6 to just under $7^1/2$ inches in length with a tail that ranges from 2 to 3 inches in length. This vole weighs from just under one to just over $1^1/2$ ounces.

The pelage of both subspecies of the long-tailed vole in western Oregon is relatively short and coarse. The back varies

DISTRIBUTION AND DESCRIPTION

Long-tailed vole, *Microtus longicaudus abditus*
total length 186-265 mm
tail 68-115 mm
weight 36-87 g

Long-tailed vole, *Microtus longicaudus angusticeps*
total length 152-187 mm
tail 51-75 mm
weight 22-47 g

from grayish brown to dark brown, with many black-tipped hairs mixed. The sides are slightly lighter, and the venter is gray to dark gray—sometimes washed with white or light brown. The tail is distinctly or indistinctly bicolored, brownish to blackish above, whitish, grayish, or brownish below. The tops of the feet vary from grayish to dark grayish or from brownish, dark brownish, to blackish.

I have always found the long-tailed vole in western Oregon to be mysterious, especially the large *M.l. abditus* because I never knew where or when I would find them. I remember vividly the first *M.l. abditus* I caught many years ago at the base of a small waterfall about two-thirds of the way up Marys Peak in the Oregon Coast Range. It was a long time before I caught another. In my experience, *M.l. abditus* is primarily associated with water, and those I have caught were always along the banks of a stream of some kind, usually one with a goodly cover of herbaceous vegetation. In contrast, *M.l. angusticeps* seems to be more of a generalist in habitat, provided there is enough grass. Grassy areas appear to be an important component of its habitat requirements.

Microtus l. angusticeps seem to be generally more abundant in than the other subspecies, but this may simply reflect ignorance of the habits of *M.l. abditus*, the name of which, you will recall, means "hidden." In addition, *M.l. abditus* is more active at night than during the day and does not readily make runways. *Microtus l. angusticeps* is active at any time and makes well-defined runways. Both of these long-tailed voles occasionally overlap in habitat with the Townsend vole, which is more aggressive and, therefore, usually more common.

Little is known about the diet of these long-tailed voles except that both subspecies eat green vegetation.

Male long-tailed voles are sexually active from February through October. Pregnant females have been captured from March through November. Litters range from two to ten offspring, but three to six young appears to be the usual size.

All I know about predation of these voles is that domestic cats catch them when and where they are available.

THIS VOLE, NAMED for the state of Oregon, was captured at Astoria, Clatsop County, Oregon, in 1836 by John Kirk Townsend.

CREEPING VOLES OCCUR in all zones (photo 102). They are absent, however, from the heart of the Willamette Valley floor of zone 2 and, in my experience at least, the higher portions of the High Cascade Mountains of zone 5.

Creeping voles range from $4^3/4$ to $6^1/4$ inches in length; their tails range from $1^1/4$ to 2 inches in length. They weigh from half an ounce to just over 1 ounce. They have tiny eyes and long, thick, soft pelages. Dorsally, the pelage varies from grayish or yellowish brown to reddish or blackish brown. Ventrally, the underfur is gray washed with whitish or light brown. Their tails are indistinctly bicolored, brownish to blackish above, dark to light gray below. The feet of most individuals are grayish on top, but some are white. Juveniles are more gray or black in overall coloration.

CREEPING VOLES ARE normally associated with coniferous forest, but are generally most abundant in areas where the forest has been removed, either by fire or by clearcutting. I have also found them in white oak woodlands along the edge of the Willamette Valley and in tanoak woodlands in southwestern Oregon, but in low numbers. Where grassy areas are surrounded by coniferous forest, creeping voles are sure to take advantage of them. Along the coast they inhabit a few of the drier meadows provided they are close to the coniferous forest and have enough dead grasses to afford cover over the voles' runways. Active at any time, they are more active at night.

Creeping voles do appear to creep when they move. These little voles, with their tiny eyes and slightly curved claws, are primarily burrowers in the soils of developing coniferous forests and appear decidedly uncomfortable when they are exposed.

Creeping or Oregon vole
Microtus oregoni

DISTRIBUTION AND DESCRIPTION

total length 120-156 mm
tail 30-52 mm
weight 14-31 g

HABITAT AND BEHAVIOR

They seem to prefer protective cover low enough that their backs are in almost constant contact with it; it is not surprising, therefore, to find them closely associated with such protective covers as large fallen trees and dense vegetation.

Although creeping voles normally construct their small round nests underground, nests are occasionally found aboveground inside large rotting fallen trees and stumps or under such things as large slabs of bark and boards. Nests are made of grass but apparently are not lined inside with finer materials.

Creeping voles can live for many generations in dense old-growth coniferous forests, where they feed primarily on truffles and such green herbaceous vegetation as they can find. They share the same source of truffles as the California red-backed voles, with whom they share this habitat, but are outnumbered by the red-backed voles.

When the old forest is either burned or clearcut, the red-backs die out within a year because their specialized food, the truffles, disappears as the trees die and the fungus quit fruiting. Left without their specialized food, red-backed voles starve to death, but creeping voles explode in numbers because they shift their diet from the truffles to green herbaceous vegetation. They can do this because their molars have open roots and grow throughout the life of the vole, which allows them to eat the more abrasive vegetation and survive. Red-backs, on the other hand, have molars with closed roots, which, like ours, stop growing at some point. The softer fungal material of their diet wears their teeth down more slowly than would the more abrasive vascular vegetation.

In this way, creeping voles, which find their optimum habitat in the early stages of forest development, can survive for hundreds of generations over the centuries, albeit in low numbers, in dense forest until once again a major disturbance to the forest makes their prime habitat available. Then inevitably the forest begins reclaiming the land. As the trees grow, the creeping vole begins at some point to decline, even as the red-backs once again begin to dominate the vole population within the maturing forest.

Although creeping voles in the Cascade Mountains have a breeding season that spans about seven months, at least some breed throughout the year along the coast and in the lower elevations of the Coast Range. Females bear four or five litters annually. A litter ranges from one to eight young but usually consists of three offspring. The gestation period is about 23 days.

At birth, creeping voles are naked, blind, and helpless. Females reach puberty at 22 to 24 days of age, whereas males do not become sexually mature until 34 to 48 days of age. After sexual maturity is reached, both sexes have a period of sterility that lasts from 5 to 14 days. Those young born late enough that they do not reach physical maturity until after July do not breed until the next year.

Owls and weasels are the main enemies of creeping voles. But snakes, marten, spotted skunks, mink, coyotes, foxes, bobcats, and domestic cats also prey on them. Humans, at least in the past, may have taken a considerable toll on these voles by spreading poisons over thousands of acres of commercially important timberlands to control various rodents, especially the deer mouse.

Gray-tailed vole
Microtus canicaudus

THE SPECIFIC NAME *canicaudus* is derived from the Latin words *canitia* ("a gray or grayish white color") and *cauda* ("tail"). The first gray-tailed vole to be taken in the name of science came from McCoy, Polk County, Oregon, in 1895. It was caught by B.J. Bretherton.

THE GRAY-TAILED VOLE (photo 103) is found in zone 2. These voles range in length from 5 to 6 inches, with tails from less than one inch to 1³/4 inches, and weigh from about half an ounce to almost 2 ounces.

Their upper parts in summer are yellowish brown (yellowish gray in a few individuals) and may be fairly bright. In winter, the upper parts are darkened with blackish-tipped hairs. The sides are slightly paler than the upper parts. The entire underside is grayish white. The tail is grayish with a usually-distinct brownish line on top. The feet are grayish.

Juveniles are rather dull grayish brown to grayish above and scarcely lighter below. The feet are gray, and the tail is gray with a blackish stripe on top.

DISTRIBUTION AND DESCRIPTION

total length 127-157 mm
tail 25-46 mm
weight 19-52 g

Photo 103. Gray-tailed vole. (Photograph by Chris Maser)

THE GRAY-TAILED VOLE probably inhabited the prairie grasslands that once covered the floor of the Willamette Valley. Today it still lives in the valley, where it is closely associated with grassy areas in and around croplands, extending into the Coast Range (zone 1) where contiguous croplands are present. The agricultural use of the valley seems to have been beneficial to these voles in creating excellent habitat, although the loss of vegetated fencerows within the last two decades has undoubtedly had a negative effect on their available habitat.

Runways of the gray-tailed vole, actually tunnels through the vegetation, can be found in fields and pastures, as well as along the banks of ditches. The runway systems are often intricate and extensive with numerous burrows. Sometimes abandoned underground tunnels of the Camas pocket gopher can be traced by the numerous holes punctured into them by these voles, which appear to use such tunnels wherever possible in lieu of making their own.

Active runways, like those of the Townsend vole, frequently have piles of feces at intersections (photo 104). Some scat piles are mounds 3 to 6 inches long by 1 to 2 inches wide. Since the voles deposit their dropping on the highest point of the mound, some mounds are 3 to 4 inches high.

Photo 104. Droppings of a gray-tailed vole; these dropping are representative in shape of the droppings of all meadow voles in western Oregon. (Photograph by Chris Maser)

These voles often build their nests of grass under old boards, bales of hay, and other materials left in the fields. Subterranean nests of grass are also constructed. Their remains, excavated by foxes and skunks, can occasionally be found on top of the ground. I have watched red fox hunting for underground nests. They walk a ways and listen, walk a ways and listen. When they hear the babies belowground, they begin digging immediately and almost as quickly excavate the nest from which they extract the babies in what seems from a distance to be a single gulp.

In some areas during the wet winters of western Oregon, the water table rises to or above the surface of the ground. If the high water is not too high or prolonged, it does not seem to bother the voles. I have observed them living in flooded fields. Their subterranean nest cavity is apparently above water while the burrows leading into it are flooded. The voles have to dive through the flooded burrows to reach their dry nests. When this happens, fresh food and droppings can be found in the flooded burrows. On occasion, however, the water rises too high, and the voles are forced to flee their nests. At such times they can be seen in moderate to great numbers on roads. Many also drown, especially the young.

The vole's food consists of the green vegetation in which they live, primarily grasses (photo 105) and clover. Wild onion and false dandelion are also eaten. Leaves of the latter are frequently cut and carried into a burrow to be eaten.

Gray-tailed voles breed throughout the year. Litters range in size from two (photos 106 and 107) to eight, usually four to six. The young may become incestuous as they reach sexual maturity while still in the nest, and they can breed at about three weeks of age. Because they are cyclical in population, like other members of the genus, there are years when gray-tailed voles are seemingly everywhere and years when one is hard pressed to find a single individual.

Owls, gopher snakes, long-tailed weasels, domestic cats, foxes, skunks, Great Blue Herons, and people all hunt these voles.

Photo 105. Clipping of grass left in its runway by a gray-tailed vole; these clippings are representative of most food refuse left in runways by all meadow voles in western Oregon. (Photograph by Chris Maser)

Photo 106. Baby gray-tailed voles in their nest of dry grass. (Photograph by Chris Maser)

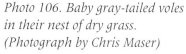

Photo 107. A close-up portrait. (Photograph by Chris Maser)

Montane vole

Microtus montanus

THE SPECIFIC NAME *montanus* is a Latin word meaning "of mountains." The first specimen of this species was caught by Titian R. Peale on October 4, 1841, at the headwaters of the Sacramento River near Mt. Shasta in Siskiyou County, California. Its scientific description appeared in 1848.

THE MONTANE VOLE (photo 108) occurs in zone 5 in only one place that I know of—the meadow below Three Creek Lake on the north flank of the mountain called Broken Top just south and east of Sisters in Deschutes County—though it is common east of the Cascades. They range in length from 5³/₈ to 7⁷/₈ inches, with tails that vary from 1¹/₄ to 2³/₄ inches. They weigh from just over 1 to 3 ounces. The back varies from light grayish, to brown, dark brown, reddish brown, sometimes blackish brown. The sides are a little lighter. The underside is whitish, sometimes washed with light brown. The tail is usually distinctly bicolored, grayish to brownish or blackish above, whitish below. The feet are grayish above. Winter pelage is grayer. Juveniles are darker and duller.

EAST OF THE HIGH Cascade Mountains, the montane vole can be found in woodlands of western juniper and big sagebrush and cultivated fields of hay and alfalfa, as well as sagebrush flats. It also occurs in subalpine and alpine meadows. Its habitat in zone 5 is a subalpine meadow.

Good swimmers, these voles like to live along water, such as ditches, streams, and rivers, or in wet meadows, as they do on Broken Top. They are active all year and throughout the 24-hour cycle; hence they can be seen at almost any time. Sociable and friendly with one another, they will fight savagely with

Photo 108. Montane vole. (Photograph by Bob Smith and Chris Maser).

anything they feel threatens them, biting quickly and severely with their sharp front teeth.

Like other meadow voles, the montane vole makes elaborate systems of runways, which can be readily seen in open or short grass, but they may not be so easily seen in the meadow on Broken Top during the height of the meadow's vegetative cover, when they are more difficult to find in the dense grasses and forbs.

Nests are made of grasses. Those of summer are subterranean, but in winter they are constructed above ground, under the cover of snow. As the snow melts in spring, the winter activities of the montane voles become visible; their abandoned runways are lined with food refuse and droppings and appear unkempt. Although summer runways may also have food refuse and feces in them, all growing vegetation will be kept neatly clipped in the runway in use.

The number of young per litter ranges from four to eight, with eight being the full complement for mature females. Although there are several litters per year, it is doubtful that breeding takes place during winter at high elevations.

These voles are cyclical, seeming to reach high populations about every four years. But sometimes they seem to almost literally overrun the countryside in an irruption that is sometimes called a "mouse plague." During such highs in population, montane voles cause a considerable amount of damage to crops, especially hay and alfalfa meadows.

Grasses seem to be the principal food for these voles, but their diet also includes sedges, rushes, and a great variety of meadow plants, including the edible parts of green leaves, stems, roots, and seeds. Food is obtained by cutting the stem or plant off at the base and then eating the desired part or parts. In the case of grass seeds, a vole cuts the stem off, pulls it toward itself, cuts off a section, then pulls it again and cuts off another section. The process is repeated until the seed head is reached, which leaves a little pile of cut stems $1^1/2$ to 2 inches long. In addition, roots and bulbs are dug up and eaten, as are the belowground runners of some plants. Normally, more than half of the vegetation cut is left as refuse, which is recycled into the soil of the meadow.

Montane voles have many predators, including hawks, owls, snakes, weasels, skunks, foxes, coyotes, domestic cats—and humans.

California vole
Microtus californicus

THIS VOLE WAS NAMED in honor of the state of California, plus the Latin suffix *cus*, which is added to a noun to denote possession. The first specimen was caught by Titian R. Peale at San Francisco Bay while on the United States Exploring Expedition from 1838 to 1842 and was described scientifically in 1848.

DISTRIBUTION AND DESCRIPTION

total length 157-214 mm
tail 39-68 mm
weight 53 g

THE CALIFORNIA VOLE occurs only in zone 3, from the California border through the Rogue River and Umpqua valleys as far north as Roseburg and Drain in Douglas County. It ranges in length from $6^1/8$ to $8^3/8$ inches; the tail varies from $1^1/2$ to $2^5/8$ inches. It weighs just under 2 ounces. This vole varies above from yellowish brown through dark brown, frequently with a reddish tinge on the middle of the back. In addition, there are long overhairs that vary from light brown to black. The sides are lighter with fewer long overhairs. The underside varies from pale blueish gray to whitish. The tail is bicolored, brownish to black above, grayish below. The feet are gray. Winter pelage is darker than summer pelage. Juveniles are duller and darker than adults.

HABITAT AND BEHAVIOR

FOUND MAINLY IN the dry upland meadows and grassy slopes of open valleys, California voles may occasionally be found in wet places, even marshes. They require less cover than regular marsh species, and their runways can be seen through the short, dry grass. In addition, small heaps of fresh soil may show where these voles have thrown it out of their burrows onto dry ground.

These voles are active throughout the 24-hour cycle. In Oregon, they appear to be restricted in both their geographical distribution and their numbers. They inhabit fertile valleys, where favorable conditions of food and cover might at any time result in an explosion of their population.

California voles breed throughout the year, and have litters that range from four to eight.

In diet they are similar to other voles, consuming grasses, rushes, young or ripe grass seeds, roots, bulbs, and tender bark.

California voles have many predators, including hawks, owls, snakes, weasels, skunks, foxes, coyotes, and domestic cats.

THE WATER VOLE IS named in honor of J. Richardson, a naturalist in North America in the early 1800s. The first specimen of this vole was captured in the vicinity of Jasper House in Alberta, Canada, and was described scientifically in 1842.

WATER VOLES OCCUR in zones 4 and 5. They are the largest member of the genus *Microtus* in North America. Adults range from about 8 inches to $10^3/4$ inches in length, of which about one third is tail. Their large hind feet are almost an inch in length. Adults may weigh at much as $4^1/4$ ounces. The pelage is long and rather coarse. The upper parts are dark reddish brown to dark grayish brown, and the underparts are washed with whitish hairs over a dark gray underfur. The tail is bicolored, dark grayish above and lighter grayish below.

THE WATER VOLE, like the water shrew, finds its favorite habitat along small, high-elevation streams that flow through subalpine and alpine meadows (photo 109). More than any other species of *Microtus*, these big, long-tailed, large-footed water voles are semiaquatic in habits. They are excellent swimmers and divers, and depend in part on the water for protection. Although they are active throughout the 24-hour cycle, their peak of activity occurs during periods of darkness, but they are often abroad on sunny afternoons, when they may be seen dashing along a runway between two burrows or swimming from one side of a stream to the other.

Most water voles live in burrows within 15 to 30 feet of the banks of a stream. Some, however, live in more outlying wet areas in a meadow, where their large, well-worn trails, usually 4 inches wide or even wider, are conspicuous during the snow-free time of the year and are often strewn with the stems of grasses and sedges they have cut for food. Their trails commonly enter the water where it is swift. The burrows, many of which are dug beneath large rocks, are about 3 inches in diameter and, constructed with no effort at concealment, often have large mounds of earth at their entrances. Freshly dug burrows may be so abundant that it seems likely that more are dug than used at any one time.

The water voles' belowground passages and nest chambers are excavated between June and late September. The voles dig their tunnels immediately below the thick network of plant roots and mosses that covers the overhanging

Water vole
Microtus richardsoni

DISTRIBUTION AND DESCRIPTION

total length 198-263 mm
tail 69-100 mm
weight 82.5-128 g

HABITAT AND BEHAVIOR

Photo 109. The stream in and around which the water voles live. (Photograph by Chris Maser)

stream banks. Belowground tunnels, 3 to 9 feet in length, and surface runways form branching travel lanes that lead to nest chambers, feeding areas, and the stream's edge.

The belowground burrows are about 3 inches in diameter; the chambers are about 4 inches high, 6 inches long, and 4 inches wide and are dug inside small rises in the microtopography, which keep the nest above the water table most of the year. Each vole fills its chamber with a large, dome-shaped nest made from short segments of leaves, grasses, sedges, and other vegetation from the meadow, and each nest is occupied by a single vole. Numerous openings in the stream's banks and onto the meadow along the stream's edge allow easy access to areas both above- and belowgound. The entrances to some of their burrows may even be under the surface of the water.

In winter the voles spread out over much of the meadow under the cover of snow, where they live from seven to eight months of each year. Their tunnels are dug through the snow along the surface of the ground, but do not lead to the surface of the snow. In fact, no vole activity will be seen above the snow once the first 3 inches have fallen without melting. Water voles make bulky nests of dry grasses under the snow. These nests, which are loosely built and fall apart soon after the snow melts, may be used throughout the winter or only during the period of snowmelt when the soil becomes saturated with water and the belowground tunnels and nests are flooded.

The voles' food consists of such things as the green leaves and stems of grasses, sedges, clovers, and numerous other meadow plants, including some of their seeds during the short mountain seedtime. Roots, bulbs, and bark of willows and other plants, such as bear grass, which is a lily, are eaten during winter.

The voles' breeding season begins in late May or early June, coinciding closely with the appearance of the first new growth of the herbaceous vegetation as melting snowbanks recede, and lasts until late September. The minimum gestation period is 22 days, after which two to ten, but usually five to six young, are born in belowground nests. Those adults that survived their first winter are responsible for most of the reproduction, because youngsters generally do not appear aboveground until after the 1st of July and only about 25 percent of them will breed in their first year. Although most voles will survive only one winter and die during their second autumn or winter, a few will survive two winters. Of the many causes of death, short-tailed weasels, long-tailed weasels, and marten account for most.

THE GENERIC NAME *Ondatra* is a North American Indian name for muskrat.

Muskrats
Ondatra

Muskrats are the largest species of voles (photo 110). Adults range in length from about 15$\frac{1}{2}$ inches to 24 inches, and they weigh up to 3$\frac{1}{2}$ pounds. They have small eyes and ears; the latter are nearly concealed in the pelage. The pelage is composed of two types of hair—underfur and guard hairs. The underfur is short, thick, fine, and very soft, whereas the guard hairs, which are interspersed throughout the underfur, are long, coarse, dark, and shiny. It is the long guard hairs that produce the dominant coloration of the back and sides, which varies from a glossy dark brown to almost black, becoming more reddish brown on the sides. The throat, chest, and belly are dominated by the underfur and are lighter in color and duller because there are few shiny guard hairs, so the pelage is light reddish brown across the chest and belly but light gray on the throat and in anal area.

Muskrats are well adapted for swimming. They have swimming fringes of short, stiff hairs along the margins of each hind foot, including the webs between the toes. These fringes increase the surface area of the feet and aid in propelling the animals through the water. Their scaly, almost hairless tails are vertically flattened and act as rudders.

These aquatic voles live in lakes, ponds, marshes, roadside ditches, irrigation ditches, streams, and rivers (including coastal tidewater rivers). They either construct houses of vegetation in open water (photo 111) or dig burrows in banks. They are expert swimmers and divers. They eat a wide variety of vegetation and some meat, such as freshwater clams and mussels.

In the northern portions of their geographical distribution, they breed from early spring to autumn, but in southern areas they breed throughout the year. Most young, however, are born from November to April. After a gestation period of 22 to 30 days, from one to eleven youngsters are born. A female breeds again while she is still nursing; several litters are produced annually. Young, sometimes called kits, are weaned when they are about one month old.

Muskrats are widely trapped for their pelts, and North American fur trappers make more money selling muskrat pelts than any other fur.

There are two species of muskrats. One species, *Ondatra obscurus*, is confined to Newfoundland. The other, *O. zibethicus*, occurs from Alaska to Labrador, south to Texas and northern Baja, Mexico, and from the Pacific coast east to the Atlantic coast, but it is absent from Florida and most of California. Although indigenous to North America, this wide ranging species of muskrat has been extensively introduced into Europe. This is the species of muskrat in western Oregon.

Muskrat
*Ondatra
zibethicus*

THE SPECIFIC NAME *zibethicus* is derived from the Greek work *zibeth* ("the civet," which is a mammal), which gave rise to the New Latin word *zibethicus* ("civet- or musty-odored"). Muskrats are so named because of the pronounced, sweet, musky odor from the secretion of glands in the anal area. The first scientific specimen of a muskrat came from "Eastern Canada" and was scientifically described in 1766.

*DISTRIBUTION AND
DESCRIPTION*

MUSKRATS OCCUR IN zones 1 and 3, and the lower elevations in the western part of 4. The physical description of our local muskrat is the same as that of the genus.

Mammals of the Pacific Northwest

THE HABITAT OF these aquatic voles is as described for the genus. When I was a boy in the 1940s and '50s, there were many muskrats in the Willamette River; today, by comparison, there are few.

HABITAT AND BEHAVIOR

Muskrats are active throughout the year. Although they may be seen at any time, they mainly are active at twilight and throughout the night. Muskrats are not particularly sociable, even with one another. They are quiet mammals, both vocally and in motion, seldom attracting attention by making noise. When startled, however, they enter the water with a loud splash and may swim a long distance under water before coming to the surface.

total length 409-620 mm
tail 180-307 mm
weight 0.541-1.575 kg

East of the High Cascade Mountains, muskrats may construct bank-burrows or houses of vegetation, depending on the depth of the water and the home range of the individual. A muskrat house is built in shallow water and is supported by a broad base resting on the bottom. Built of leaves, stems, and roots of plants, as well as mud, it may extend 3 to $4^1/2$ feet above the surface of the water. The walls of the house, although frequently a foot thick, are porous enough to maintain a sufficiently ventilated interior, even when covered with snow. A single chamber in the middle of the house just above the water, usually accessed by two or three underwater entrances, can accommodate six to eight individuals.

In western Oregon, however, muskrats seldom build the conical or dome-shaped houses of vegetation for which they are known throughout most of their geographical distribution. Instead, they dig burrows into the banks of whatever water they are inhabiting. These bank-burrows are evident where muskrats are living in tidewater near the mouths of rivers; when

Photo 112. Spadix of skunkcabbage cut and partially eaten. (Photograph by Chris Maser)

the tide goes out, many of the entrances to their burrows are exposed. A bank-burrow may be simply a tunnel leading into an enlarged chamber that serves as the living quarters, or there may be a series of chambers and tunnels. In some instances, a bank is so riddled by tunnels and chambers that it occasionally collapses underfoot.

The entrances to burrows may be from 6 inches to as deep as $2^1/_2$ feet below the surface of the water during its low level of summer. Burrows ranges from about 5 to 8 inches in diameter and from about 6 to 27 feet in length. The tunnel terminates in a spherical nest chamber that is above the level of the water and is between 12 and 15 inches in diameter. The nest is composed of a bulky, loose mass of vegetation, primarily the leaves of cattails when these are available.

Muskrats eat a wide variety of plants, such as cattails, rushes, sedges, skunkcabbage (photo 112), pondweeds, water-lilies, deerfern, and swordfern. In addition to plants, muskrats occasionally eat small pond turtles, freshwater snails, clams, and mussels, crayfish, fish, and some salamanders. In turn, muskrats are eaten by mink—their principal predators—river otters, coyotes, bobcats, and Great Horned Owls (photo 113).

Male muskrats are reproductively active from March through November, whereas females are reproductively active from May to October. The usual size of a litter is six to eight, but as many as fifteen young occasionally are born in a single litter. A female will raise at least two litters, and sometimes as many as three or four in a year. Under optimum conditions, this may translate into sixteen to twenty young being raised in one season by a single female.

The gestation period is about 29 days. At two weeks of age the youngsters can swim, dive, and eat green vegetation. Although young muskrats grow rapidly, they do not breed until the year after their birth.

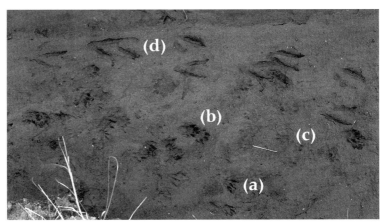

Photo 113. Tracks of a muskrat (a) along with those of its arch enemy the mink (b) interlaced with old tracks of a domestic cat (c) and those of a Canada goose (d). (Photograph by Chris Maser)

Mammals of the Pacific Northwest

THE SUBFAMILIAL NAME Murinae is derived from the Latin word *murinus* ("mouselike") and the New Latin suffix *inae*, which designates a subfamily.

Old World rats and mice range from about $4^3/8$ to 32 inches in length and weigh from a little more than one-tenth of an ounce to $3^1/2$ pounds. The tail is usually naked and scaly; in a few climbing members of this subfamily, it is semiprehensile. Members of one genus have opposable first digits on the forefeet or hind feet, with nails on some of the digits instead of claws. A few members of the subfamily have cheek pouches. Some have spiny pelages.

The habitats and habits of murid rodents vary greatly. Some are terrestrial and usually are good climbers; others burrow, dwell in trees, or are semiaquatic. A few species hop and jump. These rodents live in tunnels, hollow logs, crevices, holes in the trunks of trees, abandoned nests of birds, or buildings. Some species construct their own abodes of sticks. Murids are active day or night. They may be gregarious, living in groups or colonies, or they may live in a family group, as a pair, or alone.

Most murids eat plants and invertebrates; some include small lizards, snakes, and the eggs and nestlings of birds in their diets. The semiaquatic members of the subfamily feed extensively on mussels, crustaceans, snails, and fish, whereas those living in close association with humans eat almost anything. Although murid rodents are not known to hibernate, some store food, presumably for use during winter.

Gestation ranges from 18 to 42 days. The size of litters varies from one to twenty-two. In the warm parts of their geographical distribution, most murids breed throughout the year, often with several reproductive peaks. Most individuals in the wild live less than two years, and many less than one year. In captivity, they may live longer.

The natural geographical distribution of this subfamily is Africa, Europe, Asia (except in the extreme north), the Malayan region, Australia, Tasmania, and Micronesia. Through introductions by humans, however, the genera *Rattus* and *Mus* are nearly worldwide in distribution. There are two genera in western Oregon.

THE GENERIC NAME *Rattus* is the Latin word for rat.

The genus is difficult to define; it has more named species and subspecies than any other genus of mammals—about 570. They range in the length of their heads and bodies from $3^1/4$ to $13^1/4$ inches. The tail may be shorter or longer than the head and body and is usually scantily haired. The pelage is soft in

Old World Rats and Mice Murinae

Old World Rats *Rattus*

some species and coarse in others; in still others the hairs are enlarged and stiffened into bristles or spines.

The upper parts are black, grayish, dark brown, yellowish brown, or reddish brown, and the underparts are usually gray or white. Their appearance differs greatly. The body may be stocky or slender. Certain digits are short in members of some species but long in others. The feet of some members are modified for a terrestrial existence, others for an arboreal existence.

In most members of this genus, the gestation period varies from 21 to 30 days. Most species are prolific breeders, and at times a population may increase dramatically, at least in part because of a lack of natural enemies due to habitat alteration. Except for the black rat and the Norway rat, most species are not serious pests.

The genus *Rattus* has the largest number of species living in tropical southeastern Asia and Africa, but are found in nearly all parts of the world and in practically all terrestrial habitats. There are two species in western Oregon, both introduced.

Black rat
Rattus rattus

THE SPECIFIC NAME *rattus* is the Latin word for rat. The specimen of the black rat on which the 1758 scientific description is based came from Uppsala, Sweden.

DISTRIBUTION AND DESCRIPTION

total length 325-455 mm
tail 160-255 mm
weight 115-350 g

BLACK RATS ARE KNOWN to occur in and around human habitations in zones 1 and 2. They vary in length from 13 to $18^1/4$ inches and have tails from $6^3/8$ to $10^1/4$ inches. They weigh from 4 to about $12^1/2$ ounces.

The pelage is coarse and harsh, with long spinescent hairs. Dorsally, it varies from grayish brown to gray to black, and ventrally from whitish, to yellowish white, to gray.

HABITAT AND BEHAVIOR

BLACK RATS ARE ASSOCIATED with human habitations. Indigenous to Asia Minor and the Orient, these rats were brought to Europe during the Crusades, and are thought to have been first introduced into North America on the ships of the early European explorers. Black rats may be widely distributed in North America, but they appear to be abundant only in relatively local areas.

Black rats are expert climbers, often building their nests in vines, trees, or the roofs of buildings, hence, often called "roof rat," or "barn rat." Piles of driftwood along the banks of rivers in and about towns and cities sometimes shelter and protect colonies of these rats.

These slender rats are found on ships far more commonly than are Norway rats and have been introduced and reintroduced into most seaports. In northern seaports, the Norway rat is dominant and has evicted the black rat, but the black rat is most common in southern seaports. Where they occur together, the more aggressive Norway rat forces the black rat, the better climber, to live in the upper portions of buildings. Some think that the black rat was already abundant in North America before the introduction of Norway rats at the end of the 18th century, but that they have mainly disappeared because they cannot compete with the larger, more aggressive species.

Black rats eat almost any edible matter they can cut with their teeth.

These rats breed throughout the year and have litters consisting of two to eight young. There may be two peaks of reproductive activity, one in February and March, the other in May and June, with the period of least reproductive activity in July and August. After a gestation period of about 21 days, young rats are born naked, blind, and helpless. They mature rapidly, however, are weaned in about three weeks, and become sexually mature and capable of reproducing at about three months of age.

Large snakes, large owls, and domestic dogs and cats are predators of these rats. Humans also take a considerable toll with traps and poisons.

THIS RAT WAS NAMED after the country of Norway combined with the Latin suffix *cus*, denoting possession. The first specimen is reported to have come from England, however, and was scientifically described in 1769.

NORWAY RATS OCCUR IN and around human habitation in zones 1 through 3 and perhaps into the lower elevations of zone 4, where people live. They range in length from $12^5/8$ to $18^3/8$ inches and have tails that vary from $4^3/4$ to $8^5/8$ inches in length. They weigh from 7 ounces to one pound.

Norway rats are large and heavy bodied with tapering, scaly tails almost naked of hair and dark gray to dark brown. Their ears, usually less than an inch in length, are nearly naked, as are the soles of their feet. They have a coarse, harsh pelage, which varies dorsally from reddish brown, to grayish brown, to black. Ventrally, they vary from soiled yellowish white to a dirty gray. Young rats are dark gray above and a little lighter below. White, black, or mottled individuals are occasionally found.

Norway rat
Rattus norvegicus

DISTRIBUTION AND DESCRIPTION

total length 316-460 mm
tail 122-215 mm
weight 195-458 g

NORWAY RATS "KEPT close behind the vanguard of civilization in its progress across the continent of North America," wrote Vernon Bailey, "and were probably taken to Oregon on ships in the early part of the last century." The species was first recorded in Europe about 1553; it probably reached western Europe by ships instead of overland caravans. The first Norway rats arrived in North America about 1775. Army Lieutenant Trowbridge captured a Norway rat at Astoria, Clatsop County, as early as 1855. Today, the Norway rat is likely to occur in western Oregon wherever people live.

Norway rats inhabit so many areas in and around towns and cities that they have several common names: brown rat, house rat, barn rat, wharf rat, and sewer rat. They usually enter a new region on ships or trains and then spread rapidly as stowaways concealed in boxes, crates, or household goods being shipped about the country.

These rats are secretive, keeping under cover or in the systems of burrows they dig in banks, or under buildings, rocks, or logs. They make short forays in search of food and new areas to occupy, mainly at night, but they are also abroad during daylight hours. Their typical haunts are closely associated with the refuse of human life and they are thus common in and around many garbage dumps.

A number of years ago, just south of Yachats, on the Oregon coast, I heard snarling and squealing in the tall grasses alongside the highway. Quietly approaching the source of the sound, I witnessed an almost incredibly swift, savage fight between a large Norway rat and a large male mink. The fight itself lasted no more than five minutes but covered a fairly large area of open sand and clumps of grass. Even after killing the rat, the mink continued to worry it. Finally, satisfied that the rat was indeed dead, it disappeared.

They swim well and occupy wharves and sewers with ease, often traveling from place to place within a city through sewer pipes. I remember as a youth seeing them running about the streets of my home town whenever it rained hard enough and long enough to flood the sewer system. At such times they were commonly seen dead on the roads, casualties of automobiles. Because of improved sanitation, better construction of buildings, and a continuous campaign of eradication, the Norway rat apparently is less abundant in the U.S. today than it was at the beginning of the 20th century. Nevertheless, it has been estimated that the United States harbors between 150 and 175 million of these rats.

Scarcely a food can be mentioned that these rats will not eat, and many non-food items, such as electrical wiring, are gnawed, cut, or otherwise damaged by them in order to obtain either food or material for nesting. These rats are large enough and aggressive enough to kill and eat chickens; when Norway rats become established around poultry houses, they feed extensively on eggs and young chickens. They are also known to gnaw the hooves off of domestic livestock and even kill lambs and piglets.

Norway rats are prolific breeders, raising young throughout the year. A female normally bears six to eight litters annually but may produce as many as eleven or twelve. Although the usual size of a litter is six to eight, as many as 22 offspring per litter are known. After a gestation period of 21 to 23 days, the young are born naked, blind, and helpless. They grow rapidly and their eyes open in fourteen to seventeen days. Weaned when three to four weeks old, young rats become sexually mature at three to four months of age. Their lifespan is two to three years.

Most animals seem to have an aversion to Norway rats and kill them whenever possible though they may not eat them. Large snakes, Barn Owls, and Great Horned Owls, as well as other large owls and hawks, prey on them. Spotted skunks, long-tailed weasels, and domestic dogs often are excellent ratters. Even some domestic cats hunt rats, but most seem to be afraid of these large rodents.

House Mice
Mus

THE GENERIC NAME *Mus* is the Latin word for mouse.

There are about sixteen species of mice in the genus. Although the length of the head and body for most members of the genus is less than 4 inches, it may be as long as 5, and the length of the tails vary from depending on the species. Some African and Indian members of this genus are among the smallest living rodents; their heart beats range from 620 to 780 per minute.

The pelage may be soft, harsh, or spiny. The tail appears naked but is covered with fine hairs. Dorsally, the pelage varies from tan or pale gray through dull grayish browns, grays, to dark gray or dull brownish gray. The sides and venter may be slightly lighter than the back.

Most members of this genus are active primarily at night, but some are active any time. These mice generally are good climbers; some also swim well.

Nests are constructed of soft, shredded materials wherever adequate shelter and food are available. Food consists of a variety of plant materials, such as fleshy roots, stems, leaves, and seeds, as well as insects and meat when it is available. At times, food may be stored.

Mice of this genus breed throughout the year in the warm parts of their geographical distribution and may produce more than five litters per year. The gestation period ranges from 18 to 21 days. Litters vary from three to twelve young, but the usual size of a litter is four to seven. These mice are prolific breeders, and populations occasionally attain plague proportions.

Through human introduction, the genus is found through-out most of the world. There is one species in western Oregon.

House mouse
Mus musculus

THE SPECIFIC NAME *musculus* is the Latin word for a muscle. Apparently, it was thought that flexed muscles looked like mice under the skin. The first specimen of a house mouse, the one used in the scientific description in 1758, was caught in Uppsala, Sweden.

DISTRIBUTION AND DESCRIPTION

total length 130-198 mm
tail 63-102 mm
weight 18-23 g

HOUSE MICE OCCUR in and around human habitation in zones 1 through 3 and perhaps into the lower elevations of zone 4, where people live. They range in length from $5^1/4$ to $7^7/8$ inches and have tails that vary from $2^1/2$ to 4 inches. They weigh from about half an ounce to almost 1 ounce.

These mice are small and slender with relatively long, tapering, scaly tails. They have large ears that appear to be almost naked. The pelage is thin and coarse, varying dorsally from light brown, to grayish brown, to black. Ventrally, the pelage is light brown, brownish gray, or whitish. The tail varies from brown, to gray, to dark gray above and is slightly lighter below.

HABITAT AND BEHAVIOR

THE HOUSE MOUSE was a pest in England at the beginning of the 17th century, but it is not known when it was introduced into North America. Nevertheless, the house mouse has closely followed the advance of Europeans across the continent.

House mice are largely dependent on humans and human activities, occupying their abodes, eating their food, and raiding their stores of grain and anything else that seems edible. They have traveled long distances with humans on board ships and in covered wagons, trains, and trucks. Once established, they multiply quickly and spread out with the protection afforded by buildings. Eventually becoming too numerous for the buildings, they spill over into the surrounding fields and meadows, where they live under the cover of grasses and other plants, finally overrunning the most fertile parts of arable valley bottoms. Here they burrow into banks, under rocks, walls, and piles of wood or dig safe retreats of their own and exhibit more skill than most indigenous mice in avoiding enemies, including humans. They are good gnawers and diggers and thus difficult to prevent from getting where they wish to go. A house mouse is quick, as anyone knows who has tried to catch one on the run, which at maximum speed is about eight miles per hour.

House mice sometimes seem to completely overrun buildings because they are not confined to one runway or one corner. Once they have become established, they remain within a small, general area, seldom traveling more than 50 feet from their home base, but can appear to be everywhere when in large numbers. Although usually keeping out of sight, their presence is announced by their persistent gnawing and scampering between the walls of a house or in the attic, as well as their characteristic little black fecal pellets.

Although largely nocturnal, they appear to see equally well in light because they are often about voluntarily in daylight hours searching for food. When they are foraging, they wander slowly here and there to sniff this and that, all the while searching for something to eat. They are adept at climbing rough surfaces and can easily jump upward 8 inches, making them ideal little thieves of human food items. These mice are essentially omnivorous, consuming almost anything that is deemed edible. They even drink milk, fresh or sour, and eat soap. Because they seem to survive on little sustenance, house mice have generated the phrase "poor as a church mouse."

House mice in western Oregon are sometimes trapped away from human habitations, while using the runways of other small mammals and in the lodges of dusky-footed woodrats. Their nests, usually well hidden, are composed of whatever materials are handy, such as paper, cloth, grasses, and leaves. The inner nest is lined with finer materials than those used for the outer portion. When sites and materials for nesting are scarce, house mice may share communal nests.

In the warm portion of their geographical distribution, such as western Oregon, house mice breed throughout the year, but there appear to be slight peaks in reproductive activity in April and May and again in August and September. The gestation period is 18 to 21 days. Litters range from three to sixteen young, but four to seven is the usual size, and a female's first litter is normally smaller than subsequent litters. House mice are naked, blind, and helpless at birth but in ten days are covered with hair and in fourteen days open their eyes. They are weaned and begin to disperse in three weeks and may begin to breed at 35 days of age. In captivity, they have lived as long as six years, although their lifespan in the wild is undoubtedly much shorter.

Because house mice are prolific breeders, their populations sometime explode.

House mice are controlled to some extent by snakes, owls, hawks, shrikes, weasels, skunks, foxes, and other predators. Spotted skunks and striped skunks are excellent mousers, as are domestic cats in and around human habitation. Humans also take a huge annual toll with traps and poisons.

Birch Mice and Jumping Mice Zapodidae

THE FAMILY NAME Zapodidae is derived from the Greek words za (an intensive meaning "very") and podos (foot), combined with the Latin suffix *idae* (which designates it as a family).

These small mice, with elongated hind limbs and very long tails, are modified for jumping. The length of the head and body ranges from 2 to 4 inches in length, and the tail, which is used for balance while jumping, varies from $2^5/8$ to $5^3/4$ inches in length. Most species have narrow, long hind feet, ranging from 1 to $1^1/2$ inches in length. The mice weigh from one-fifth of an ounce to just over 1 ounce. Zapodids gain weight in the autumn because of the thick layers of fat they accumulate before hibernation, so they are the heaviest just before entering dormancy, where they spend six to eight months of the year in warm subterranean chambers lined with plant materials.

These mice inhabit forests, thickets, meadows, swamps, and bogs. Most members of this family can jump well over 6 feet when startled. They generally do not make runways but may climb into low vegetation. Some are good swimmers. Most of these mice are nocturnal and either are solitary or associated in pairs. They seek shelter in subterranean burrows, which they either dig or appropriate from some other animal; these burrows are inconspicuous because they do not have loose earth around the entrances. Shelter also may be sought under newly fallen or rotting trees.

Members of the family Zapodidae eat berries, seeds, certain fungi, and small invertebrates (mainly insects); they are not known to store food.

Gestation periods of birch and jumping mice vary from 18 to 23 days. One to two litters are raised annually. Litters range from one to eight young. The youngsters become independent in about one month but are not sexually mature until the following year.

Mice of the family Zapodidae occur in northern and eastern Europe and middle and eastern Asia. In North America, they are found from subarctic Canada south to about latitude 35 degrees N. There is one genus in western Oregon.

North American jumping Mice Zapus

THE GENERIC NAME *Zapus* has the same derivation as the familial name and refers to the exceptionally large hind feet of these mice.

Jumping mice are small and brightly colored. Their pelage is somewhat stiff, almost brittle in appearance. Dorsally, the pelage is grayish brown to brown; the sides are yellowish to orangish brown, and the venter is clear white. The shades of overall coloration vary somewhat in different species. The length of head and body ranges from 3 to almost $4^1/2$ inches and the tail from 5 to just over $6^1/2$ inches.

The rear portion of the body is much heavier than the forepart, and the hind limbs are much larger and more powerful than the forelimbs. These mice have internal cheek pouches. Their upper incisors are narrow, and each has a longitudinal groove the full length of the outer surface.

Jumping mice prefer moist areas; some select forests close to streams; others frequent grassy places. When alarmed, they can leap to almost 8 feet, using their long tails for balance. These mice are excellent swimmers.

During the summer, one or two adults may occupy a compact, globular nest made of grasses woven together, with an entrance on the side. Summer nests are usually located on the ground or in vegetation near the surface. Throughout the remainder of the year they are snug in nests of mosses, grasses, and plant fibers in chambers from a few inches to about 6 feet under the surface of the ground. They become fat as autumn approaches and enter hibernation at the onset of freezing weather. When hibernating, each mouse curls into a ball, with its nose and feet tucked on its abdomen and the tail curled around its body.

North American jumping mice eat seeds, acorns, and certain fungi, as well as insects, particularly beetles and the larvae of butterflies and moths.

Litters of three to eight young are born from May to September, but usually early in the season. Late litters may represent second families.

North American jumping mice occur throughout most of the forested parts of North America as far south as North Carolina, Missouri, New Mexico, and California.

There is one species of jumping mouse in western Oregon.

THE SPECIFIC NAME is derived from the Latin word *trinus* ("triple") and the Greek word *notos* ("the back"), combined with the Latin suffix *atus* (meaning "provided with"), and alludes to the distinctly tricolored pelage of this mouse. The first specimen was collected on Lulu Island, mouth of the Fraser River, British Columbia, on May 32, 1892, by Samuel N. Rhoads and described scientifically in 1859.

Pacific jumping mouse
Zapus trinotatus

PACIFIC JUMPING MICE (photo 114) occur in zones 1 through 5. They are about $9^1/2$ inches long, most of which is a long, slender, sparsely haired, tapering tail that is distinctly bicolored—brown above and white below. Their fur is composed of stiff, bristly hairs that lie close to the body and are strongly tricolored. Their backs are brown with an infusion of many yellowish-tipped

DISTRIBUTION AND DESCRIPTION

total length 205-252 mm
tail 126-160 mm
weight 12-38 g

Photo 114. Adult Pacific jumping mouse. (Photograph by and courtesy of Robert M. Storm.)

hairs. Their sides are yellowish-orange with many scattered brownish- and blackish-tipped hairs, and their undersides vary from clear white to white washed with light orange. Their feet are long and narrow with long, slender toes, and their ears are long and narrow.

PACIFIC JUMPING MICE are associated with areas of grass and herbs along the edge of the coniferous forest, along streams, in thickets of salmonberry, in marshy areas (especially with skunk cabbage), and in early successional stages of coniferous forests.

Pacific jumping mice occasionally walk on all four feet, but normally progress in short hops in an upright position, solely on their hind feet (photo 115), steadying themselves with their long, strong tails. Their tails also act as counterbalances that compensate for the vigorous thrust of their long hind limbs. They are also good swimmers.

When pursued, they propel themselves through the air in long leaps, covering almost 6 feet in a bound. After a few rapid leaps, they stop suddenly, crouch slightly, and remain motionless. If pursued further, they take flight in earnest. At the height of a jump, a mouse turns its head down, arches its back, and dives headlong into vegetation. Even though it may strike thick vegetation, it lands on its feet. Landing on its forefeet, then bringing its long hind legs well forward beneath its body, it leaps again.

Although jumping mice are usually silent, they do vibrate their tails rapidly against some resonant body, such as dry leaves, and produce a drumming sound. They also squeak when disagreeing with one another and are noisy as they rummage in vegetation; their rustlings can be heard for several feet. They are particularly noisy when they forage for salal berries up in

Photo 115. Note the elongated heel and long toe marks of the jumping mouse mixed in with the rounded, somewhat stubby tracks of deer mouse. (Photograph by Chris Maser)

the bushes, often 6 inches or more above the ground. The dense, shrubby nature of salal thickets, as well as the long hind feet and long tails of the mice, allow the mice to move freely in or on a thicket. When startled, they either dive headfirst into the thicket or escape by leaping across the surface of the springy top of the thicket. When resting on top of a salal thicket, the mice normally have their tails braced across the upper surface of the broad, stiff leaves.

During summer, jumping mice construct their well-hidden, fragile, spherical or dome-shaped nests on the ground (photo 116); some are in slight depressions dug by the mice. These summer nests are composed of loosely interwoven coarse or broad-leaved grasses, and are about 6 inches in diameter and 4 inches high. Nests in marshy areas are made of mosses and lined with grasses or sedges. Each nest has a single opening in the side and appears to belong to one individual.

Jumping mice eat the seeds of the skunk cabbage, parts of some mosses, the seeds of grasses, and some of the belowground-fruiting, mycorrhizal fungi. As summer progresses, however, and the fruits of salmonberry, thimbleberry, trailing blackberry, huckleberries, and stink currant fall to the ground, the mice eagerly concentrate on them.

Most male jumping mice become sexually mature in June, but occasionally a sexually mature individual may occur in late May. Some individuals still have enlarged, descended testes as late as the first half of August.

Jumping mice are born pink and naked in a well-hidden nest of grasses and moss amongst the July grasses. Their eyes are closed and their ears are folded over and deaf. They have short, stubby heads, relatively long tails, and weigh less than one-twentieth of an ounce.

Photo 116. Summer nest of the Pacific jumping mouse made of grasses; the entrance is in the side of the nest, just visible in the lower right hand side of the photograph by the violet leaf. (Photograph by Chris Maser)

As autumn approaches, the mice begin to accumulate layers of fat under the skin, over the muscles of the body, and throughout their body cavity. Although some individuals will begin to accumulate fat as early as the latter part of August, most will not begin until the latter part of September. With these accumulations of fat to sustain them, they will enter hibernation in their warm, dry, belowground nests in late October or early November. During hibernation, as the winter winds howl and the snow accumulates, the jumping mice are rolled up in little furry balls and appear to be quite dead. If, however, a mouse were to be given warmth, it would be fully awake in half an hour, but remove the warmth and it would again doze off. Jumping mice remain in hibernation until released by the warmth of the late May sun.

Barn Owls, Great Horned Owls, Long-eared Owls, and Northern Spotted Owls prey on jumping mice. In addition, bobcats eat these mice when they are available, as do domestic cats. Some snakes, marten, mink, weasels, spotted skunks, foxes, coyotes, other species of owls, and hawks also catch them.

THE FAMILIAL NAME Erethizontidae is derived from the Greek word erithizon ("to provoke") combined with the Latin suffix *idae* (which designates it as a family).

New World porcupines are large, heavy-set rodents. The length of the head and body is 12 to 37 inches and the tail 3 to 18 inches. Adults weigh from about 7 to 40 pounds. Their feet are modified for an arboreal life: the sole of each foot is wide and has only four functional digits, each with a strong, curved claw. In members of one genus, the tail is prehensile. The limbs are fairly short. Most of the body and tail hairs are modified into short, sharp spines or quills with overlapping barbs; depending on the species of porcupine, spines vary in distribution on the body. There are long, bristlelike hairs on the back. The pelage is marked with blackish to brownish, yellowish, or whitish bands.

Of the four genera of New World porcupines, only the habits of the North American porcupine have been documented. This genus occurs from the treeline of arctic North America south throughout Canada and the United States (except the southeastern portion) into Sonora, Mexico; members of the other three genera occupy southern Mexico, Central America, and South America on the eastern side of the Andes.

THE GENERIC NAME *Erethizon* has the same derivation as the familial name.

Since there is only one species in this genus, refer to the species description and distribution.

THE SPECIFIC NAME *dorsatum* is the Latin word for "the back." The first specimen of this species was collected in eastern Canada and was described scientifically in 1758.

THE NORTH AMERICAN PORCUPINE occurs in all zones, although more commonly in some areas than in others (photos 117, 118, and 119). They are infrequently found in the floor of the Willamette Valley, for example, but are relatively common in some parts of the Cascade Mountains. They range in length from 26 to 37 inches and their tails vary from 6 to 12 inches. They weigh from approximately 7 to 40 pounds.

The porcupine is the second largest indigenous rodent in North America and one of the most distinctive. It is a heavy-set animal with short legs and a relatively short, but very strong tail. The ears are small and hidden by the body hairs. It has small eyes and a soft, furry nose. The porcupine's feet are unique among all North American mammals in that the oval soles are

New World Porcupines Erethizontidae

North American Porcupine
Erethizon

North American porcupine
Erethizon dorsatum

DISTRIBUTION AND DESCRIPTION

total length 648-930 mm
tail 148-300 mm
weight 3.5-18 kg

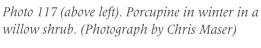

Photo 117 (above left). Porcupine in winter in a willow shrub. (Photograph by Chris Maser)

Photo 118 (above right). Business end of a porcupine; note the quills and the short, strong tail used to deliver them into the flesh of an enemy. (Photograph by Chris Maser)

Photo 119 (right). Hind foot of a porcupine showing the fleshy knobs that are characteristic of the species. (Photograph by Chris Maser)

covered with small, fleshy knobs (photo 119). The claws are sharply curved and heavy.

The pelage in winter consists of a thick, somewhat woolly coat of long, soft, black underfur beneath an armor of stout, sharp, barbed quills from 1 to 4 inches long. The quills, which cover the back, sides, and tail, are partly concealed on the back and side by a loose outer coat of stiff, erect, yellow or yellow-tipped guard hairs, 6 to 10 inches long. The underside is covered beneath with underfur and coarse, short guard hairs. The under surface of the tail is covered with rigid bristles that serve as a prop when the animal climbs.

During the summer, the pelage is composed primarily of naked quills and long outer guard hairs but little underfur. When erect, the blackish-tipped whitish quills are fully exposed (see photo 118). The young are blackish in overall coloration.

NORTH AMERICAN PORCUPINES in western Oregon are associated primarily with coniferous forests They are essentially solitary mammals, active mainly at night. They are cautious, slow, deliberate, and peaceful.

HABITAT AND BEHAVIOR

Porcupines are deliberate, but excellent climbers, spending much of their time in trees, though they appear to be equally at home on the ground. They are active throughout the year but may hole up during inclement weather. They walk flat-footedly, often leaving their lines of oval tracks and the drag marks of their tails in dusty trails, or mud, or snow (photos 120 and 121).

In western Oregon, porcupines usually retire to the comparative safety of a tree to rest and sleep, but occasionally may select a hollow log or stump, or a pile of logs or debris from logging operations. In other parts of their geographical distribution, they have regular retreats in large crevices and small caves, among boulders and talus, and in particular, large, limby trees. Such regularly used retreats can be identified by an accumulation of feces. A retreat usually houses a single individual, but during the winter more than one animal occasionally occupies an especially favorable den.

Porcupines are calm, methodical animals that would rather retreat passively under cover or up a tree than confront an enemy. They do not attack and cannot shoot or throw their sharply barbed quills. If confronted in the open, their only defense is to erect their quills, giving the appearance of a gigantic, living pincushion. The combination of sharp, barbed, loosely attached quills and a strong, flexible tail gives a porcupine ample defense in most instances. Should an adversary ignore a porcupine's weapons and approach too closely, the

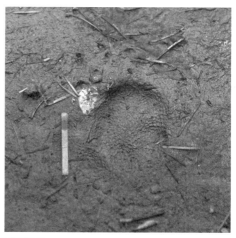

Photo 120. Distinct track of a porcupine in mud; note indentations made by the fleshy knobs. (Photograph by Chris Maser)

Photo 121. Track of porcupine in mud in which the indentations of the fleshy knobs are indistinct (a) but the drag marks of the quills on the tail (b) are distinct in the lower right hand corner of the photograph. (Photograph by Chris Maser)

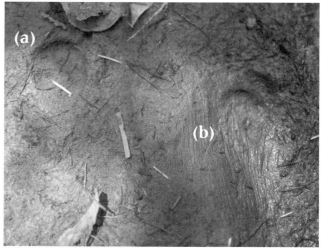

porcupine flicks its tail vigorously from side to side and up and down, driving the quills into the assailant.

Mature quills, like mature feathers, are dead structures. The great reduction of the surface area of a quill's root, and the correspondingly reduced tension the follicle exerts on the root, is an important factor in the ease with which a quill becomes detached from the animal's skin. Furthermore, each barb or scale along the tip of a quill acts as a small anchor, holding it firmly in the tissue of the adversary. The backward-directed, overlapping barbs also increase the penetration of the quill because, when they engage the enemy's muscle, the pulling action draws them farther into the tissue, at a rate of about 2/50 or more of an inch per hour.

Although a porcupine's complement of quills is usually thought of as purely a defense mechanism, it also serves other purposes. The backward orientation of the quills, as well as the body hairs, facilitates passage through underbrush; moreover, the general pelage prevents the penetration of rain and snow, effectively screens the sun's burning rays, and in varying degrees, conserves body heat.

In addition, the long, outer guard hairs, extending well beyond the end of the general coat of hair and quills, arise from sensitive follicles and thus keep the individual aware of the conditions in its immediate surroundings. They enable a porcupine to maintain an acute, peripheral sensitivity to touch 6 inches or more beyond the surface of the skin; in fact, the follicles are so sensitive that even a gentle touching of these hairs on a resting or sleeping animal stimulates a prompt erection of the quills so that their effective points are prominently exposed in all direction. If a porcupine is in a deep sleep, however, it takes stronger stimulation to bring about this reaction. The vibrissae or whiskers are also extremely sensitive to touch and, together with the long guard hairs, effectively guide their owner safely through night wanderings and dark hiding places.

Although porcupines seem to wander aimlessly, there is some indication of seasonal migrations. They move to localized shelters during long periods of inclement weather and disperse again during fair weather. During their wanderings, they usually cover an area—their home range—of about almost a mile to almost 3 miles.

Porcupines, often thought to be silent or even voiceless, make a variety of squeaks, moans, whines, wails, grunts, and coughs. Some of these sounds are termed "songs" and are thought to be "love songs" since they are heard most often during the mating season.

Porcupines appear to be strictly vegetarians. They eat a wide variety of plants and exhibit definite seasonal preferences; they also consume tremendous quantities of food, as much as three pounds at a time. During spring and summer, most of their diet is low-growing vegetation, such as skunk cabbage, sweet coltsfoot, clovers, lupines, geraniums, asters, sedge, and grasses, as well as the twigs and leaves of a variety of shrubs (photo 122). During autumn and winter, they depend on the inner bark of trees, which they expose by scraping off the rough, dead, outer bark. They eat the inner bark of such trees as ponderosa pine, lodgepole pine, sugar pine, whitebark pine, Douglas fir, western hemlock, western redcedar, and Port-Orford-cedar, which is called "white cedar" by local residents along the central Oregon coast, where the species grows (photo 123).

The breeding season begins in September and may last into January, but the peak of reproductive activity occurs during

Photo 122 (above). Soft dropping of a porcupine on spring and summer diet of succulent herbaceous vegetation.

Photo 123 (below). Hard droppings of a porcupine on a coarse winter diet of twigs and the bark of trees. (Photographs by Chris Maser)

October and November. A female loses interest in the male four to six hours after breeding; she may even become antagonistic toward him. After a while, each porcupine goes its own way.

After a gestation period of about seven months, a single young is born, usually in May or June. The young may nurse for four months or longer. Babies weigh from 12 to 20 ounces at birth, and their coats are generally long, thick, and black or gray, being longest in the northern climes. Their quills are a quarter of an inch to an inch or longer and cover most of the back on the head, body, and tail; they are functional as soon after birth as they dry out and even sometimes while still wet. Their eyes are open and thoroughly functional in locating and responding to movement within their range of vision. The erection of quills as a defense can be perfectly executed. The reaction of turning the head away from and the tail toward a moving object is displayed within minutes to hours after birth. The ability to flick the tail in such a way as to drive quills into an adversary is performed well within the first hours. The crowding reaction, in which a porcupine withdraws its feet and rolls on to one side, thus pushing side quills into an enemy, is well timed and very well done. Further, the reaction of protecting the head by sticking it into a corner, or under something, is well developed and functional within minutes to hours after birth.

Male porcupines may attain sexual maturity at sixteen months of age; females probably are not sexually mature until their second year. Porcupines are fairly long lived, even in the wild; one female is known to have lived at least eleven years.

Despite their armament of quills, porcupines are preyed on by a number of animals, most notably the fisher. Consequently, 24 fishers were imported into Oregon and released in 1961 as part of a porcupine control program. Pumas also prey relatively heavily on porcupines, eating them quills and all without apparent deleterious effects. In addition, bobcats and coyotes prey on them to some extent.

Mammals of the Pacific Northwest

Domestic dogs frequently have painful encounters with porcupines, usually resulting in a face full of quills, a bad temper, and occasionally an increased respect for porcupines—but not always. Quills are best pulled out of a dog's face with a quick, straight jerk using the fingers or, preferably, a pair of needle-nose pliers (but any pliers will do), gripping the quill as close to the skin as possible. Be extremely careful not to break the quill off and leave part of it in the wound because the embedded portion will work continually deeper into the tissue. Allow wounds to bleed as a method of self-cleansing, provided the bleeding is not excessive, and then cleanse with soap and water.

Humans undoubtedly kill more porcupines with poisons, guns, traps, clubs, and automobiles than do predators. When a porcupine is struck by an automobile, the quills may embed themselves in a tire, work their way in, and cause the tire to go flat. It is wise, therefore, to stop immediately and withdraw any quills while they can still be easily pulled out.

Coypus
Myocastoridae

THE FAMILIAL NAME Myocastoridae is derived from the Greek words *mys* ("mouse") and *kastor* ("beaver") combined with the Latin suffix *idae* (which designates it as a family). The name probably alludes to this animal's mouselike appearance and beaverlike habits.

Since there is only one living genus and species in the family Myocastoridae, refer to the species description.

Coypus are indigenous to Central America and South America, but they have been widely introduced into North America and Eurasia.

Coypus
Myocastor

THE GENERIC NAME *Myocastor* has the same derivation as the familial name.

Since there is only one living species, refer to the species description. The geographical distribution of the genus is the same as that of the family.

Coypu or
Nutria
Myocastor coypus

THE SPECIFIC NAME *coypus* is the South American indigenous name for this rodent. Although nutria is the name most widely used in North America, coypu is probably a better name, because nutria means otter in Spanish. The coypus in the Pacific Northwest came originally from Argentina. The species was described in the scientific literature in 1805.

DISTRIBUTION AND DESCRIPTION

total length 671-1400 mm
tail 300-440 mm
weight 2.3-11.4 kg

COYPUS OCCUR IN zones 1 through 3, and may also exist in the western lowlands of zone 4. A number of coypus kept by a fur farmer in Tillamook County, in western Oregon, were liberated from their pens during a flood. These became established at or near Garrison lake on the northern coast of Lincoln County, the county just south of Tillamook County, where the liberation took place. Although definite data are lacking, a "wild" coypu is known to have been caught in a marshy area in Portland, Oregon, in 1936 or 1937. One or more were also trapped during the autumn or winter of 1938 along the Nestucca River in Tillamook County, and were still doing well in 1943.

Coypus are large rodents, ranging in length from 26 to 56 inches and have tails that range from 12 to just over 17 inches. They weigh from 5 to 25 pounds. They have long, round, sparsely haired, scaly, ratlike tails. They have small eyes and ears, the latter being almost naked of hair. The incisors are dark orange and protrude noticeably beyond the lips. The tip of the muzzle and the chin are definitely white. The hind feet of coypus are much larger than their front feet and have four of the five toes connected by a web of skin that aids in propelling them through water. The fifth or outer toe is free of the web and may be used in combing the pelage.

The pelage is composed of thick, soft underfur that is largely obscured by the long, coarse, stiff, protective guard hairs that give these rodents a shaggy appearance. The overall coloration is darkest on the back and lightest on the belly. Depending on the coypus' environment and ancestry, the dorsal pelage varies from light yellowish brown, light reddish brown, dark reddish brown, to black. Coypus living in densely vegetated marshes, croplands, or swamps are darker than those in open areas. Some may also have patches of white fur scattered over their bodies; this is an apparent carry-over from domestication and breeding for fur.

COYPUS IN WESTERN Oregon live in coastal lakes, tideland rivers, inland rivers and sloughs, and even in some foothill rivers. They are almost strictly nocturnal, but occasional individuals may be seen foraging along the edge of protective cover close to water. Normally, however, daylight hours are spent resting, grooming, and playing.

HABITAT AND BEHAVIOR

Coypus are at home in water, but appear slow and cumbersome on land, usually ambling here and there with a high-rumped, waddling movement. When disturbed or frightened, however, they move rapidly in a waddling, bounding, or low-bellied, creeping motion. Coypus are reasonably efficient climbers, able to negotiate steep banks, entanglements of roots, gently sloping trees, and even wire fences.

They have an acute sense of hearing, but their senses of sight and smell are poorly developed. When frightened while on shore, they enter the water with a loud splash and swim for cover under water or simply remain submerged for several minutes. They can maintain perfect buoyancy with little or no body motion and can stay well hidden under very sparse vegetation with only their noses and eyes or the tops of their heads above the water. If undisturbed, they are methodical, unhurried swimmers, with most of the head and back above water and the tail floating freely at the surface. During long-distance swims or when mildly disturbed, only the head is visible above the surface.

Wild or feral coypus generally are phlegmatic, unwary rodents. In recent years, however, people have eliminated many of the fearless coypus, leaving the wariest individuals to reproduce, though some individuals remain indifferent toward humans. For some reason we do not understand, coypus exhibit periodic changes in behavior; their nonchalant indifference gives way to extreme alertness, or viciousness, or both, only to change again to phlegmatic indifference. Nevertheless, when cornered or captured, they can be aggressive and can inflict serious injury.

Because they are aquatic mammals, it is vital that coypus keep their pelages clean or they will lose their insulation against the water. Consequently, they almost ritually scratch and groom themselves. Their forefeet are used for both these chores, whereas the hind feet are used only for scratching. Coypus, unlike beaver, do not have split toenails or combing claws on the hind feet; instead they use the fifth, outer toe, which is free of the other four, webbed toes, for combing themselves and for cleaning their ears.

Coypus often use platforms of vegetation for feeding, resting, nesting, and to escape danger. Most platforms appear to result from piles of uneaten portions of plants at favored feeding sites, rather than being purposefully built structures. Platforms, which may be more than 6 feet in diameter, often serve several generations. They vary in size and density, depending on the coarseness of the vegetation and appear to consist of the most available vegetation, rather than of a preferred type of building material.

Ordinarily, coypus living in marshes are not extensive burrowers but make their homes among dense vegetation on top of the ground. They also appropriate the burrows of other animals, as well as the lodges of beaver and muskrats. Burrows constructed by coypus are usually situated in steep earthen banks offering a protective cover of rank vegetation. They are normally located adjacent to watercourses, but now and then may be a half mile or more away. Burrows vary from simple structures with one entrance and short tunnels to very complex units with several multilevel entrances, tunnels, and living compartments, which are variously used, depending on the level of the water. Tunnels usually extend from 4 to 6 feet into a bank but may extend from 50 to 150 feet. Living compartments range from small ledges, about 12 inches in width, to large family units, 3 feet or more in width; they are covered either with soil or with plant debris, apparently food refuse.

Most coypus have a definite home range within which there is a central area, where an individual spends most of its time, not leaving it for days or even months. Coypus living in agricultural areas move much farther during a lifetime than those living in marshes.

Coypus are almost exclusively vegetarians, daily consuming several meals amounting to a total of just over 1 to $3^1/2$ pounds of food. They eat while on land, on floating objects, or in the water. On land, they either graze on grasses or clip tall vegetation and manipulate it with their forefeet. Their usual method is to hold the vegetation with their front feet and lean forward, resting on their elbows, or to sit upright, using their tails as braces. Because of their remarkable buoyancy, they also can float freely in the water for long periods, all the while putting floating plants into their mouths with their forefeet. Coypus often pull up underwater vegetation with their forepaws or cut it off with their front teeth, but seem always to bring it to the surface to be chewed and swallowed. They sometimes wash their food, but this usually does not happen until feeding has progressed well into the night.

The coypus' sense of touch plays a major role during mealtime. Their forefeet are both sensitive and dexterous, enabling them to locate food items, to pick up a goodly quantity of plants or grains, or to manipulate a single kernel of grain or an individual stem of clover. It is evident that coypus use their sense of smell to locate feeding sites that have been used by other individuals, and although they seem to be continuously sniffing their food, neither their sense of smell nor taste appears to play much of a role in the final selection.

These rodents generally eat the soft, succulent, basal portions of plants, but they also eat several different parts, or an entire plant. Coarse plants, such as cattails and rushes, are commonly consumed when they are available, but soft, floating plants, such as duckweed, are eaten also. Though coypus often exhibit definite seasonal variation in their food habits, they depend on cattails and sedges for their dietary staples, and where these plants are abundant, they seem reluctant to greatly alter their diets, regardless of the season.

Many factors, such as the kind and amount of available food, the weather, and the prevalence of enemies and of diseases, influence the reproductive capacity and the survival of coypus in the wild. Female coypus usually come into estrus every 24 to 26 days, remaining in this condition from one to four days, and most come into estrus one or two days after giving birth or after having aborted. Males, on the other hand, can breed at any time.

Courtship is common just before the onset of estrus; it consists of vocalizations by both sexes, chases on land or in the water, playful fighting, wrestling, and biting. A male occasionally will squirt urine or seminal fluid on the female during courtship. Once estrus commences, courtship is brief or disregarded. Breeding is prompt and takes place almost anywhere, in or out of water. A female may breed with one to several males during each estrus; inbreeding is also common.

Gestation lasts about 130 days. On average, only about 60 percent of the embryos survive to be born; many are either aborted or absorbed within the uterus. Litters range from one to eleven offspring, but the usual size is four to six. Babies weigh from 6 to 8 ounces at birth, depending on the abundance and nutritional value of the mother's diet. The size of a litter also shows a correlation with the quality and quantity of food—the better the food, the larger the litters.

Females about to give birth, and for about four to six weeks thereafter, frequently appear to be in a tranquilized state. They freely allow themselves, as well as their young, to be handled, making no effort at defense. They will not tolerate the intrusion of a strange young coypus, however, and become vicious. After they repulse a stranger, their tranquillity returns and normally remains until the young are weaned.

Youngsters are born with their eyes open and are fully haired; they are ready to swim shortly after drying off. A female's teats are situated so high on the side that the babies can nurse even while in the water or while the mother lies on her stomach. Females may begin producing milk as early as one to two weeks before giving birth and may continue for six to eight weeks thereafter. Most young are weaned at about five weeks.

When the supply of high-quality food is good, sexual maturity is reached at about four months, but when the supply is poor, it may not occur until a youngster is about five or six months old. Regardless of the age at which sexual maturity is reached, most coypus do not breed until they are about eight months old. As with many rodents, a female's first litter is usually smaller then subsequent litters.

Coypus have lived in captivity up to twelve years, but their lifespan in the wild is probably much shorter. Because of their large size, adult coypus probably have few enemies other than parasites, diseases, and humans. Since most females readily abandon their young when approached by people and domestic dogs, it seems likely that the same would happen with wild predators. If so, and if the abandoned young are small enough, such predators as mink, long-tailed weasels, otter, raccoons, bobcats, coyotes, and large owls and hawks could add young coypus to their diets. Humans, however, with traps, guns, and poisons, are the main enemy of coypus.

Flesh-eating Mammals
Carnivora

THE ORDINAL NAME Carnivora is derived from the Latin words *carnis* ("flesh") and *voro* ("to devour"), but this term is somewhat misleading, because many members of this order are really omnivorous. Two common household pets—dogs and cats—are classified as carnivores.

Some carnivores, such as dogs and cats, walk on their toes; others, such as bears and raccoons, walk on the soles of their feet with heels touching the ground. Intermediate conditions also exist. Members of this order have at least four digits or toes on each foot. On those that have five, the first digit is sometimes smaller and located higher on the foot than the others. All digits terminate in claws. Cats, except for cheetahs, are the only carnivores with fully retractable claws that can be retracted into sheaths.

Carnivores have heavy skulls and strongly developed facial muscles. They have well-developed strong canine teeth, which are elongated and curved, with pointed tips, adapted for seizing and holding prey. The last upper premolar and the first lower molar are often specialized shearing teeth called "carnassials," and are most highly developed in members of this order with almost strictly carnivorous diets, such as the cats; they are least developed in those with omnivorous diets, such as bears.

The order Carnivora is highly variable. Some members are primarily arboreal, some semiaquatic, some truly marine, but most are terrestrial. Most carnivores are good climbers; in fact, two genera have prehensile tails. Some are strictly nocturnal; some are strictly diurnal; others are abroad day or night. They find shelter in caves, burrows, hollow trees, and hollow logs. Although some bears are dormant during winter, most carnivores are active all year. The dark summer pelages of some species become white in winter. Some carnivores are adapted to arctic regions, some to deserts, and others to humid jungles.

Carnivores have a variety of methods for securing food. Cats stalk and then pounce unexpectedly on their prey. Dogs give swift pursuit to their prey, and weasels engage their quarry in a tireless, bounding chase. Some species, such as martens, follow their prey swiftly through the trees; others, such as river otters, are adept at catching prey under water; and some species, such as badgers, dig their prey out of burrows. Although some carnivores live almost solely on freshly killed animals, many are scavengers.

Members of this order generally have one or two litters per year, but a few have three. Gestation ranges from 49 to 113 days; however, the delayed implantation of the fertilized egg in bears and some members of the weasel family makes gestation considerably longer. Litters range from one to thirteen offspring. The young are usually born blind and helpless but with some hair on their bodies. Babies are normally cared for by their mothers, but both parents share the responsibility in some species. Sexual maturity is usually attained by one to two years of age in small to medium-sized species—later in large species.

Distribution of carnivores is worldwide except for Antarctica and some islands, such as New Zealand, Melanesia, Polynesia, and Taiwan. Humans have introduced carnivores on some islands where they were not indigenous. There are five families of carnivores in western Oregon.

Dogs and Foxes Canidae

THE FAMILIAL NAME Canidae is derived from the Latin word *canis* ("dog") combined with the Latin suffix *idae* (which designates it as a family).

Most members of this family have lithe, muscular, deep-chested bodies, with long, slender limbs, and bushy tails. Their muzzles are long and slender (except in some domestic varieties), and the ears are relatively large and erect. The claws are blunt, and the first digit on both the forefeet and the hind feet is smaller than the others.

The average male is larger than the female. The length of their heads and bodies ranges from $13^1/2$ to 54 inches and their tails from $4^3/8$ to $21^5/8$ inches. Adult canids weigh from just over 2 to 174 pounds. Although most members of the family are either uniformly colored or speckled, one species has stripes on the sides of its body and another is covered with blotches.

In the northern portions of their geographical distribution, these alert, crafty mammals are active at any time, provided they are not molested by humans; in the tropics, however, they are more active at dusk and dawn. They use burrows, caves, and hollow logs and trees for dens. With the exception of the raccoon-dog, which sleeps most of the winter, members of this family are active throughout the year. The coyote may be the swiftest wild dog, reaching speeds of 38 to 40 miles per hour. Gray foxes often climb trees.

Some canids hunt in family groups or in packs, sometimes forming a relay system to bring down prey; others hunt singly. Although hunting is done mainly by scent, canids have acute hearing and good sight. Prey normally is subdued by an extended chase in open areas. The habits and habitats of canids are varied, extending from the arctic, throughout forests, prairies, and deserts, into the tropics. Most canids are opportunistic, eating whatever is available. Some however, are primarily hunters; some rely on carrion; others are mainly omnivorous. Food habits, on the whole, vary with season and locality. A population of canids fluctuates with the increases and decreases of its prey.

In members of most genera, gestation ranges from 49 to 70 days, with an average of 63 days; but canids of at least two genera are reported to have gestation periods as long as 80 days. There usually is one litter, sometimes two, per year, consisting of 2 to 13 young. Babies, usually born in a burrow, are blind and helpless at birth but well covered with fur. They are weaned in 6 to 10 weeks and are cared for solicitously until they can fend for themselves. In members of some genera, more than one female may raise their young in one den. Young mature in one, two, or three years in the wild and live 10 to 18 years.

The family Canidae is worldwide in distribution, except for New Zealand, New Guinea, Melanesia, Polynesia, the Moluccas, Celebes, Taiwan, Madagascar, the West Indies, and some other oceanic islands. Domesticated dogs, however, have been introduced by humans in many areas. There are three genera in western Oregon.

THE GENERIC NAME *Canis* is the Latin word for dog.

Dogs, Coyotes, Wolves
Canis

The genus consists of eight species. The length of the head and body ranges from 22 to 55 inches and the length of the tail from 9 to $22^3/8$ inches. These animals weigh from about 15 to 174 pounds.

Members of this genus are intelligent and, to some degree, social creatures. They live and hunt in family groups, packs, or pairs. They occur from the Arctic throughout the temperate region, to the open savannah country of eastern and southern Africa. Their diets are as varied as their habitats and habits. There is a relatively close relationship among the various species within this genus as evidenced by successful interbreeding among all the wild species and the domestic dog.

The genus *Canis* occurs throughout North America and Mexico to Central America, throughout much of Europe, Asia, the Middle East to Turkestan, and throughout northeastern, eastern, and southern Africa, as well as Australia. Humans have introduced the genus into places that it had not previously occupied. There formerly were two species within this genus inhabiting western Oregon—the wolf and the coyote. Today, only the coyote remains.

Coyote
Canis latrans

THE SPECIFIC NAME *latrans* is derived from the Latin word *latro* ("to bark"). The first scientific specimen of a coyote came from Engineer Cantonment, about 12 miles southeast of the present town of Blair in Washington County, Nebraska, on the west bank of the Missouri River. It was described scientifically in 1823.

DISTRIBUTION AND DESCRIPTION

total length 1052-1320 mm
tail 300-394 mm
weight 6.4-20.9 kg

THE COYOTE IS FOUND throughout all zones in western Oregon; in fact, it is found throughout the entire state (photos 124 and 125). Coyotes range in length from 42 to almost 53 inches and have tails that vary from 12 to $15^3/4$ inches. They weigh from 14 to 46 pounds.

Coyotes have erect, pointed ears, bristling manes on the back of the neck, and a triangular cape of long hair behind the shoulders. There is an elongated gland on the top of the tail near the base, covered with long, black, bristly hair. The winter pelage is long and relatively soft. The upper parts are light brownish gray over tannish underfur; the long, coarse, outer hairs are heavily tipped with black on the mane, cape, and back, as well as the top and tip of the tail. The backs of the ears are light reddish brown, but the muzzle, crown, outsides of the legs, and the lower surface of the tail are bright yellowish brown. The underparts, except the grayish throat band, are tannish to tannish white. The summer pelage is thinner, coarser, darker and more brown than the winter pelage.

HABITAT AND BEHAVIOR

COYOTES, BEING ONE of the most adaptable of North American mammals, occupy almost every conceivable habitat, including the suburbs of large cities, such as Los Angeles, California, and small towns, such as Corvallis, Oregon. Sometimes called "brush wolves," they are primarily active at night but are occasionally seen during the day. More often heard than seen, these graceful, easy travelers leave their tracks and droppings throughout most of Oregon (photo 126). Their characteristic howl, as much a symbol of the American West as the cowboy, is given most often at twilight and at dawn; it may be heard at any time during the night but seldom during the day. Coyotes also utter short, high-pitched barks or, in distress, a shrill screaming howl.

Compared with some other predators, coyotes are not physically strong, but their wits, craftiness, and courage more than make up for lack of physical strength. Furthermore, their capacity to withstand and overcome injury, pain, physical handicaps, and other hardships, is almost incredible. Their

Photo 124. Close up of a coyote's face. (USDI Bureau of Land Management Photograph by Robert Kindschy.)

Photo 125. Coyote traversing a field. (USDA Forest Service Photograph.)

senses of sight and hearing are exceptionally well developed, but their sense of smell is the most remarkable.

Coyotes are sociable; they may change mates yearly, but for the most part, pairs remain together for several years or for life. Even as adults, they have a capacity for playfulness with one another, together with a reputation for adaptability, wariness, intelligence, and independence.

Whereas the geographical distribution of numerous other North American animals is steadily shrinking, the coyote has increased not only its numbers but also its geographical distribution. At least until the 1890s, the coyote inhabited about

Photo 126. Part of the art of reading animals' sign is keeping track of the weather so as to be able to age the tracks. (A) Note the square pad across the back of the track; this is a characteristic of tracks of the genus Canis, and distinguishes their tracks from the tri-lobed margin of a cat's track. (B) Tracks of a coyote in soft soil. Note that the soil was not only soft enough for the weight of the front foot to cause a crack along the right side of the back pad but also has dried some since the coyote passed, as evidenced by the crack caused as the soil shrinks that is just entering the left front toe on the track of the front foot. (C) Coyote tracks in soft mud. Note that the tracks were made while the mud was soft enough to be displaced outward along the right margin of the front foot. Note also that the mud has dried enough since the coyote passed that the a crack caused by drying has entered the back pad or the hind foot. (D) Old coyote track made in soft mud, which has since been rained on for a week. Note the fir needles that have fallen into the track and the way the larger particles of soil have settles on the surface of the track, indicating substantial rain.(Photographs by Chris Maser)

three million square miles of open spaces in the American West. Even in his 1936 publication on the mammals of Oregon, Vernon Bailey depicted the coyote's distribution as east of the High Cascade Mountains, as well as southwestern Oregon and the Willamette Valley. Today, however, the coyote is found throughout western Oregon, right to the shore of the Pacific Ocean, primarily because extensive clearcut logging of the forests has created suitable passages and open habitat.

Coyotes usually rest during the day. In central and eastern Oregon, they frequent rimrock, canyons, and brushy areas, where they can lie either in the sun or in the shade. Sometimes they sit with their backs close to a western juniper tree and simply watch the world as they sun themselves. Along the coast, they spend the day in thickets at the forest's edge, or around piles of logging debris, or fallen logs in clearcut areas. Underground dens are also used and may be appropriated from other animals and remodeled or dug entirely by the coyote itself.

In western Oregon, coyotes normally travel along the tops of ridges, as well as on logging roads, visiting saddles between hilltops and old landings, where logs were loaded onto trucks for the trip to the mill. They may travel great distances, following the migrations of their prey, such as deer, into the high mountains in spring and summer and back to lower elevations in autumn and early winter. They do have home ranges, but these are not well understood.

Although the coyote's role as a predator of livestock and game animals gets it into trouble with some people (photo 127), it also helps to control rabbits, hares, and rodents, which are also considered detrimental to various human endeavors, such as raising domestic sheep by competing for forage. Coyotes are opportunists, both as hunters and scavengers. They eat any small mammal they can capture, as well as porcupines, deer, pronghorned antelope, domestic livestock, indigenous birds, and domestic poultry. They also eat great amounts of carrion, as well as occasional lizards and snakes. Insects, such as grasshoppers and crickets, are included in their diet. Grasshoppers are especially important to the young-of-the-year coyotes as they learn to hunt on their own. Coyotes are fond of fruits in season and partake of a wide variety, such as blackberries, cascara berries, strawberries, and salal berries. They occasionally raid gardens to get melons, earning them the name of "melon wolf."

Because of its method of hunting, normally relying on an extended chase, the coyote is basically an animal of fairly open country, where it can run down its prey instead of ambushing

Photo 127. Part of the art of reading animals sign is understanding what their droppings tell you. (A) Coyote dropping that is full of digested meat and hair of a mule deer. (B) Coyote dropping that is the remains of a meaty meal in which the coyote has cleaned itself out by eating grass. (C) A place where a coyote has twice deposited feces; first it cleaned itself out; note twisted grass mixed in with the ball of deer's hair. The second visit is represented by the dark dropping in the lower right, which is full of digested meat and hair of a mule deer. (D) The dropping of a coyote full of cascara berries. (Photographs by Chris Maser)

it as do most cats. When hunting small animals, however, a coyote may stalk and pounce. But it chases rabbits and hares, taking short cuts until it is close enough to seize its prey. A pair of coyotes or a family group may pursue prey, such as deer, in relays.

Sperm production begins in males over one year old in November; coyotes are capable of breeding by mid-January and remain sexually potent for about two months. The mated pair will have their dens, situated wherever the pair has a sense of security, usually on a south-facing slope, prepared and ready by the time the pups are born. Coyotes usually have several dens cleaned out and other suitable areas located. When a den or its surroundings are disturbed, particularly by humans, the coyotes move to another den. Even if they are undisturbed, coyotes tend to move periodically. Such movements probably have advantages for their survival by minimizing an enemy's potential detection of the young, as well as avoiding excessively high populations of parasites, such as fleas, and accumulations of urine, feces, and food refuse. If coyotes are unmolested and their den undisturbed, they may use it the following year.

Pups are born from late March through early May, after a gestation period of 60 to 63 days. Litters range from 4 to 19 young, but the usual litter consists of five to seven pups. A single family is usually raised per year. Pups weigh about eight ounces at birth, but their weight doubles in about eight days. Although their eyes are closed at birth and do not open until they are 10 to 12 days old, the fuzzy, grayish or dark brown pups can crawl around in two to three days and can walk in 8 to 10. By the time they are a month old, they can run fairly well. Pups are fed only milk during their first two weeks; nursing continues for at least three months, sometimes considerably longer. At about two weeks of age, their deciduous teeth become visible, and the mother may begin to regurgitate small amounts of meat for them. Their permanent teeth replace their milk teeth when they are four to six months old, but feeding by regurgitation continues until the pups can tear food apart by themselves. Young coyotes usually attain sexual maturity by the next breeding season. Wild coyotes have lived 14 years, but these undoubtedly were very old individuals.

Humans, with traps, guns, and poisons, appear to be the coyote's only real enemies. Parasites and disease, as well as periodic shortages of food and occasional prolonged periods of severely inclement weather, probably are the main nonhuman factors that control coyote populations.

Red foxes
Vulpes

THE GENERIC NAME *Vulpes* is from the Latin word *vulpis* ("a fox"), but it also means "cunning" or "craftiness."

Foxes of the genus *Vulpes* are medium to small members of the family Canidae. They have large pointed ears, elongated sharp muzzles, and rounded bushy tails, which are usually as long as the head and body. The pupils of their eyes are generally elliptical in strong light. Most, if not all, foxes have a definite foxy odor arising from glands situated under the tail near the anus. They usually have six teats.

The length of the head and body ranges from 15 to $34^1/2$ inches and the length of the tail from 9 to 22 inches. Adults weigh from $6^1/2$ to 20 pounds. They vary from grayish, tannish yellow, pale yellowish red, deep reddish brown, to black.

Foxes are intelligent. Although not naturally wary, as a result of contact with humans, they have developed a cunning that now connotes the name "fox" with the acme of cleverness or craftiness. They have keen senses of sight, smell, and hearing, as well as considerable endurance.

Red foxes live in many different habitats, from the northern arctic coast of Alaska and Canada, through the temperate brushy woodlands, to the Sahara Desert of north Africa. Members of some species generally rest during the day in a sheltered spot; some species are nocturnal, but others are active at any time. Red foxes tend to be omnivorous, eating such items as small mammals, birds, eggs, fruits, and insects.

Over most of their geographical distribution, they breed during late winter. The gestation period ranges from 49 to 56 days. Litters range from 4 to 10 young, but the usual size is about five. The pups are born in a den and are cared for by both parents. Red foxes apparently mate for life; their lifespan is about 12 years.

Red foxes occur throughout most of North America, Europe, including the British Isles, most of Africa, north and south of the Sahara, and most of Asia. They have also been introduced into Australia. There is one species in western Oregon.

Red fox
Vulpes vulpes

THE FIRST RED FOX of this species was taken in the state of Virginia and was described in the scientific literature in 1820.

DISTRIBUTION AND DESCRIPTION

total length 900-1120 mm
tail 330-461 mm
weight 3.1-6.8 kg

THE RED FOX OCCURS in the northern half of zone 1 and throughout zones 2-5 (photo 128). They range in length from 36 to $44^3/4$ inches and have tails that vary from $13^1/4$ to $18^3/8$ inches. They weigh from almost 7 to 15 pounds.

Red foxes are slender and light in build. They have large, erect ears and long, bushy, white-tipped tails. Their feet are small and furry. Their pelages are very long and soft. The normal red fox varies from pale yellowish red to deep reddish brown above, with white or light to medium grayish underparts. The lower portion of the legs is black. There are, however, other color phases.

The "cross fox" is reddish brown above, and gets its name from the cross formed by a line of blackish hairs down the middle of its back intersected by another line of blackish hairs across the shoulders. The cross is most readily visible after an animal has been skinned and its pelt has been stretched out flat. The "silver fox," most prized of all foxes for its pelt, varies from silvery to nearly black, the general effect depending on the proportion of white or white-tipped hairs to black hairs. The "black fox" in reality is a silver fox that has only a meager number of white or white-tipped hairs. In all color phases, the tip of the tail is normally white.

THE RED FOX OCCURS in just about every terrestrial habitat except deep forest and cities, even occupying the edge of suburbia.

HABITAT AND BEHAVIOR

According to Ernest Thompson Seton, the red fox was all but absent from the eastern United States during the early period of our history. European fox-hunters introduced the European

Photo 128. Red fox. (USDA Forest Service photograph.)

Dogs & Foxes • Canidae

red fox into the eastern United States in colonial times; thus the eastern red fox in times past may have interbred with the European red fox. The red fox is indigenous, however, in northern and northwestern North America.

Red foxes are creatures of open areas, such as meadows, interspersed with patches of brush and timber. They are resident wherever they are found and seldom emigrate to other areas unless forced to by shortages of food or other severe circumstances. Red foxes are mild tempered and often frolic at twilight or on moonlight nights, but in no sense are they gregarious or sociable, except within the family group. These animals are intelligent, cautious, cunning, swift, alert, and quick to take alarm.

A red fox crossing a meadow on a foggy morning is indeed a beautiful sight. Its graceful, fluid motions make it appear to float through the fog without touching the ground. Vernon Bailey once wrote of the red fox: "Few animals are so quick and agile or so light and graceful in motion….Usually the long lines of delicate tracks in the snow, the prints of the narrow, furry feet in dusty trails, or the pungent almost musky odor greeting one's nostrils in the dewy morning, furnish the only evidence that a fox has passed along in the night."

Red foxes generally are quiet. Their principal sound has been variously described. To some, it is a rather high-pitched howl that consists of a few short yaps or barks followed by a long gurgling *Ya-a-a-a-r*, sometimes ending in a near scream or screech. To others, it is a short, sharp, little bark, like that of a small dog, only more rapid and prolonged.

In addition to helping a red fox maintain its balance when forced to turn or dodge swiftly, the bushy tail is also used to keep its nose and the pads of its feet from becoming frost-bitten when a fox sleeps during severe winter weather in some parts of its geographical distribution—the colder the climate, the larger the tail. A fox may also use its tail to ward off attacks either in a fight with another fox or by aerial predators, such as Golden Eagles.

Data suggest that the size of the home range of an individual red fox is similar to the size of the territory of the family, about one to three square miles. Red foxes appear to have innate minimum and maximum spatial requirements, which are manifested in their territoriality. As a red fox population declines, the remaining individuals increase their territories to the maximum size; if the population continues to decrease and all the remaining individuals have established maximum-sized territories with adequate supplies of food, then areas of suitable habitat will remain unoccupied.

> Joseph S. Dixon, in 1933, described the following encounter between a fox and a Golden Eagle:
>
> *I was fortunate enough to witness an attack by a golden eagle upon a cross fox … in Alaska on July 8, 1932…. Seeing me he started off down the road at full speed, but before he had covered 100 yards there was a "hiss of wings," and the eagle flew at him like a thunderbolt. Just as the eagle struck, the fox jumped nimbly to one side and thus evaded the eagle's talons. As the eagle descended the second time, the fox fluffed out his tail and stuck it straight up in the air over his back, so that it served as a protecting foil at which the eagle struck, and again missed the fox. The third attack of the eagle was frustrated by a dive into a narrow crack in the solid rock, where I found the fox with his nose just sticking out.*

The red fox hunts by stealth, creeping low, stopping to look and listen, stretching its head high to sight its prey, then, when close enough, pouncing with its forefeet on the vole, mouse, or rabbit, biting the back of the head of large prey or merely nipping it if it is small. It does not run its prey any distance and seldom waits by a trail unless it senses prey close by.

Although red foxes have a bad reputation with farmers because they eat domestic poultry, on the whole their diet is beneficial to farmers. The red fox is an opportunistic hunter and is omnivorous, eating mammals, birds and their eggs, reptiles, fish, insects and other invertebrates, carrion, as well as many fruits and other parts of plants. In many areas, their main prey consists of voles, rabbits, and hares but also includes shrews, moles, mice, pocket gophers, ground squirrels, chipmunks, tree squirrels, a variety of birds, and whatever else may be available.

The breeding season for the red fox begins in January and lasts through February, although occasional individuals may breed as early as the last of December. There usually is a single litter per year, which is born in a clean den after a gestation period of about 53 days. Litters consist of 2 to 10 pups, but five to six pups is the normal size. Yearling females produce smaller litters than do older females, generally called vixens, and sizes of litters vary from year to year, depending on the available supply of food for the vixens, as well as social stresses.

At birth, the young are lightly covered with fine, fuzzy, dull gray, dark brown or blackish fur, but with a distinctive white tip on the tail. Their eyes open in about nine days; they grow rapidly, beginning to play around the entrance of the burrow in about three weeks. Morning and evening are the principle playtimes. Either parent, usually the mother, watches or both parents watch the youngsters as they play. At about two months, pups are weaned and food is then brought to them by their parents. One pair of red foxes observed near Corvallis, Oregon, in 1943 brought the following food items for their seven pups: eight Beechey ground squirrels, several black-tailed jackrabbits, one brush rabbit, one domestic turkey, one domestic chicken, and one rufous-sided towhee; the remains were scattered in and around the den.

The pups usually abandon the family den shortly after four months of age, at which time they are nearly three-fourths grown and their permanent teeth are replacing their deciduous teeth. The young are essentially mature when they are about six months old, and they may breed the following winter. They begin to disperse from the familial territory—that of their parents and siblings—during October when they are about seven months old.

Although red foxes seldom use dens except to house and rear their young, there is usually more than one den in a family's territory. The foxes probably excavate most dens themselves, but they may renovate the burrow of some other animal. When possible, they construct dens on or near the tops of south-facing slopes and take advantage of the warming sun in spring. Dens ordinarily are 13 to 20 feet long, sometimes 40 feet or more. Living quarters are at least 3 feet below the surface of the ground. The entrances are 12 to 15 inches high, 8 to 10 inches wide. There are usually at least two entrances to each den, opening in different directions, affording a measure of safety. Even though the entrance to a den is usually littered with food refuse and other waste materials, the inner living quarters are often bare but always clean and tidy.

Although humans are undoubtedly the red foxes' worst enemy, they may be preyed on by wolves, lynx, fishers, wolverines, and occasionally a coyote. Some also die from the effects of porcupine quills after either an accidental or a purposeful encounter with the owners, and Golden Eagles may get some.

Gray Foxes
Urocyon

THE GENERIC NAME *Urocyon* is derived from the Greek words *oura* ("a tail") and *kyon* ("a dog").

There are two species of gray foxes. The length of the head and body ranges from 19 to 27 inches and the tail from about 5 to 17 inches. Adults usually weigh from just over 5 to 15 pounds.

Gray foxes have coarse pelages. The face, upper part of the head, back, and sides, and most of the tail are gray, whereas the throat, sides of the neck, lower flanks, and underparts of the tail are bright reddish brown. The hairs along the middle of the back and the top of the tail are heavily tipped with black, creating the illusion of a black mane. Black lines also occur on the legs and face of most individuals. The throat, insides of the legs, and underparts of the body are white.

Gray foxes are often called "tree foxes" because they frequently climb trees. They use caves, cavities and crevices in rocks, and hollow logs and trees for dens; they may also dig their own dens or appropriate those of other animals. They are mainly nocturnal, resting in a secluded place during the day. Gray foxes have a varied diet, including small mammals and birds, as well as many fruits and other portions of plants.

A single litter, consisting of two to seven—usually three to four—young, is born each year in the spring after an average gestation period of 63 days. The pups, born in a den, weigh about four ounces and are blackish at birth. They begin eating solid food in about six weeks and begin to disperse in the late summer or early autumn. The parents may remain together throughout the year.

Gray foxes occur only in the Americas: in the eastern, southwestern, and extreme western United States, south of the Columbia River, including certain islands off the coast of southern California, through Central America into northern South America. There is one species in western Oregon.

Gray fox
Urocyon cinereoargenteus

THE SPECIFIC NAME *cinereoargenteus* is derived from the Latin words *cinereus* ("ash colored") and *argenteus* ("silvery"); the name refers to the overall coloration of the backs and sides of these beautiful little foxes. The location of the first specimen of the gray fox is given as "Eastern North America." Its scientific description dates back to 1775.

DISTRIBUTION AND DESCRIPTION

THE GRAY FOX occurs throughout most of the five zones, although its geographical distribution may be spotty. Individuals range in length from 32 to 45 inches and their tails vary from 11 to $17^3/4$ inches. Adults weigh from 7 to 15 pounds.

total length 800-1125 mm
tail 275-443 mm
weight 3.2-7.04 kg

Gray foxes are relatively small; they have rather short legs, and fairly sharp, well-curved claws. Their pelages are short and coarse, and their tails appear somewhat flattened vertically. The face, upper part of the head, back, sides, and most of the tail are gray because of the numerous white- and black-tipped outer guard hairs that obscure the brownish underfur. The throat, sides of the neck, lower flanks, and underparts of the tail are reddish brown. The hairs along the middle of the back and along the top of the tail are heavily tipped with black, creating the illusion of a black mane. The tip of the tail is black. Black lines also occur on the legs and face of most individuals. The throat, insides of the legs, and underparts of the body are white. The summer pelage is brighter than that of winter, with lighter gray and more orangish on the sides and legs. A nursing female has pinkish hairs on her belly.

HABITAT AND BEHAVIOR

GRAY FOXES ARE ESSENTIALLY inhabitants of wooded areas, particularly mixed hardwood or mixed hardwood-conifer forests. Although primarily nocturnal, they have been seen during the day, apparently foraging for food. These little foxes are less easily observed than red foxes, except for the sign they leave, such as tracks and droppings (photo 129). They ordinarily sleep throughout the day in a hollow log or tree or some other sheltered place.

Gray foxes lack the speed and fluid grace of the red fox; nevertheless, they are exceedingly quick at dodging and turning and, of necessity, take advantage of rocky or brushy cover. If adequate cover is lacking, but trees are at hand, a gray fox will take refuge up a tree. It is the only member of the dog family in North America that regularly climbs trees, not only to escape enemies but also to rest. They have even been found curled up in abandoned nests of hawks in the tops of trees. When tree limbs are close to the ground, a fox simply hops from branch to branch as it ascends the tree. When necessary, they can also climb cat-like up the limbless trunk of a tree.

Photo 129. Close-up of fresh gray fox droppings. (Photograph by Chris Maser)

The voice of the gray fox is a sharp little bark, less sonorous, lower in pitch, and not as loud as that of the red fox, but similar in some respects. It apparently does less yapping and barking than the red fox. When contented, a gray fox may utter a purring grunt, but it is usually a quiet animal.

Vernon Bailey in 1905 wrote of the gray fox: "Strange as it may seem, these foxes go up the trunk of a tree with almost cat-like ease. I have found them looking down at the dogs from 20 to 40 feet up in the branches of nut pines and live oaks, and have known their climbing a yellow pine…where 20 feet of straight trunk over a foot in diameter intervened between the ground the first branch. More often they take to a live oak or juniper, where the lower branches can be reached at a bound, and then, squirrel-like, hide in the swaying topmost branches."

The home of a gray fox is a den well concealed in brushland or woodland; it may be in or under a hollow log or stump, in the base of a hollow tree, in a natural crevice among rocks, but rarely, if ever, in a burrow in an exposed meadow or field. One den of a gray fox that I found three miles south of Corvallis, Oregon, was in a small grove of white oak trees. Dug by the foxes themselves, the den had three well-worn trails leading to its single entrance concealed under a clump of poison oak. It was occupied by two adults and four juveniles.

Inside the den may be a scanty nest composed of leaves, grasses, fur, and any other available soft material scratched together by the fox. There is little evidence of sanitation around the den, although the entrance is usually not as messy with food refuse as in the case of the red fox, and unlike the den of a red fox, the gray fox does not usually have a platform adjacent the entrance where the young play.

Gray foxes are hunters of small animals, but depend on fruits and other portions of plants for food more than other foxes do. In Oregon, they feed on such small mammals as mice, gophers, kangaroo rats, woodrats, ground squirrels, chipmunks, and brush rabbits, as well as whatever small birds and unprotected domestic poultry they can catch. They also eat grasshoppers, beetles, berries of manzanita, juniper, and cascara, and any accessible cultivated fruits, such as blueberries, grapes, figs, plums, cherries, and apples. Unwanted food may be left with little attempt at concealment; food, as far as I know, is neither hidden nor covered for storage.

Mating occurs from the middle of February through March. After a gestation period averaging 63 days, a litter of two to seven—usually three or four—pups are born in a den. At birth, pups weigh about four ounces; they are blind, helpless, and nearly naked. Their eyes open in about nine days, and they are soon covered with grayish or blackish fuzz.

The mother apparently cares for the young by herself until they are about two or three weeks old, at which time the father may begin bringing food. At about three months of age, the pups begin to fend for themselves, but they do not leave the family until late summer or early autumn, when they are nearly full grown. The parents, on the other hand, may remain together throughout the year. Gray foxes in the wild usually live less than six years; in captivity, they usually live less than eight years.

Although few predators probably molest adult gray foxes, the bobcat, coyote, Great Horned Owl, and possibly some of the larger hawks may prey on the pups. Almost none are killed by automobiles. In some places, the population of gray foxes

As to disposition, Ernest Thompson Seton in 1928 described well the gray fox: "If we mix equal parts of Red-fox, Coon [raccoon], and Bobcat, and season the combination with a strong dash of Cottontail Rabbit, we shall have the Gray-fox's disposition synthetically produced. He is shy, he is cunning, he is a desperate fighter when at bay, he loves the trees and yet rejoices in the briar brush, he can run for hours and is adept at trick-trailing, but will hide in a burrow or up a tree; and the places he frequents are the places where any one of these animals also may be found."

decreases as the coyote population increases, and vice versa. People are the principal enemy of gray foxes, killing many of them with guns, traps, and poisons—some for pelts, some for elimination, and some for sport.

Gray foxes are susceptible to tularemia, a bacterial disease that is highly infectious to people. They also carry rabies. During 1960 and 1961, gray foxes in the Willamette Valley of Oregon all but disappeared and did not come back until 1965 or 1966—an outbreak of rabies was thought to have been responsible for this sudden and prolonged decline. A sick fox is best left strictly alone.

THE FAMILIAL NAME Ursidae is derived from the Latin word *ursus* ("bear") combined with the Latin suffix *idae* (which designates it as a family).

Bears have large compact bodies, big heads, short powerful limbs, and short tails. They have small eyes and small, round, erect ears. Their feet terminate in five digits, each with a strong claw for tearing, digging, and in some cases, for climbing. Bears walk flat-footedly. Those that are mainly terrestrial have hairy soles, but those that climb a lot have naked soles. Males average about a fifth larger than females. The length of the head and body ranges from about 3 feet to about 9 feet and the tail from 3 to 5 inches. Adult bears weigh from 60 to 1,715 pounds. They have long, shaggy coats; the pelage generally is unicolored, usually some shade of black, brown, or white.

Some bears are active primarily during the evening and night, others during the day. Bears that inhabit steppes and barren areas often dig dens in hillsides; the usual den, however, is a cave, hollow log, hollow base of a tree, or the hollowed-out base of a stump.

Bears are related to the dogs and wolves (family Canidae), but the exact line of descent is uncertain. Bears have poorly developed senses of sight and hearing, but their sense of smell, although not on par with some of their canine cousins, is excellent.

Bears are usually peaceful, inoffensive animals who try to avoid conflict, and it is not surprising that they are usually most numerous in areas remote from civilization. But if they must defend themselves, their young, or their supply of food, they are formidable and dangerous.

In temperate and colder regions, bears, except the polar bear, accumulate fat before winter and, with the onset of cold weather, they enter a den prepared earlier in the year and sleep throughout the most inclement weather. Bears are not true hibernators, however, because their body temperature is not reduced and their bodily functions are not completely slowed down; they can be awakened easily and sometimes arouse during mild weather, which is not usually the case for creatures that really hibernate.

Bears inhabit the arctic, temperate, and tropical regions. Their diets are as varied as their habitats and habits. Chiefly omnivorous, they consume grasses, leaves, roots, and fruits, as well as insects, fish, mammals, and carrion.

Except when breeding or when accompanied by their young, bears are solitary. A female has a single litter of one to four young. The gestation period ranges from six to nine months in members of most genera as a result of delayed fertilization and implantation of the ova. Most babies are born while the mother is asleep during the winter, between October and March. Young are very small at birth, weighing from eight ounces to one pound. They become sexually mature at two and a half to six years of age and usually live fifteen to thirty years in the wild.

Bears inhabit the Northern Hemisphere and northern South America; they do not occur in Africa, Madagascar, Australia, various oceanic islands, or the Antarctic. With the grizzly bear long extinct in the state, there is only one genus of bear represented in western Oregon.

North American Black Bear
Ursus

THE GENERIC NAME *Ursus* has the same derivation as the family name.

Since there is only one species of black bear, refer to the species account.

Black bear originally inhabited practically all the wooded areas of North America north of central Mexico. Today, however, they have been displaced from much of their former geographical range.

North American black bear
Ursus americanus

THIS BEAR WAS NAMED after America. The location of the first scientific specimen of the black bear is given as: "Eastern North America." It was described scientifically in 1780.

Photo 130.Cub black bear. (Oregon Department of Fish and Wildlife photograph.)

DISTRIBUTION AND DESCRIPTION

THE BLACK BEAR can be found in any suitable habitat in all five zones, although it is least likely to appear in zone 2 (photo 130). These bears range in length from 50 to about 70 inches with tails that vary from 3 to 5 inches. They weigh from 50 to more than 200 pounds.

The black bear is one of the largest mammals in Oregon, exceeded in size only by some of the hoofed mammals, and the largest carnivore. Black bears are massive, with strong, heavy bodies and moderately sized heads; their facial profile is rather straight. They have small eyes and small, round, erect ears. The tapering nose has a broad pad and large nostrils. The broad feet terminate in five digits, each with a strong claw. Compared with those of other species of bears, the claws are relatively short and those on the forefeet sharply curved. The short tails are hairy and inconspicuous.

total length 1270-1780 mm
tail 80-125 mm
weight 22.7-90.9 kg

The black bear has a number of color phases—black, blue, brown, reddish brown (often referred to as cinnamon), and white. Different phases may occur in the same litter. Black bears occasionally have some white on the throat and chest. Only the black, brown, and reddish brown phases occur in Oregon.

BLACK BEARS IN WESTERN Oregon are primarily animals of forested areas mixed with openings, although they periodically show up in places one would not normally expect them, such as the residential part of town. These bears were much more common when I was young than they are today, in large part due to loss of quality habitat. They were even more common in the days Vernon Bailey was studying the mammals of Oregon, as he wrote in 1936:

HABITAT AND BEHAVIOR

> In the densely forested range of these west-coast black bears there was and still is a wealth of food and cover to support large numbers, and probably nowhere in North America were bears originally more numerous. In 1909, on a trip down the coast of Oregon the writer [meaning Bailey himself] found them still abundant all along the way, although that part of the State had been well settled for many years. On one sheep ranch on the Chetco River [which is west of Brookings] in southern Curry County more than a hundred bears had been killed within the year without much impression being made on the general supply. In 1929 the Forest Service credits the Siskiyou National Forest in that same general section as harboring 910 black bears, a fairly generous supply for even this extensive area of wild rough land.

Black bears are probably still more numerous in the Oregon Coast Range and along the coast than elsewhere in the state, probably because of improved habitat resulting from forest fires and extensive clearcut areas. Fires and logging have removed most of the virgin forest that once covered the land and have created relatively open, brushy country with a general abundance of food for bears.

Photo 131. Tracks of a black bear crossing sand along a stream. (Oregon Department of Fish and Wildlife photograph.)

The setting sun usually brings the bears out to forage. Normally remaining active throughout the night and into the early morning, they may be seen occasionally during the day. They may appear awkward and comical; a bear lumbering along a trail in full flight, its head swinging from side to side and its hind feet stretching past its forelegs, may look amusing, but a bear's speed is deceptive. While it may appear slow, a bear can easily outrun a human.

The average length of the home range of a female (called a sow) may be around one and a half miles, whereas that of a male (called a boar) may be in the neighborhood of four miles (photo 131). Home ranges of adult males may overlap with those of adult females, but there is minimal overlap between the home ranges of two adults of the same sex, though home ranges of subadults may overlap considerably with those of adults. As an individual matures, its home range may increase in size.

Black bears construct their winter dens in a variety of places, the most common being at the base of a hollow trees (photo 132). Other sites include the undersides of fallen trees (logs), in rock caves, under buildings, or holes either dug in the ground by the bears or enlarged by them. Although all bears tend to make hollows in which to lie, only about a third of them seem to move bedding materials into their dens. Occasionally a bear will lock the entrance to its den with bedding material in a manner similar to that of true hibernators. A few individuals rearrange their beds during the winter.

Late autumn and early winter weather greatly influence the onset of winter dormancy. Bears usually enter their dens by late October, but the onset of dormancy may be delayed into the beginning of November. The first heavy snowstorm generally sends bears into their winter quarters, where, along the Pacific coast, they may remain for nearly three months, even during relatively mild winters. From late February until the end of March, there may be a period of mixed activity and inactivity, with complete arousal occurring during the first part of April. At higher elevations, such as in the High Cascade Mountains, emergence may be delayed until the middle of May.

In contrast to winter dens, summer beds or shelters are merely concealed places scratched in the ground among dense

Indians of several tribes thought that bears ate the skin off their feet during winter and thus had sore feet in spring when they emerged from dormancy. The early European colonists also believed this. An old saying that I often heard used by backcountry folks when I was a young man was: "He or she is as angry as an old sow with sore feet!" I was not the only one who heard this notion; in 1974, L.L. Rogers, aware of such beliefs, examined the feet of dormant black bears. He found that they do indeed shed the hairless pads of their feet during winter dormancy. The newly forming pads are tender until they harden; thus, some bears do in fact have tender feet upon emergence from their dens in spring. Droppings near the entrances of the bears' winter dens had portions of their foot pads in them—not surprisingly, the Indians had been correct.

vegetation, by a rock or under the branches of a fallen tree. A resting place may be lined with bedding materials and is fairly neat even though its owner does not habitually clean it, primarily because the bear goes some distance to defecate or urinate.

As with most animals of solitary habits, black bears do not have any known regular call notes; they simply growl, bawl, and bellow when things are not to their liking. A mother, on the other hand, frequently vocalizes to her cubs. She calls them with a whimpering sound and warns them with a loud "woof-woof." When unsure of the disturbance, a bear will rise onto

Photo 132. Winter den of a black bear dug into the base of a hollow, living tree. (Photograph by Chris Maser)

its haunches, sniff loudly, and look around. Although a black bear's sight is poor and its hearing only moderately acute, its sense of smell is highly developed.

These bears are good, surprisingly fast, climbers, though they cannot climb small trees. When starting to climb, a bear stands full length at the base of a tree, reaches up with its forefeet and grasps the tree with its claws and legs in a series of bounds.

The black bear is also a good swimmer and seems to enjoy water. In hot weather, a bear often bathes and wallows in shallow pools. Through regular use, such pools become know as bear wallows. I have seen bears wallowing in muddy pools in the High Cascade Mountains. I watched one large mother as she wallowed for three hours on an extremely hot summer afternoon, but she did not share it with her three cubs. Every

Some years ago, I was asked to examine black bear damage to trees in the Mt. Hood National Forest. It was the most severe damage I had ever seen.

We spent most of the day in the field looking at 20- to 30-year-old plantations of Douglas fir, where black bears had been eating the inner bark off the trees. These plantations were on densely forested steep slopes, on gentle slopes, and on flat ground with more widely spaced trees. The bears had climbed the trees on the steep slopes and had eaten the cambium on the open, downhill side about two-thirds of the way up the trees. They had hooked their claws under the bark of trees on the gentle slope and pulled it off in great strips (photo 133). They had then scraped off the cambium with their lower, front teeth (photo 144). On flat ground, the bears had to work hard to get around the low limbs and reach open areas of bark.

After four or five hours in the field, the wildlife biologist asked what they could do to control the damage. "Cut trees differently," I said. "That will stop most of the bear damage. Let's go back to the office and I'll show you."

Once in the office, he got out all the maps and aerial photographs of the area. I then showed him that all the clearcuts were within one-quarter mile of one another and were all about the same age. The way the sale units had been laid out created much open bear habitat within an otherwise dense old-growth forest. The bear population had responded, as would be expected, to the boon of new habitat. As the plantations grew up, however, the expanded bear population suddenly found itself with a rapidly shrinking supply of food, and thus supplemented its diet with the cambium of plantation trees. A bear population also existed outside the plantations; the population in the plantations could not therefore move out of the plantations to find alternative foods without having to compete with other, already-established resident bears.

With this understanding, a solution to the bear problem was simple: take their habitat needs and habits into account when planning future timber sales and make appropriate adjustments. Bear habitat can either be created or minimized as desired.

Photo 133. Bark stripped off a young Douglas fir by a black bear. The bear hooked the claws of its front foot under the bark and pulled it off. Arrows indicate some of the bear's claw marks on the inside of the bark. (Photograph by Chris Maser)

Mammals of the Pacific Northwest

time a cub tried to enter the water, she would throw it out with a single swipe of her forepaw. Although I have found no such wallows along the Oregon coast, they may exist.

Black bears are opportunistic in their feeding. They eat a great variety of green vegetation, fruits, and fungi. In addition to plant material, they eat insects, as well as other invertebrates, mammals, birds, and carrion. In western Oregon, they also eat the cambium of coniferous trees.

Testes begin to enlarge before the males arouse from winter sleep and contain mature sperm before and after the period in which females are receptive, but by the time a male enters dormancy, the testes have again shrunk drastically and are nonfunctional. Females normally breed and produce litters only every other year. They come into estrus at least by the last week in May and remain in estrus until about the second week in August. The peak of the breeding season, however, is usually from mid-June through mid-July. A female remains in estrus either until she is bred or until her ovaries become reproductively nonfunctional.

Ovulation is induced through copulation. After fertilization, the development of an embryo either ceases or is considerably slowed, and it does not become implanted in the wall of the uterus until about the first of December. It develops rapidly during the six to eight weeks before birth in late January or early February, after a gestation period of about 220 days. Litters

Photo 134. Once the outer bark is removed, the bear scrapes the inner bark (cambium, the living tissue) off the stem with upward vertical strokes of its lower incisor teeth. (Photograph by Chris Maser)

Photo 135. A very young black bear cub; note its decidedly fuzzy appearance. (Oregon Department of Fish and Wildlife photograph.)

range from one to four young, rarely five; two or three constitutes the usual litter.

Cubs, born while their mother is dormant, are about 6 to 8 inches long and weigh 7 to 12 ounces at birth. They are born with closed eyes and are sparsely covered with fine, stiff hairs. Development is slow the first few days after birth, but if sufficiently nourished by their mother, the young grow rapidly thereafter. They open their eyes in about 25 days, at which time they are covered with short, fuzzy brownish hairs, even those that will be black adults. A mother is very attentive and protective of her young. If unduly disturbed, she may move to a new home and will carry each baby to the new den by the nape of its neck.

Before they are three months old, the cubs commence playing by themselves, but they are carefully watched and guarded by their mother (photo 135). At six months, they are able to fend for themselves, but remain with their mother for a year or more, usually sleeping either with her or near her during their first winter. By the time they are a year old, they weigh between 40 and 75 pounds. A family normally breaks up during the breeding season. Sows have their first litters when they reach $6^1/2$ to $8^1/2$ years of age. The potential longevity of black bears in the wild is about thirty years, but they normally live about twenty-five years.

Humans have been and still are the main "predator" of black bears, killing many of them annually either for sport oo for elimination. Other than humans, black bears appear to have few enemies.

THE FAMILIAL NAME Procyonidae is derived from the Greek prefix *pro* (meaning "before") and the Greek word *kyon* ("a dog") combined with the Latin suffix *idae* (which designates it as a family).

The family Procyonidae, although containing only nine genera, is a diverse one and difficult to describe in general terms. Members of this family are close to the primitive dog-bear (canid-ursid) stock, ranging in length of the head and body from 1 foot to almost 5 feet, with tails varying from $4^3/4$ to $27^5/8$ inches. Adults weigh from almost 2 to 297 pounds, males averaging about one-fifth larger and heavier than females.

Procyonids have short, broad faces and short ears that are hairy and erect. The tail is prehensile or semiprehensile in some members of the family. Each foot terminates in five digits, of which the third is the longest. The claws are short, compressed, well curved, and in members of some genera can be partly withdrawn into a sheath. In members of several genera, the soles of the feet are hairy. The pelage varies from gray to rich reddish brown. Facial markings are often present, and the tail is usually ringed with light and dark bands.

Most procyonids become active in the evening, but some are active at any time. Members of this family take shelter in a variety of places, such as hollow trees, crevices in rock, and occasionally in buildings.

Some procyonids walk on the soles of their feet with their heels touching the ground. Others walk partly on the soles and partly on the digits; their movement is usually bearlike. Members of this family are good climbers; in fact, one species spends nearly all of its life in trees. Although most procyonids travel in pairs or family groups, members of one species travel in bands of as many as two hundred individuals but usually in groups of five to forty. Procyonids are mainly omnivorous.

Litters range from one to six offspring. Young are born in the spring and generally weigh an ounce or more at birth. There may be one or two litters per year. Females of most genera begin to breed during their first year or shortly thereafter, but males do not begin to breed until they are about two years old.

The procyonids inhabit the temperate and tropical regions of the Americas and the mountainous portions of south-central Asia. There are two genera in western Oregon.

THE GENERIC NAME *Bassariscus* is derived from the Greek word *bassara* ("a fox") and the Greek suffix *iskos* (diminutive form) and probably alludes to this small mammal's foxlike features.

Since there is only one species within the genus *Bassariscus*, refer to the specific description.

Ringtails, Raccoons, and Allies
Procyonidae

Ringtail
Bassariscus

The ringtail occurs from southwestern Oregon east as far as central Colorado, and south through Texas to southern Mexico.

Ringtail
Bassariscus astutus

THE SPECIFIC NAME *astutus* is derived from the Latin word *astus* ("craft, dexterity") and alludes to this alert little mammal's quickness and agility. First captured near the City of Mexico, Mexico, which is probably Mexico City, it was described in the scientific literature in 1830.

RINGTAILS OCCUR IN the southern hakf of zone 1. They range in length from $24^3/8$ to $32^3/8$ inches with tails that vary from $12^3/8$ to $17^1/2$ inches. They weigh from almost 2 to $2^1/2$ pounds.

A ringtail is about the size of a small house cat but more slender, with larger ears and a long, bushy tail marked with eight blackish rings, including the tip, and seven whitish rings. The blackish rings are not completed on the whitish underside of the tail. The face is brownish gray with whitish spots both above and below the eyes and in front of the ears. The color of the pelage is about the same throughout the year; the back is dark yellowish gray and the underside tannish white or light yellowish white.

The ringtail is also called "ring-tailed cat," "miner's cat," and "cacomistle." "Ring-tailed cat" is not a good name because this mammal is not closely related to the cats. "Cacomistle" is a name more generally, and perhaps better, applied to the genus *Jentinka*—also in the family Procyonidae—the Central American cacomistle. The name "miner's cat" came about in earlier years when much prospecting was going on in the western United States. Miners tamed these mammals and kept them around their cabins, instead of house cats, because ringtails are excellent mousers.

HABITAT AND BEHAVIOR

RINGTAILS ARE PRIMARILY nocturnal and seldom seen during the day. Vernon Bailey was so taken with the ringtail that in 1905 he penned: "… most of all, the big, soft, expressive eyes give a facial expression of unusual beauty and intelligence."

Throughout much of their geographical distribution, they show a preference for rough, rocky, broken terrain, with or without trees. In western Oregon, they are most common along the rock cliffs near rivers, such as the Rogue River in Curry County. A ringtail descends a vertical surface headfirst. With the aid of fully extended claws, it can quickly ascend or descend vertical surfaces, even the edge of a smooth, narrow slab, such as a board or metal door, a narrow rod, or a smooth branch. An individual's strength and the friction of its naked soles, as well as its ability to rotate the hind feet either $90°$ or $180°$ and then clasp them tightly together, make such feats possible. Rotation of the hind feet for grasping also occurs during play-fighting when an opponent is embraced and drawn close by all four feet. Grasping is further aided by a limited amount of opposability of the first digits of the forefeet.

Ringtails use a mountaineering technique called "chimney stemming" to negotiate in any direction a crevice between two closely spaced nearly vertical or vertical surfaces devoid of footholds. It may press all four feet against one wall and its back against the other, or it may use any combination of its four feet distributed between the two walls, except crossing the limbs. When using this technique to descend a crevice headfirst, a ringtail does not necessarily rotate its hind feet; it can maintain sufficient control by applying adequate pressure to both walls. A frightened ringtail may seek a narrow crevice-like retreat, where it wedges itself into the most inaccessible portion. Even while playing or exploring new places and objects, it crawls into any space that will accommodate its body.

Another amazing trait of ringtails is their ability to rebound or ricochet from one surface to another when rapidly ascending, descending, or running. This technique may be used to negotiate crevices that are too wide to allow use of the chimney technique and too smooth to climb directly. During an ascent, a ringtail appears to push, gaining momentum at the point of the ricochet; when descending the point of ricochet seems to slow the momentum. The ricochet technique enables a ringtail to exert some control over the direction and speed of its travel, even permitting changes of direction from 90^0 to 180^0.

When leaping, a ringtail propels itself at a specific landing spot with considerable force; this is a deliberate, accurate launch at a particular location regardless of whether the direction is up, down, or horizontal. This power leap is frequently used even to jump only 1 to 2 feet, occasionally resulting in a resounding thump.

A ringtail moving along a narrow ledge less than 2 inches in width can reverse direction at least two ways. It may raise the forequarters upward, climbing in a semicircle with the pressed abdomen against the wall, dropping down facing in the opposite direction; or it may keep the head and torso aimed toward some object away from the ledge and walk the hind quarters up the wall in a semicircle over and down the other side while its tail hangs down over its back.

Ringtails begin to forage well after dark. They move swiftly and quietly while hunting in thick vegetation and shadows; when they have to cross moonlit places, they arch their long tails over their backs, making them appear much larger.

With clear admiration for the ringtail, Ernest Thompson Seton wrote in 1928:

"Snug in his high aerial castle dwells the Ringtail, and sleeps placidly through the hours of brightness. Not even at sundown will he venture forth, as do so many of the gray-light creatures. Night, the very blackest, must be on woods and hill, before he unwinds his famous tail from his elegant body, peers with big, bright eyes from his high doorway, and commits himself to the adventures of his daily night.

Long before dawning, he is fed, back, and upcurled again in his home hole; and the only proof that one commonly finds of his life and presence, is the endless chains of catlike tracks in the dust."

In April 1972, 10 miles east of Brookings, along the coast in southwestern Curry County, I heard a slight noise outside the cabin I was using as field headquarters while studying the mammals of that region. It was a dark rainy night, and as I quietly opened the door, flashlight in hand, there was a blurr of motion and the lingering odor of a ringtail. Although I did not clearly see the animal, the odor was positive identification of this late-night visitor.

When frightened, ringtails secrete an amber fluid from their anal glands, which has an unmistakable sweet and musky odor. The fluid spreads over the anal area but apparently is neither forcibly expelled nor wiped or rubbed on anything.

The most commonly described vocalization emitted by ringtails is an explosive bark. Other sounds include a piercing scream and a plaintive, long, high-pitched call.

Ringtails eat a variety of foods, including small birds and mammals, snakes and lizards, insects and an assortment of other invertebrates, as well as various fruits.

The breeding season seems to be relatively short, after which a single litter is produced. Litters of one to five, but usually three or four, young are normally born during May and June. Nursery dens may be in crevices in rocky bluffs or in hollow stumps in the vicinity of such bluffs. There is little or no apparent effort to construct a nest within the den.

Three or four days before giving birth, a female becomes antagonistic toward the male, driving him off. Baby ringtails weigh an ounce at birth. They are pink, fuzzy, and helpless, and their eyes and ears are closed. They have blunt muzzles, no teeth, and pigmented bands on their tails. Newly born ringtails vocalize by squeaks and move in an awkward crawling, wriggling motion. At 26 to 29 days, their ears are open; and by the time they are 31 to 34 days old, their eyes are also open. The young become fully haired in a fuzzy pelage that is lighter than that of the adults but similarly marked. At this age, the ears droop, the muzzle is long, and the canine teeth and the outer incisors are through the gums. By the time they reach 134 days of age, they are subadults, capable of adult movements, and have the explosive bark of an adult; the testes of males, although immature, have descended.

Large owls and bobcats are probably the main enemies of ringtails. In some states, ringtails are hunted and trapped for their pelts. They are occasionally killed by automobiles.

THE GENERIC NAME *Procyon* is derived from the Greek prefix *pro* ("before") and the Greek word *kyon* ("a dog").

The length of the head and body ranges from 16 to 24 inches and the tail from 8 to 16 inches. Raccoons weigh from 3 to 49 pounds. Behind the pointed muzzle, the head is broad. There is a black "bandit mask" across the face and eyes; there are also five to ten complete black rings on the hairy tail, including the tip. The pelage is predominantly grayish to almost black. The toes are not webbed and the claws cannot be retracted. The front toes are long and can be widely spread allowing the forefeet to be skillfully used.

Raccoons, although primarily active during the late evening and throughout the night, are occasionally active during daylight hours. They are excellent climbers and swimmers. These mammals inhabit timbered and brushy areas, frequently near fresh or salt water. They are omnivorous and eat such things as frogs, fish, crayfish, marine invertebrates, small mammals, and birds, as well as various seeds, nuts, and fruits.

Breeding takes place from January through June. The gestation period ranges from 60 to 73 days but averages 63 days. There may be two litters per year of one to seven young, but three or four offspring is the usual size. The young travel with their mother when they are ten weeks old and become independent at about one year. Females may breed when one year old, but males generally do not breed until their second year. Raccoons may live more than ten years.

There are currently seven recognized species of raccoons. They occur from southern Canada south throughout most of the United States, Mexico, and northern South America and occupy various islands off the coasts of North America, Mexico, and Central America. There is one species in western Oregon.

THE SPECIFIC NAME *lotor* ("a washer") is a New Latin word derived from the Latin word *lotus* (" washing"). The name refers to the raccoon's habit of manipulating its food in water before eating it The raccoon was described scientifically in 1758 from a specimen taken in Pennsylvania.

Raccoon
Procyon lotor

RACCOONS OCCUR IN zones 1, 2, 3, and the lower elevations of zone 4. They are heavy-bodied, compact mammals that reach about 3 feet in length and weigh up to 49 pounds. They have pointed noses and short faces. Their ears are erect and they have black "masks" across their faces, through their eyes.

They walk completely flat footedly on naked soles (photos 136 and 137). Their well-curved claws aid them in climbing. The soft underfur is largely obscured by the long, coarse, outer

DISTRIBUTION AND DESCRIPTION

total length 603-1050 mm
tail 192-473 mm
weight 1.8-22.2 kg

guard hairs. The pelage on the back is a dark, coarse gray, appearing grizzled. They have a narrow whitish band across the forehead and cheeks and three whitish spots covering the chin and the sides of the nose. The tips of the ears are whitish. The round, bushy tail is completely encircled with six or seven black rings, including the tip, and six or seven gray rings. The tops of the feet are grayish. The pelage of the underside is light brown, more or less silvered by the scattering of long, whitish hairs. The throat is brownish to brownish black and the chin is whitish. The nosepad and the soles of the feet are black.

HABITAT AND BEHAVIOR

ALTHOUGH RACCOONS PRIMARILY are active during the late evening, throughout the night, and into the early dawn, they may be abroad during the day. During the spring, adults become active when the sun sets, but during the autumn, they may not commence their nightly activities until two hours after sunset. Regardless of when their activities commence, raccoons sleep most deeply around midday. Most raccoons, except females about to give birth, sleep on or near the ground during the day. A few rest in trees. Although some resting sites may be used more than once, most individuals shift to new resting places daily and do not follow a predictable pattern. Habitat seems to exert some influence on an individual's selection of a resting place, because most of those living in low-lying swampy areas rest on or near the ground, whereas most of those living in upland, wooded areas rest in trees. Raccoons generally are restless sleepers. When disturbed, a raccoon will leave its resting place and seek another in a similar situation.

I have seen a male shortly after dawn between 60 and 70 feet up in an ash tree; he was curled up in a natural bed, where three branches separated from the main limb. I also watched a male one sunny October afternoon as he draped over the broken top of a young Douglas fir tree about 50 feet up. Such resting sites seem to be fairly common in areas of young Douglas fir trees from 25 to 60 years of age, as evidenced by raccoon hair stuck to the bark. The tops of the trees are frequently broken and either permanently bent over or completely removed by wind and snow. With the leader gone, the smaller, secondary branches take over, forming a well-concealed area in which to sleep. On one occasion, I found a female raccoon and her four offspring sleeping about 30 to

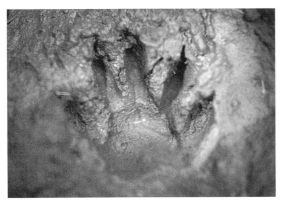

Photo 136. Track of a raccoon's front foot in soft mud. (Photograph by Chris Maser)

Mammals of the Pacific Northwest

40 feet off the ground in the crotch of a large Oregon ash tree. The female was wedged into the crotch, and her young were all piled on top of her. Other resting sites I have found in trees include the nests of squirrels, dusky-footed woodrats, Oregon red tree voles, Cooper Hawks, and nest boxes made by humans for wood ducks.

Photo 137. Tracks of a raccoon along a stream. Lower track is front foot; upper track with blunt toes is hind foot; note the almost complete register of the raccoon's heel due to its flat-footed walk. (Photograph by Chris Maser)

Raccoons sleeping in trees sleep in compact positions—lying on their sides, tightly curled—during the early morning; but by midday or later, they tend to sprawl with one or more legs, as well as the tail or head, hanging over the supporting branches. Hollow trees are used for sleeping also. Entrances to dens in trees vary from about 5 feet to 40 feet above the ground, and sleeping quarters in most dens are well below the entrance. One feature common to all den trees is the absence of a ground-level entrance. Hollow trees are most frequently used during the late autumn and early winter, when more than one raccoon, usually siblings or a mother with her young, may occupy a den during severe weather.

The size of a raccoon's home range as well as the nightly distances traveled vary greatly and depend on the habitat they encompass. Marsh-dwelling raccoons, for example, spend the majority of their time each night in the vicinity of shallow water, moving an average of about 520 yards per hour. They move a considerable distance during a brief period, followed by a long period spent in areas of shallow water. During autumn, the greatest movement occurs between three and four hours after sunset, followed by three to four hours of limited movement. Movement then gradually increases until one or two hours before sunrise when it suddenly decreases. During spring, this pattern of movement is essentially the same, but the total distance covered is more evenly distributed throughout the night. Generally speaking, individuals with large home ranges travel farther per night than do those with small home ranges.

During the summer and autumn, raccoons accumulate a layer of fat under the skin. Throughout the winter and early

Young raccoons are cute and would seem to make delightful pets because they are intelligent and curious, but they are wild animals and not meant to be pets, as I found out as a youth.

I got a young raccoon one day from a boy at school whose parents wouldn't let him keep her, because she got into everything. His mother threatened to throw him out of the house if that "damn coon" got into her flour again and tracked it all over the kitchen. The final straw, it turned out, was when she killed a couple of the people's young chickens, even though she was inside her cage. She had apparently waited quietly until they came close enough to grab, and she then proceeded to pull them through the chicken-wire side of her cage a piece at a time. With that, the raccoon was out of their house for good.

Although the young raccoon was very cute, I quickly learned just how sharp her teeth were. She was also extremely curious; nothing was spared her close scrutiny. She was exceedingly fond of ice cream—any kind of ice cream—and lots of it. If anyone was eating ice cream and she discovered it, she was not to be denied, which often meant nipped fingers and ice cream-coated raccoon tracks all over. But it was her second adventure that I remember most, an episode of such magnificence that it ensured her return to the wilds.

My mother was hosting a bridge party of several ladies in the living room. I found the whole notion of cards boring and therefore thought the ladies would enjoy an interlude to visit with my cute pet. So I let the raccoon loose in their midst. And I was disappointed, to say the least, with their abrupt, nervous silence. Some even developed a catatonic state. The raccoon, however, wasted no time, heading straight for the wax fruit on one of the ladies' hats, eliciting lengthy high-pitched screams, bulging eyes, heavy breathing, and general pandemonium.

But the worst was yet to come. As I have already said, when hunting for food in a stream, a raccoon doesn't watch what it is doing but instead feels around under the water with its sensitive forefeet, appearing to gaze absent-mindedly into space while doing so. This same behavior takes place whenever a raccoon is searching in any dark place into which it cannot see, and that was the coup de grâce.

After sampling the fruit-basket hat, the raccoon walked along the back of the couch until she came to a very heavy, buxom lady wearing a low-cut black dress—an invitation that was impossible to resist. So she immediately climbed onto the woman's shoulder and began exploring the depths of her cleavage as though searching for a frog or a crayfish amongst the rocks of a stream, all the while looking off into space. The woman in turn emitted small, strangling sounds, turned purple, and fainted.

Mammals of the Pacific Northwest

That was the final straw, even for my mother who was exceedingly patient with me and my animals, including the rattlesnakes, scorpions, and black widow spiders I kept in my room to ensure some privacy. The next day my raccoon was returned to the river from whence she had originally come, and I never saw her again.

---

spring, sometimes as late as May, the fat is used. Average daily weight loss by adults may be about two percent; by spring, they may have lost over 50 percent of their weight the previous autumn.

Raccoons are omnivorous. Those that live along the coast may feed entirely on tidewater and mudflat animals, such as mussels, shrimp, fish, and other marine organisms. Inland from the beach they eat such things as crayfish, freshwater mussels and clams, frogs, earthworms, and young brush rabbits; they also rob the nests of birds, squirrels, and other tree-dwelling animals. Raccoons also eat such fruits as evergreen huckleberries, salal berries, blue elderberries, cascara berries, and salmonberries, apparently shifting toward a diet of available fruits during the autumn and early winter.

When hunting for food in a stream or along a river, a raccoon does not watch what it is doing but feels around under the water with its very sensitive forefeet, appearing to gaze off into space while doing so. When something of interest is encountered, it is manipulated briefly in the water with the forefeet and then inspected with the nose and the mouth. The habit of manipulating food in water before eating it may have an adaptive value; sand and grit are washed off before the food is consumed, preventing unnecessary wear on the teeth. Nevertheless, adult raccoons often do have badly worn and sometimes decaying teeth that may result from sand eaten with their food.

Lengthening days partly influence the commencement of breeding, which may start in April. But low temperatures immediately before the onset of the breeding season may so restrict the movements of the females that they miss being bred in the first estrous cycle; therefore, they may not be impregnated until they experience a second estrus four to five months later.

The gestation period ranges from 63 to 65 days. As the time of birth approaches, pregnant females select hollow trees and may begin to sleep there one to three days before the young arrive. They remain in their tree-dens for one or two nights before giving birth. Two to three nights after the young arrive, they spend short periods outside the dens—less than six hours, sometimes less than two hours.

Litters range from one to seven, but the usual size in western Oregon appears to be two to three. Raccoons weigh about three ounces at birth; their eyes are closed. The mask on the face is present as pigmented skin that is sparsely haired, but the tail-rings are present only as pigmented skin. The backs and sides are uniformly colored, and young of different colors—black, dark gray, dark brown, or light brown—may occur in a single litter. Hair on the backs and sides of newborns is relatively sparse, but at one week of age they are well covered, and by the time they are seven weeks old, their pelage is similar to that of an adult. The eyes of young raccoons open when they are about twenty days old. At four weeks they can walk shakily but can neither run nor climb; at seven weeks they can walk, run, and climb. They begin to eat solid food when they are nine weeks old and are weaned at about twelve weeks.

The mother removes her young from the den and places them in a bed on the ground when they are between seven and nine weeks old. She may move them several times before they are old enough to begin accompanying her. When sixteen weeks old, the young begin to follow their mother, and by seventeen to eighteen weeks they become semi-independent. In late November, raccoons begin their winter inactivity either as a family or at least close to one another. During this period of semi-independence, youngsters operate within their mother's home range. Although females, and occasionally males, may become sexually mature during their first year of life, most do not breed until their second year.

Although bobcats and Great Horned Owls prey on raccoons to some extent, humans undoubtedly are the raccoon's main enemy—hunting them with dogs for sport, trapping them for their pelts, and trapping and poisoning them for elimination. They are also killed by automobiles. I have, on more than one occasion, put a badly injured raccoon out of its misery, thanks to a hit-and-run driver.

If inclined to hunt raccoons for food, be sure their flesh is thoroughly cooked before it is eaten, because they may carry trichinosis. They can also carry leptospirosis, tularemia, rabies, and Chagas' disease.

THE FAMILIAL NAME Mustelidae is derived from the Latin word *mustela* ("a weasel") combined with the Latin suffix *idae* (which designates it as a family).

This is a highly diverse family of mammals, consisting of 25 genera and nearly 70 species. The length of the head and body ranges from $5^1/_2$ inches to almost 5 feet. Adults weigh from just under an ounce to 79 pounds. Males average about one-fourth larger than females. The pelage may be uniformly colored, spotted, or striped; some species of weasels turn white in winter in the northern regions of their geographical range. Members of most genera have long, slender bodies, but skunks (which some authorities now put in their own family, Mephitidae), wolverines, and badgers are stocky. Ears are short and round or pointed. Short limbs terminate in five digits bearing compressed, curved, nonretractile claws. Badgers have large, heavy, relatively long, straight claws for digging. Otters have webbed feet, which aid in swimming. All mustelids possess well-developed anal glands.

Members of this family are nocturnal, diurnal, or both. North American badgers sleep throughout most of the winter. The same is true for striped skunks in the northern regions of their geographical distribution. Winter sleep may be interrupted by trips outside the shelter during warm weather. Members of several genera are agile climbers; others are skillful swimmers. Mustelids travel singly, in pairs, in family groups, or in bands of up to thirty individuals representing several family groups. Some members have a home range of about $12^1/_2$ acres; others may range over many square miles. Sea otters, which seldom leave the water, may range over an area of about 105 miles.

Members of many genera, particularly otters and European badgers, are extremely playful, especially when young. But they, like most other mustelids, can fight viciously if they must. Many emit pungent secretions from anal glands as a defensive mechanism. Members of some genera, such as skunks, are black and white; this combination of colors is thought to represent a warning coloration associated with the fetid secretions of the anal glands, indicating that the animals should not be molested. In some species, this contrasting coloration is exposed and emphasized by particular movements of an individual's body when it is alarmed.

Mustelids are primarily carnivorous. Although they hunt by scent, their senses of hearing and sight are well developed. Members of a few genera are more or less omnivorous; wolverines are scavengers, and otters feed mainly on aquatic life. Members of some genera store food.

Martens, Weasels, Skunks, Otters, and Allies
Mustelidae

The gestation period in members of many genera may range from 39 to 65 days; in others, with a prolonged delayed implantation, it may be as long as $12^1/2$ months. Most mustelids produce only one litter per year, but one species that does not have delayed implantation may bear up to three litters per year. Litters range from one to thirteen young. With the exception of sea otters, mustelids are blind at birth. Most young are able to fend for themselves at about two months of age and are sexually mature in about a year. The lifespan of most wild mustelids ranges from five to twenty years.

Mustelids occur throughout the world, except for Australia, Madagascar, the Antarctic, and most oceanic islands. There are six genera in western Oregon.

Martens and Fishers
Martes

THE GENERIC NAME *Martes* is the Latin word for a marten.

Members of the genus *Martes* are inhabitants of forests, both coniferous and deciduous. The length of the head and body ranges from $13^5/8$ to $24^5/8$ inches with tails that vary from 6 to 17 inches. Adults weigh from $1^1/2$ to 18 pounds.

Martens vary dorsally from yellowish golden brown to almost blackish, usually darkest on the feet and the tip of the tail; they vary from white to yellow to orange on the throat and chest. Fishers, on the other hand, have dark brown or grayish foreparts and blackish rumps, tails, and legs.

Members of this genus are generally solitary and may be abroad at any time. They are active throughout the year, though during severely inclement weather, they may remain in their dens. In mountainous regions, they may descend to lower elevations in winter. These agile and graceful mammals are more or less arboreal. They eat a variety of foods, such as mice, voles, squirrels, porcupines, and fruits.

Most species have delayed implantation. Martens are born after a gestation period that ranges from 220 to 290 days, whereas fishers have a gestation period of 338 to 358 days. Litters, ranging from one to five young, are usually born in a den, such as a hollow tree, stump, or log. Individuals may live more than 17 years. Members of this genus have been highly prized for their pelts, and were nearly exterminated in many areas. Through rigid measures of protection, these animals have regained a little of their former abundance in some areas.

In the New World, martens inhabit Alaska, and both martens and fishers inhabit Canada and those parts of the United States generally corresponding to the boreal coniferous forest. Only martens occur in the Old World, from the limits of tree growth south to the Mediterranean and across Europe and Asia into the Malay Archipelago. There are two species within this genus in western Oregon.

THIS SPECIES IS NAMED for North America. The site of the first scientific specimen of the North American marten is given as: "Eastern North America." It was described scientifically in 1806.

North American marten
Martes americana

Photo 138. Wind-thrown trees, which generally constitute areas of prime habitat for marten; photo taken near the Middle Sister in the High Cascade Mountains of Oregon. (Photograph by Chris Maser)

ALTHOUGH THE MARTEN has been found in all zones in western Oregon in the past, today it is most likely confined to the more remote areas of coniferous forest. Marten range in length from $20^1/2$ to $29^3/8$ inches, and their tail vary from $6^5/8$ to $9^5/8$ inches. They weigh from $1^1/2$ to just over 3 pounds.

The marten has a long, slender body, a short, pointed nose, small eyes, and prominent broad and rounded ears. The bushy, cylindrical tail accounts for about one-third of the animal's total length. It has short legs, and each foot has a densely haired sole with five toes that terminate in semiretractable, slender, sharply curved claws. The pelage is thick, long, fine, and silky in winter but thinner and somewhat more coarse in summer. The coloration is about the same throughout the year. Dorsally, the pelage varies from light to dark yellowish brown or more reddish brown; the color may be slightly paler on the head and shoulders but darkens to blackish on the tail and feet. The throat, chest, and sometimes the lower abdomen vary from yellowish to yellowish orange to rich orange.

DISTRIBUTION AND DESCRIPTION

total length 513-735 mm
tail 165-240 mm
weight 0.68-1.45 kg

MARTEN ARE PRINCIPALLY inhabitants of dense coniferous forests, but they also travel extensively throughout the more open subalpine forests up to and beyond timberline. In general, martens are most closely associated with heavily forested east-

HABITAT AND BEHAVIOR

I remember a marten that I once "treed" in the High Cascade Mountains. The tree, a small fir near timberline, was so far from its neighbors that the marten could not jump from it to another tree. Being cornered, the marten threw a temper tantrum. It ran up and down the trunk of the tree and out onto a branch toward my face, chattering, hissing, and snarling. Near the end of the limb, it would stop and bounce up and down on its front feet, all the while clearly communicating its displeasure and indignation. When I hissed or snarled in return, the marten would swap ends in midair and disappear up the tree, returning almost immediately. After twenty minutes or so, I left the hissing marten bouncing on the limb.

and north-facing slopes on which there are numerous windfalls (photo 138). They avoid areas that do not have the overhead protection of living trees and are therefore frequently characterized as inhabitants of the deep forest.

According to local trappers along the Oregon coast, martens have a particular affinity for cedar swamps, wet areas where the dominant trees are western redcedar or Port-Orford-cedar. Marten also tend to be ridgetop travelers. Since they do not tolerate the destruction of their habitat, they are extremely scarce in the Oregon Coast Range but still abundant in some parts of the Cascade Mountains, as well as elsewhere within their broad geographical distribution. They also show little inclination to remain in areas of concentrated human use. The last real stronghold of the marten in the Coast Range appears to be in the vicinity of Loon Lake in Douglas County.

Although martens are active during the night, chiefly at dusk and dawn, it is not unusual to see an individual abroad during the day. They appear to be as much at home in the trees as they are on the ground. When I have startled martens in the wild, they have almost invariably escaped by bounding up a nearby tree and then traveled from one tree to another, frequently leaping between trees with agility and grace. Martens are alert, wary animals that can disappear quickly when they so choose. But if an observer remains quiet and motionless, a marten may at times approach closely, apparently unaware of the person. Nevertheless, a marten's senses of sight, hearing, and smell are well developed, as is its curiosity.

On the whole, a marten is a quiet animal, making almost no noise as it roams its forested haunts. When with its young, however, it may utter a soft purrlike grunt or a cooing sound. When captured or cornered, on the other hand, a marten may chatter, snarl, growl, hiss, or even screech or scream sharply.

Mammals of the Pacific Northwest

Martens are basically solitary and do not seem to be particularly communicative vocally. Before and during the breeding season, however, both sexes establish scent posts by frequently rubbing the abdominal scent glands on such objects as the branches of trees.

A marten uses whatever den it can find wherever it happens to be—hollow logs, holes in stumps, hollows in trees, and so on. Although there seems to be little regularity in their use of dens, the proximity of food is an important consideration.

The home range of a male is about one square mile, whereas that of a female is about one-quarter of a square mile. The home ranges of adult martens of either sex tend to be relatively evenly distributed with little overlap. Martens of the opposite sex appear quite tolerant of one another; the home range of a male often includes the home ranges of one or more females. Juveniles establish their home ranges without regard to those already established by the adult. The boundaries of some home ranges coincide with various features of vegetation or topography, such as the edge of a large meadow or a recently burned area. Since martens seem to avoid entering such open areas, they apparently act as psychological rather then physical barriers. On the other hand, a flowing stream may be a physical barrier to the expansion of a home range during the summer; if the stream freezes solidly during winter, a marten may readily cross it.

Martens prey on a wide variety of small mammals, such as shrews, mice, voles—especially the red-backed vole—woodrats, pikas, rabbits, hares, mountain beaver, chipmunks, squirrels, and even an occasional bat. Although small mammals are their most important food, they also eat insects, birds, and berries, such as Oregon grape, snowberries, strawberries, serviceberries, and currants, as well as carrion.

Martens pursue their prey either on the ground or through the trees; they can go almost anywhere a tree squirrel can go. They also spend time on large rock slides, searching among the boulders for available prey, such as pikas. Prey is initially bitten on the head and neck; once secured, it is taken to a sheltered spot to be eaten. Since martens consume the hair, bones, feathers, and entrails as well as the flesh of small animals, little evidence of their kill is left. When large prey, such as a squirrel, rabbit, or hare, is eaten, the feet and large bones are discarded as refuse.

Martens breed from June through August, during which time males and females become playful with one another. As the breeding season terminates, however, they revert to their solitary lives. Because of delayed implantation, the gestation period ranges from 7 to possibly $9^1/2$ months. Embryos, usually implanted in the uterus during February, have an active growing period of 27 days before birth. Litters, ranging from one to six (usually two to four) young, are born in April, May, or June.

The young are born in a nest within a hollow tree, stump, or log, which is lined with leaves, grasses, mosses, or other vegetation. Young, weighing about 1 to $1^1/2$ ounces, are naked, blind, and helpless at birth. They develop rapidly, and by the time they are three months old, they weigh almost as much as the parent. Young martens stay with their mother until about September, after which they begin their solitary lives. Males become sexually mature at one year of age, females at two years old.

Marten are preyed on by bobcats, lynx, fisher, Great Horned Owls, and Golden Eagles. Their greatest enemy, however, is humans who not only trap them for their beautiful pelts but also destroy their habitat with clearcut logging.

Fisher or Pekan
Martes pennanti

DISTRIBUTION AND
DESCRIPTION

total length 716-
1184 mm
tail 300-422 mm
weight 2.1-6.8 kg

THIS MAGNIFICENT MAMMAL was named after Thomas Pennant, a British naturalist. The species was first described in 1777 from a specimen taken in "Eastern Canada."

The name "fisher" is really a misnomer—the fisher does not fish. The Abenaki Indian name of "pekan," by which this mammal is also known in eastern North America, is perhaps the best and most distinctive name.

Fishers are known to have occurred in the southern portion of zone 1, potentially most of the forested areas of zone 3 (where a few may yet remain), and the southern half of zones 4 and 5, and it is likely that they once occurred throughout zones 1, 4, and 5. In 1936, Vernon Bailey cited fishers caught in the Coast Range of Oregon in Lane, Douglas, and Curry counties in 1913 and 1914. Two fishers are known to have been accidentally trapped in Curry County as recently as 1968. Early records in western Washington indicate that fishers were probably more abundant in that state than in Oregon.

Fishers range in length from $28^1/2$ to $47^3/8$ inches, and their tails vary from 12 to $16^3/4$ inches. Adult fishers weigh from $4^1/2$ to 15 pounds. The fisher is similar in structure and body proportions to the marten but is nearly twice as large, four times as heavy, and a different color. The head is broad and flat, narrowing to a pointed face and nose. The ears are round, low, and broad. The tail is moderately long, tapering, and bushy. Fishers have strong, moderately large feet with hairy soles. Each foot has five toes that terminate in sharply curved claws.

The general tone of the pelage is very dark brown to blackish brown, but darkest on the lower back and rump. The tops of the head, neck, and shoulders are grizzled or frosted because of white- or pale gray-tipped guard hairs. The underparts are dark brown, occasionally with one whitish spot to a few whitish spots on the throat, and frequently with a white blotch or blaze on the belly. Females have finer and commercially more valuable pelages than do the males. The nose, feet, and tail are blackish.

LIKE THEIR SMALLER COUSIN, the marten, fishers appear to be strongly reluctant to travel through areas lacking overhead cover; hence they are definitely forest dwellers by nature. They avoid crossing open areas, such as meadows and frozen lakes; they will travel a mile or more around a frozen lake rather than cross it. Fishers seldom follow roads; when they do, they most frequently cross them near ridgetops covered by coniferous forest.

HABITAT AND BEHAVIOR

Although they clearly prefer coniferous forests, they also live extensively in hardwood forests in eastern North America, though they seldom bypass even small stands of conifers.

Normally nocturnal, they are also active in the daytime. If an individual has recently gorged itself, however, it is likely to go into a deep sleep and may not awaken for 48 hours.

Throughout most of the year, they tend to travel in more or less direct routes, but they commonly deviate from their general direction of travel to investigate dense thickets of coniferous trees or tangles of fallen trees. During the peak of the breeding season, however, several fishers in an area sometimes follow one another's tracks. When so engaged, they often separate, circle, rejoin, and backtrack so many times that it may be impossible to determine if the same individual is being followed. Such meanderings usually occur in late winter in areas of less than three square miles. Even in the breeding season, when proceeding from one definite location to another, fishers travel in a relatively direct manner.

Although fishers are agile climbers, their tendency to climb seems to vary considerably among individuals and perhaps time of year. They can jump at least as far as 9 feet from one tree to another. Trees are normally descended head first, but in winter, a fisher may descend from a tree by leaping into the snow from heights of 15 to 20 feet. Fishers are also good, strong swimmers, entering water voluntarily.

During winter, fishers use temporary dens in snow, piles of brush, under fallen trees, or under the upturned roots of trees blown down by wind. Temporary dens in snow are seldom used more than a few days. They are simple, consisting of a single tunnel dug into the snow for 3 to 4 or 5 feet beneath the surface. An occasional burrow in the ground is used, but whether a fisher digs such burrows or appropriates them is not known.

Home ranges of fishers appear to encompass from $1^1/2$ to $4^1/2$ square miles. In some areas at least, there undoubtedly are more than one fisher per $1^1/2$ square miles. Even though their home ranges overlap, fishers are not very tolerant of one another. They are predominantly solitary animals, except during the breeding season and the time that the young are dependent on their mothers.

Although a fisher's method of hunting is not, as yet, well understood, it seems to be largely an opportunist, neither lying in wait nor pursuing prey for long distances. Most of a fisher's hunting consists of investigating sites in which small mammals are likely to be found, and some effort may be made to dig at the dens or nests of small mammals. The only prey the fisher apparently seeks out and hunts deliberately is the porcupine.

Fishers kill all prey—except the porcupine—by biting the back of the head. To kill a porcupine, a fisher circles it, making repeated, swift, rushing attacks at its face and head. Such attacks may last half an hour or more, after which the porcupine dies from the cumulative effects of the wounds inflicted by the fisher's teeth and foreclaws. A fisher can kill a porcupine that weighs a little more than twice the weight of its own body.

Once killed, porcupines are eaten in one of two ways, both of which minimize contact with the sharp, barbed quills. One method is to begin eating on the unprotected underside; the other is to begin eating at the face, head, and neck, proceeding toward the tail in the same direction as the quills lie. Both methods normally leave a porcupine's skin, with quills attached, flattened and remarkably clean of flesh and most of the bones. One porcupine may afford a fisher a supply of food for several days.

As expected, fishers may sustain some injury from the quills of their prey, but even though quills sometimes penetrate a fisher's viscera, where they soften, they seldom cause inflammation, infection, or even serious damage to the animal. M.W. Coulter wrote the following account in 1966 about the quills he found in fishers in Maine:

> Quills were found embedded in the tissues of 127 of 365 (35 percent) [of the] fishers examined.... Counts of more than 100 quills were made in some individuals. They were found in all regions of the body, although it was usual to find more of the barbed structures about the heads, necks, and foreshoulders. In some specimens

the quills had pierced the stomachs or intestines, or were found lodged in the mesenteries [tissues] supporting the above organs. Quills were common throughout the larger muscles. One animal had a single quill in a kidney.

In addition to porcupines and snowshoe hares, the fishers' other primary prey, they eat a wide variety of small mammals, such as shrews, moles, rabbits, hares, squirrels, mice, and voles. They also eat a variety of birds, such as jays, nuthatches, woodpeckers, chickadees, thrushes, sparrows, grouse, and owls. A few amphibians and reptiles are consumed, as are some insects, nuts, and fruits. In addition, large quantities of carrion, such as deer, are eaten.

Fishers normally eat the hair, feathers, and bones of small prey, but the tufts on the ends of squirrels' tails, as well as the stomachs of squirrels and larger mammals, are discarded. Many of the bones of hares and porcupines are broken open and eaten, but only the softer ends of the ribs of deer are gnawed. Fishers also eat birds' eggs by opening them near one end and consuming their contents.

Fishers are probably polygamous in their breeding habits. They are decidedly solitary during the three or four months before the onset of the breeding season, which lasts from late February until at least mid-April, but toward the beginning of March they commence traveling together. Not all individuals, nor even the majority, associate at any one time.

During the second and third weeks of March, fishers may stop as often as sixteen times per mile and establish scent stations marked by the fluid from their anal glands. Such behavior, occurring only in late winter, probably serves as a means of communication among individuals approaching readiness to breed. The odor of the fisher's anal gland secretion is not as strong as that of weasels and mink. Furthermore, although the latter often emit fluid from their anal glands when frightened or hurt, fishers apparently do not do so.

Since fishers are predominantly solitary, males likely follow females in estrus by their scent. Females come into estrus and breed within six to eight days after giving birth, and once a pair has mated and the female is no longer in estrus, she is probably no longer attractive to the male.

Litters, ranging from one to four (usually three) young, are born after a gestation period of about 51 weeks. The gestation period is so long because of delayed implantation. Young are born from late February through the first of April, but the majority arrive during March. Babies are born in dens, which, so far as is known, are situated in hollow trees as high as 40 feet above the ground.

Fishers are born blind, helpless, and only sparsely covered with fine, light gray hair on the middle of the back. They can utter short, high-pitched cries. Growth and development are rapid in the first few days. In three to four days, babies are almost entirely clothed in fine brownish gray fur, and they weigh from 1 to $1^1/2$ ounces. They try to avoid direct light shortly before their eyes open, which is around 52 to 54 days of age, and for about ten days thereafter. They can crawl awkwardly at about two months, at which time they weigh close to a pound.

At ten weeks, young fishers become playful, and by twelve weeks, they exhibit extreme curiosity with frequent physical contact in the manner of hunting activities. When about two months old, they eat meat brought to them by their mother. They begin drinking water at 10 or 11 weeks and are weaned by the time they reach seventeen or eighteen weeks of age, in late June or July. By this time, they can leave the nest, and they begin to attack small prey. Although young fishers do not attain their full growth during their first year of life, some males and most females breed when still yearlings. Females, therefore, give birth to their first litters at two years of age.

Humans are the only creatures known to regularly kill fishers, which are hunted and trapped for their beautiful and valuable pelts.

Wolverine
Gulo

THE GENERIC NAME *Gulo* is the Latin word for "glutton."

The single species of wolverine, *Gulo gulo*, inhabits the taiga and forest-tundra around the high northern latitudes of the world.

Because there is a single species, refer to the account of the species for a description.

Wolverines still occur in the western United States; there are still a few in western Oregon.

Wolverine
Gulo gulo

THE FIRST SCIENTIFIC specimen of the wolverine caught in North America came from Hudson Bay, Canada, and was described in 1780, but this species was later recognized to be circumpolar, so the original specimen, the one on which the scientific description is based, is from outside North America.

DISTRIBUTION AND description

total length 820-1130 mm
tail 170-260 mm
weight 14-27.5 kg

WOLVERINES OCCUR IN the higher elevations and more remote areas of zones 4 and 5. They range in length from $32^3/4$ to 45 inches, and their tails vary from $6^3/4$ to $10^3/8$ inches. Their weight usually ranges from 30 to 60 pounds.

Males are larger and darker than females. Wolverines are robust with short, wide bodies and powerful heads, necks, and legs. Their ears are low, their legs short, their tails short and bushy. They have short toes and sharp, curved claws for climbing.

The fur is long, quite thick, and coarse. The short, dense underfur is wholly concealed by the blackish brown guard hairs of the back. A broad yellowish to light-brown band extends from the shoulder to the rump along each side of the body and join across the base of the tail. The shoulders may be yellowish brown or gray. The head, behind the eyes, may also be gray. The face, nose, feet, and tail are black. Ventrally, the fur is dark brown to blackish except the throat and chest, which are usually heavily mottled with white or pinkish orange.

WOLVERINES ONCE INHABITED the boreal forest across the northern part of North America and southward down the Rocky Mountains to Colorado and down the Cascade Mountains of Washington and Oregon, into California. Adolph Aschoff reported wolverines around Mt. Hood in 1896, and George Moody caught one in the upper McKenzie Valley west of the Three Sisters in 1912. Today, they are rare in the Cascades. I reported finding the tracks of a wolverine at Santiam Lake in March 1958 while camping there on 15 feet of snow. No one believed me until a hunter shot one in 1965, close to where I had reported the tracks.

HABITAT AND BEHAVIOR

Although wolverines belong to the weasel family, they are in build and habits more like little bears. Usually solitary and active throughout the year, they tend to have a cycle of activity in which they are active for three or four hours and then rest for an equal amount of time. Although mainly terrestrial with a gait that is a short, loping gallop, they can climb trees with considerable speed. Often portrayed to be great wanderers, they are not nomadic, but occupy definite but vast areas, which for a male may be 750,000 acres.

Although wolverines seem to prefer carrion, they also eat the eggs of ground-nesting birds in summer, as well as the larvae of wasps, and berries in autumn. In winter, they prey mainly on animals, including porcupines, when the cover of snow enables them to travel faster than their prey. Wolverines are even strong enough to kill deer and elk that become hampered by snow and have been known to drive both adult bears and pumas off their own kills. Caches of prey or carrion are covered with earth or snow, or sometimes wedged in the fork of a tree.

Although wolverines have delayed implantation, the gestation period, to the best of my knowledge, is unknown. A litter may range from two to five, but two or three young are the normal size. Babies are born from February through May. They nurse for 8 to 10 weeks, and remain with their mother

It is a misconception that wolverines are aggressive and prone to attack. I met a Canadian woman some years ago whose father had been in the Royal Canadian Mounted Police stationed in a remote area of northern Canada. This woman, and a friend of hers whose father was also in the Mounties, befriended a wild wolverine one winter by putting food out. She told me that it was not at all aggressive, which corroborates other observations.

for about two years, when she drives them out of her territory. Wolverines become sexually mature at four years of age.

Eskimos and other people of the far north wear parkas on which the outermost trimming around the hoods is made of wolverine fur, because it retains less frozen moisture from one's breathing than do other available furs.

I am unaware of anything that preys on wolverines, other than disease and old age.

Weasels, Minks, and Allies
Mustela

THE GENERIC NAME *Mustela* is the Latin word for a weasel.

Members of this genus are long-bodied, lithe carnivores that range in length from 5 to 21 inches and weigh from one-fifth of an ounce to $3^1/_2$ pounds. Dorsally, pelages vary from yellowish brown, light brown, reddish brown, and dark brown, to black. Ventrally, they vary from white, yellowish, and light reddish brown, to black. Several members of this genus have distinct black tips on their tails, and one species also has a black mask across the eyes and black feet. A few species turn white in winter and brown in summer.

Members of this genus occur from the arctic to the tropics; their habitats and habits are correspondingly varied, but they are for the most part solitary mammals. Some are good climbers, others good swimmers; all are persistent in the chase. Their long, cylindrical, supple bodies allow them to pursue prey almost anywhere the prey can go; they can enter any burrow into which their heads fit.

Gestation periods vary tremendously, depending on whether a species has delayed implantation. Litters vary from three to thirteen young. Born in a nest, babies become self-sufficient in about five or six weeks. Most members of this genus probably live less than two years in the wild.

There are about fifteen species within the genus *Mustela*. Members of this genus occur throughout North America, Mexico, and Central America, south to northern South America. They are also found throughout Europe and Asia, south to northern Africa, Java, Sumatra, and Borneo. There are three species in western Oregon.

Short-tailed weasel
Mustela erminea

THE SPECIFIC NAME *erminea* is from the New Latin word *ermineus*, which in turn is derived from the Old French word *ermine*. The scientific description of this little weasel was published in 1758. The species is circumpolar; the first specimen is from somewhere outside of North America.

Mammals of the Pacific Northwest

Short-tailed weasels occur in all zones. They range in length from 7³/₈ to 13⁵/₈ inches and have tails that vary from 2 to 4 inches. They weigh from 1 to 3 ounces.

DISTRIBUTION AND DESCRIPTION

total length 186-341 mm
tail 50-100 mm
weight 30-84 g

These small, cylindrical, carnivores have long, slender, lithe bodies, short legs, and relatively short, hairy, slightly bushy tails. Their heads are small, horizontally flattened, and taper to a blunt nose. The ears are prominent, round, and hairy. The feet terminate in five toes, each of which has a small, curved, sharp claw.

The pelage is short, moderately fine, but not thick. In zones 1 through 3 and the lower elevations of zone 4, these little weasels retain their dark coats throughout the year. Those in the high elevations of zones 4 and 5, however, turn white in winter.

The upper parts are uniform, ranging from brown to dark brown to reddish brown, being darkest on the face and nose. The white of the underparts extends from the lips or chin to the crotch and down the insides of the hind legs onto the tops of the feet; there often is some white on the tops of the feet also. Although the undersides are white, the amount varies. The tail is brown with a distinct black tip.

I have always found short-tailed weasels to be associated with forested or wooded areas. Also called ermines or stoats, these little weasels are active primarily during the night, but daytime activity occurs as well. I have both seen and caught these lithe carnivores on bright sunny midsummer days in the middle of the afternoon. Active throughout the year, they do not migrate but may move their home base if confronted by a shortage of food or other defects in the habitat. Although they are not abundant anywhere, these weasels occur from sea level up to and beyond timberline. In fact, a mummified female short-tailed weasel was found on June 20, 1956, on the North Ridge of Mt. McKinley, Alaska. It was at the 14,860-foot level, 9,906 feet above the highest vegetation on the mountain, having traveled two vertical miles above timberline, where she apparently starved to death.

HABITAT AND BEHAVIOR

Photo 139. The droppings of a short-tailed weasel, the shape of which is characteristic of all weasels, as far as I know. (Photograph by Chris Maser)

Short-tailed weasels in western Oregon seem to have a strong affinity for protective cover and seldom venture far from fallen trees or thick vegetation. But I have watched several of them climb as high as 30 feet up into fir

*All weasels
have distinctively
twisted droppings
(photo 139). I
remember, for
example, having
my base camp set
up in May 1967
at Phulung
Ghyang, Newakot
District, Nepal,
between 11,000
and 12,000 feet
above sea level as
I studied the
Nepalese
mammals in
relationship to
tick-borne
diseases. One day,
as I walked up the
narrow, rocky
trail along the
edge of the fir
forest, I came to
an old, fallen tree
alongside the
trail, and there,
on a flat rock
were the
unmistakable
droppings of a
weasel. I therefore
immediately set a
trap, and caught
a most beautiful
Himalayan
weasel that very
night.*

trees, excursions that appeared to have been made out of curiosity—an attribute with which weasels are generously endowed. Nonetheless, their small size undoubtedly makes them wary about exposing themselves unnecessarily; therefore these secretive little predators are seldom seen and are little known by the people of western Oregon. Even though they may reside in proximity to human habitations, they leave little sign and their presence usually goes undetected.

Although individual weasels are captured, I have most often caught what appear to be family groups, consisting of four to six individuals. Such groups, occurring in a small area within two or three days, appear suddenly and then just as suddenly disappear. It is thought by some that these weasels customarily travel in pairs at times other than the breeding season.

When in motion, a short-tailed weasel has a bounding gait. Because this long-bodied mammal brings its front and hind feet together, its back arches with each bound. The hind feet land almost, if not exactly, in the impressions left by the front feet; however, the right front and hind feet lag slightly behind the left feet, causing a diagonal pattern of tracks. The normal distance between bounds is about 2 feet, but when traversing open spaces, the short-tailed weasel may use long, graceful bounds up to almost 6 feet in length. Whatever it does, it does quickly, and it is a master at dodging.

Short-tailed weasels normally appropriate the nest of one of their victims and then use the hair of their prey to line it.

Prey is frequently eaten within the nest, resulting in an accumulation of food refuse. As victim after victim is brought to the nest, the weasel plucks out the hair and places it here and there, pushing and nudging it against the side of the nest cavity, which forces the structure to expand. The longer a short-tailed weasel occupies a particular nest, the larger and better insulated it becomes. Although such behavior appears to be most typical in winter, these weasels seem to line their nests, at least to some extent, throughout the year. A weasel undoubtedly derives an advantage for its survival from this behavior because, in addition to its high expenditure of energy, the general structure of its body is such that there is a disproportionate amount of surface area and, therefore, a great loss of body heat.

The nest is kept clean of fecal material because the weasel defecates in a particular place outside the abode; such behavior is evident both in the wild and in captivity. Although most nests are discovered on the surface of the ground and prove to have been the renovated nests of victims, they are also located belowground. Underground nests are composed of some vegetation with the hair of the prey interwoven into it.

Short-tailed weasels emit a variety of vocalizations, including a repetitive, low chatter; when excited or alarmed, they vent a shrill, sharp note—almost a screech, or a series of sharp, soft barks. The sound most often heard, however, is a distinct hiss.

Short-tailed weasels are superbly adapted carnivores, preying mainly on small mammals, such as shrews, shrew-moles, baby brush rabbits, chipmunks, deer mice, voles, house mice, and jumping mice. But they also eat small birds, occasionally frogs and small fish, and sometimes even earthworms. It is possible that from time to time they may also include some insects in their diets. Their long, muscular, supple body is well suited for the chase, and their small size allows them to enter small burrows. In addition to exceedingly quick reflexes, they have acute senses of sight, hearing, and smell. With their senses always keen, they spend much of their time foraging about fallen trees, banks of small streams, thickets of vegetation, and rockslides. When hunting, they are in almost constant motion; no nook or cranny escapes their careful scrutiny.

A weasel uses all of its senses during the hunt. Although the sense of smell appears to be its primary hunting tool, when an object or sound has attracted its attention, but cannot be fully understood, a weasel often stands on its hind legs and looks the situation over. Dropping down, it may advance toward the object and stand up again; this procedure is usually repeated until the weasel ascertains what is going on.

When prey is located, a weasel's excitement is often portrayed by its tail, which suddenly appears to be twice its usual size as all the hairs stand straight out; the weasel may also lick its lips a time or two before, in a blur of motion, it attacks. The rapid dash usually terminates with the weasel's fangs sunk deeply into the back of its prey's skull. At times, however, a swift chase ensues before contact with the prey can be made. Occasionally during such a chase, a weasel misses the head and grabs the prey's back, legs, or even tail. At such times, the weasel clutches the prey with its forefeet, while at the same time, it quickly jerks the prey downward between its forelegs with its mouth, releases its grip and attempts to secure another hold closer to the head. If the prey does not escape, the weasel continues these maneuvers until at last its fangs are securely fastened in the base of the prey's skull.

Should the prey perchance escape, the chase is on; a hunting weasel is the epitome of single mindedness. Once its teeth are sunk into the base of the skull, the weasel hangs on, periodically clenching together its jaws. If the prey is small, such as a deer mouse, it is drawn toward the weasel by the latter's forelegs and feet while the claws of the hind feet are used to tear at the

Toward the end of March 1965, I caught a pregnant short-tailed weasel in a live trap. As soon as I saw "Weasel," as I came to call her, I knew she was different from her kin with whom I had so often had encounters over the years. I say this because weasels are not only feisty but also prone to hissing and try immediately to escape. But Weasel did neither of these things; she was calm, seeming somehow centered within her being, and simply blinked at me. So, going against my basic principle of not keeping wild animals in captivity unless it is absolutely necessary for study, I put her in a large cage and made her as comfortable as I knew how. I gave her a live deer mouse to eat, a normal item in her diet.

By morning, Weasel, who was about 8 inches long and, were she not pregnant, would have weighed about an ounce, had made a nest of the cotton and fine, dry grass I had put in her cage, and she had interwoven some of the deer mouse's hair into it. I gave her clean water and some ground beef and left her alone.

I visited her again in the evening, when she was out and about. I was surprised that she watched me calmly without trying to hide, or escape, or try to bite me when I opened the cage and put in food. She did not even spit or hiss. Each day I put my hand into Weasel's cage. On the third day, she came up and sniffed my fingers. I fully expected to be bitten, as every weasel I had ever known would have done, but nothing happened. By the end of a week, she was eating out of my hand, and I felt a deep love for her. We developed a bond of trust, the likes of which I have never known with a weasel of any kind.

Then I opened her cage one morning while she was asleep. She awakened, blinked her dark eyes at me, yawned, and stretched her head toward my hand. So I began, gingerly at first, scratching her on the top of her head. She liked it; so each day I scratched her head a little longer. After a few days, she moved her head to where she wanted it scratched, frequently around her ears. Our wonderful trust grew continually stronger, and my love for Weasel exceeded all bounds.

As Weasel began to look like she was nearing term and would soon give birth, I became as nervous as an expectant father. Then one day Weasel did not raise her head and yawn at me. She did not lick my finger. She did not even move. My heart stopped, and I got a cold, sinking feeling throughout my body.

I gently opened the top of her nest, and there she lay. Hesitantly, I touched her. She slowly lifted her head and laid it in the palm of my hand. I knew something was drastically wrong as I petted her. She lay still, looking up at me, and I watched the light slowly fade from her bright, beautiful eyes. Then her eyes closed gently, and her head sank a little deeper into my hand as her spirit winged its way back from whence it came.

I was sick at heart. Our time together had been so very short, yet full of love. I stared at her for a long time with tears running down my face. What happened? Had I done something wrong? Instead of babies, there was only the cooling body of my beloved Weasel. Her babies died with her.

I could not bring myself to bury Weasel's body. I had to know if somehow I had been the cause of her death. This was particularly important to me since I had knowingly violated my basic principle in caging Weasel, a wild animal—just to be with her because I had felt such an instantaneous bond. Had I been wrong? Yes! But I still needed to know why she had died. So I froze her body and sent it to Murray Johnson, a friend of mine who was a student of mammals by avocation and a surgeon by profession.

The wait seemed interminable. At last I heard from Murray. Weasel had died of an internal hemorrhage caused by a tiny piece of mouse's bone that had gotten wedged into and had finally penetrated the side of her colon.

Weasel had given me a kind of love I cannot to this day explain. She seemed somehow ethereal, as though she held within her tiny being a wisdom one seldom finds in life. She touched my life for a brief moment and in that moment touched me forever. Would she have lived to deliver her babies, I wonder even today, if I had honored my basic principle of not caging wild animals? I will never know.

---

prey. If, on the other hand, the prey is large and more difficult to handle, such as a Townsend chipmunk, the weasel flops onto its side, periodically clenching its jaws, driving its fangs deeper into the prey's skull.

When an animal is thus bitten, its legs and feet work wildly for a time. By lying on its side, the weasel tends to keep the prey's legs off the ground, allowing the kicking to take place in the air. Such behavior undoubtedly makes the prey easier to handle and also conserves the weasel's energy. If more than one prey individual is present, the first is swiftly dispatched and the second is attacked. When a lone prey animal is killed, however, the weasel may dawdle over its meal.

Once prey is secured, it is usually, but not always, carried to the nest, or at least under cover, to be eaten. The weasel normally chews through the victim's skull and eats the brains (photo 140). When more than one individual has been killed, the weasel may store the excess in a pile near the nest.

As far as I know, the breeding season for short-tailed weasels occurs during July and August. I have trapped males with enlarged testes in western Oregon in March, April, and May,

Photo 140. A deer mouse with its brains eaten out by a short-tailed weasel. (Photograph by Chris Maser)

and males with maximum-sized testes in July. Implantation is delayed in short-tailed weasels; young are born from the middle of April to early May, indicating a gestation period of about 255 days. Embryos implant and develop for about eight weeks before birth. Litters range from four to thirteen young, but six or seven seems to be the usual size. There apparently is a single litter per year.

Short-tailed weasels are born in underground nests, which are frequently located under the roots of trees, under or in stumps or fallen trees. There may be many tunnels and runways leading to and from a nest.

Babies are born helpless, blind, and sparsely covered with short, fine, white hair. They are pinkish with long necks. One-day-old short-tailed weasels average about 6/100 of an ounce. At fourteen days of age, a heavy brown mane stands out markedly along the top of the neck in contrast to the rest of the body, which is still sparsely covered with white hair. The babies' eyes open when they are 35 days old, and by the time they are 45 days of age, the brown hair on the back obscures any trace of the mane. At this time, the young are active and playful. They are cared for by both parents. Females become sexually mature during their first summer and usually mate, bearing their first litters the following spring. Males probably do not become sexually mature until the year following their birth.

No mammals that I know of, other than humans, persistently prey on short-tailed weasels, although domestic dogs and cats kill some. Owls, such as the Barn Owl, Great Horned Owl, Northern Spotted Owl, Barred Owl, and Snowy Owl, capture these weasels. Some hawks, such as the Rough-legged Hawk and Goshawk, also obtain a few. Ernest Thomson Seton wrote in 1928 of a Bald Eagle that had been shot with the bleached skull of a weasel, thought to be a short-tail, clinging in a "death grip" to the eagle's throat.

THE SPECIFIC NAME *frenata* is derived from the Latin word *frenum* ("a bridle") combined with the New Latin suffix *ata* (denoting the use of the word as an adjective). The name refers to the white markings on the heads of some individuals of this species of weasel, which resemble a bridle—hence the subspecies *frenata* is known by the common name of "bridled weasel." The first long-tailed weasel, on which the 1831 scientific description is based, came from Ciudad Mexico, Mexico.

LONG-TAILED WEASELS OCCUR through all zones. They range in length from $11^3/8$ to 22 inches with tails that vary from $4^3/8$ to $7^3/4$ inches. They weigh from 4 to $11^1/2$ ounces.

Long-tailed weasels are one of the smaller carnivores in western Oregon. They have long, cylindrical, slender, lithe bodies. Their legs are short, and their hairy, slightly bushy tails are relatively long. Their heads are small and horizontally flattened, and taper to a blunt nose. The ears are prominent, round, and hairy. The feet terminate in five toes, each of which has a small, curved, sharp claw.

The pelage is short, moderately fine, but not thick. Summer pelages vary from brown to yellowish brown on the back, sides, and the outsides of the legs, including the feet, but are darkest on the face and tail; the latter terminates in a distinct black tip. There may be some white hairs on the face. The throat, chest, belly, and insides of the legs vary from yellowish white, light yellowish, dark yellowish, to orange. Although the long-tailed weasel in the lower elevations west of the Cascade Mountains retains essentially the same pelage throughout the year, those in the high elevations are all white in winter with the exception of the black tip on their tails. Occasionally, as winter progresses, the underside and the tail may show some sign of a yellowish stain.

THE LONG-TAILED WEASEL is not restricted to forested areas or protective cover as is the smaller short-tailed weasel. Moreover, the long-tailed weasel is frequently active during the day in areas that are almost devoid of protective cover. Regardless of where an individual lives, it is active throughout the year.

Long-tailed weasels do not appear to be abundant in western Oregon; they are, in my experience at least, most easily caught in the proximity of the burrow systems of mountain beaver. I do not know whether this habit of frequenting the burrows of

Long-tailed weasel
Mustela frenata

DISTRIBUTION AND DESCRIPTION

total length 285-550 mm
tail 115-196 mm
weight 110-326 g

HABITAT AND BEHAVIOR

Although some people think long-tailed weasels are poor climbers, in the summer of 1969, I watched a domestic cat chase a male up a large cottonwood tree. The weasel bounded straight up the trunk for 20 feet before reaching the first branch, where it turned and faced the cat, which, thinking better of the chase, departed quietly. There was no hesitancy on the part of the weasel at any point during its climb. Ernest Thompson Seton wrote in 1928 that he watched a long-tailed weasel give swift chase to a red squirrel almost to the topmost branch of a large hemlock tree.

mountain beaver in any way influences their distribution or the routes and distances traveled. It is not uncommon for eight or nine long-tailed weasels to suddenly appear in a mountain beaver's system of burrows.

I have on several occasions tracked long-tailed weasels in snow for more than a mile in open country without finding them. They seem to travel farther in open country than in brushy country. These weasels, at times at least, travel routes repeatedly; thus, when tracks are located and a trap is set, the weasel is often caught on its next trip through the area.

Long-tailed weasels, like short-tailed weasels, take over and renovate the nests of their prey, both aboveground and below. I examined three such nests in central Oregon in October 1971. The nests had been those of montane voles; the weasel had enlarged the nests and lined them with the fur of its prey, the voles, making the nests snug and warm. Long-tails also use underground nests, often situated in the burrows of mountain beaver on which they prey. Such nest are made of mosses or other vegetation and kept clean because the weasels defecate in specific areas, including special chambers that act as latrines, away from their nests.

When a long-tailed weasel is irritated, its tail becomes twice the usual size as all the hairs stand straight out; it may also stamp its front feet and voice an explosive hiss. If further irritated, it forcibly emits some musk from its anal glands and may attack.

The long-tailed weasel is a predator well adapted for either an open chase or pursuit of its prey in any burrow system wide enough for its head to enter. It is so quick that J. M. Edson, after being nipped on a finger, commented: "If lightning is any quicker than a weasel the margin is of microscopic breadth." Coupled with its speed and precision of movement and judgment, a weasel also uses its body and feet to advantage in securing prey. When, for example, it is within striking distance of small prey, such as a chipmunk, it seizes the handiest portion of the victim's body with its teeth and quickly throws its own body in a loose, snakelike coil over that of the prey. Such a maneuver effectively subdues the prey's frantic struggles, allowing the weasel to shift its initial grip to the back of the head or neck for the killing bite, while arching its body around that of the prey, holding it with all four feet.

Large prey, such as rabbits and pocket gophers, are rendered helpless by a bite at the base of the skull. A long-tail can accomplish this without even tearing the victim's skin. In the wild, in an open chase, a long-tailed weasel can kill young rabbits; but under the same circumstances, adult rabbits can

Many people tend to think of weasels, especially the long-tailed weasels with which they are most familiar, as highly nervous, energetic, fearless, and blood-thirsty creatures. But Joseph C. Moore observed in 1945 that inexorably coupled with the long-tail's seemingly limitless curiosity is its propensity toward playfulness:

Hand in hand with curiosity went playfulness. This was first noticed on the third day of the weasel's captivity when it was observed sprawled on its back in the doorway of its nest box toying with the door. For minutes at a time the weasel scratched at it, bit it, and tugged it up and down with its teeth. After it had become accustomed to its environment, the animal invariably played with its food, leaping upon it from every angle, biting it, rolling over and over, wrestling with it, tossing it about, and "killing" it again and again. Whenever fresh bedding…was provided in the cage, the weasel rolled and romped in it and wormed its way through and through it.

When its playground widened to include my table top, many more playthings became available. It pushed ink bottles about, rolled vials back and forth, wrestled with an electric extension cord, essayed to climb the goose-necked table lamp, jumped in and out of its cage, and wrestled with the large feet indecorously propped on the table. From one thing to another it dashed, pausing and posing charmingly at the end of each short rush, a superb picture of grace and alertness. Attracted by my hand on the table the weasel crept up to it and sniffed it. If the hand stayed there, the weasel nipped it gently and ran away, and then presently returned to sniff and nip and run again. This continued with the nip growing harder each time, eventually provoking a shout and a vigorous cuff which always missed the quick rascal. Lurking furtively behind objects at the far side of the table until the scolding tone went out of the voice which addressed it, the weasel then came forth once more to start another game of some kind.

When hunting pikas in and out of the boulders, a long-tailed weasel may secure its meal, but its success in doing so is not a foregone conclusion. I watched one hunting pikas in the talus at Multnomah Falls, along the Oregon side of the Columbia River, in 1966. The weasel pursued one pika into the talus but came out following another individual, only to disappear again into the talus.

Some years ago I shot a rabbit and, going over to pick it up, found that a long-tailed weasel had already claimed it, with not the slightest intention of sharing it with anyone or anything. Every attempt to retrieve the rabbit was repulsed by a stamping, sputtering weasel with a stiff, oversized tail. The weasel's attitude was an unmistakable communiqué that it would take even more drastic action if necessary. The courage of this weasel earned it not only my respect but also my rabbit.

kill a weasel by kicking it with their powerful hind legs and sharply clawed hind feet. The story is different, however, if an adult rabbit is confined to close quarters, which hampers its maneuverability but not that of the weasel.

Although long-tailed weasels do pursue their prey aboveground, they are really masters of subterranean burrow systems, where they hunt pocket gophers, ground squirrels, and mountain beaver. I have watched them hunt ground squirrels, popping up in the entrance of one burrow after another, often many feet apart.

In hunting, long-tailed weasels seem to use the senses of smell, sight, and hearing, which undoubtedly aids them in capturing a wide variety of prey under varied circumstances. Though this weasel depends primarily on mammals for food, some insects, reptiles, and birds are also captured and eaten.

The testes of male long-tailed weasels begin to enlarge in late March and reach their maximum size in April. Adult males have sperm in their testes and are fertile from April through August. The testes shrink rapidly during August and the first of September, reaching their minimum size by mid-September. The increase and decrease in the size of the testes correlates with the onset of the spring and autumn molts. Most females are probably bred during July.

These weasels have delayed implantation. The gestation period ranges from 205 to 337 days, averaging 279 days. The embryo is not implanted until 27 days before birth, after which time development is rapid. A female comes into heat between 65 and 104 days after giving birth. She will remain in heat for several weeks if she is not bred.

One litter is produced annually, ranging in size from four to nine babies, but usually six to eight. The babies are born between mid-April and mid-May. At birth, they are pink and wrinkled with a few sparse, rather long, white hairs on the head and back. They have very long necks, resembling those of their parents. Babies average about a tenth of an ounce when one day old. By two weeks, they are covered with silky, white hair, which is longest on the back of the head, neck, and over the shoulders, but they do not develop the mane that characterizes the young of short-tailed weasels. Young begin to eat meat when 21 to 25 days of age, but their eyes do not open until they are 36 or 37 days old. They are weaned in about five weeks and have their permanent teeth by the time they are 75 days old. They attain the peak of their growth by the tenth week, after which they grow more slowly. It is thought that both parents care for the young, at least until they are weaned.

Females are sexually mature by the time they are three months old, and are bred by adult males during their first summer of life. Young males, on the other hand, are not sexually mature until the following spring, when they are 10 to 11 months old.

Long-tailed weasels are killed by Great Horned Owls and probably by other large owls and hawks. Foxes also kill some, as occasionally does a large or poisonous snake. Domestic dogs and cats kill long-tailed weasels, as do automobiles, but humans take the largest annual toll, killing them largely for their white winter pelts.

THE SPECIFIC NAME *vison* is either an Icelandic or a Swedish word meaning a kind of marten or weasel, and is derived from the Danish and Swedish word *vissen*, meaning "withered" or "shriveled." The specimen on which the 1777 scientific description is based came from "Eastern Canada."

Mink
Mustela vison

MINK OCCUR IN ZONES 1 through 5. Mink are long-bodied, muscular, cylindrical carnivores. A large male may reach 28 inches in length and weigh up to $2^{1}/3$ pounds. They have short legs and a hairy, moderately bushy tail that comprises about one-third of their total body length. The head is horizontally flat, tapering to a blunt nose. The ears are low, wide, round, and hairy; the eyes are relatively small. A mink's pelage is composed of coarse, glossy guard hairs overlying a thick, soft, fine underfur. In summer, the pelage varies from dull, reddish brown to light reddish brown; in winter it varies from rich, dark brown to nearly black. The pelage is uniform in color except for variable streaks or spots of white on the chin, chest, or belly, all of which remain the same throughout the year. Mink have well-developed anal glands through which a startled or frightened animal may secrete an unmistakable, pungent odor that often hangs in the air long after the animal has vanished.

DISTRIBUTION AND DESCRIPTION

total length 418-720 mm
tail 127-245 mm
weight 0.68-1.64 kg

MAINLY AQUATIC IN HABITS, mink are most commonly associated with freshwater streams, rivers, and lakes; along the coast, however, they also frequent the brackish water of estuaries, the mouths of rivers, and salt marshes, as well as occasionally visiting rocky points jutting into the sea.

Though they are primarily active during the night, it is not uncommon to see mink abroad during the day. Activity normally begins after sunset and during the summer is most closely correlated with the onset of darkness. Cessation of activity is less affected by sunrise. The amount of nightly activity is higher and more concentrated during short summer nights;

HABITAT AND BEHAVIOR

during long winter nights the amount of activity is lower at any one time and is spread out over a longer period. There also is a direct correlation between activity and temperature. Mink increase their activity as the temperature drops; thus the colder the night, the longer the duration of activity, presumably because of the need for a greater amount of food.

The size and shape of home ranges of mink vary in accordance with the immediate topography as well as the availability of food. It tends to be long and narrow, approximating the shape of the body of water along which it lives (photo 141). Although an individual covers its entire home range over a few days or two weeks, such coverage is not evenly distributed.

A mink tends to concentrate its activity in a certain area, usually about 325 yards long, the location of which seems to be dictated by the availability of usable dens, normally two to five. During its period of activity, a mink continually moves back and forth within this area, but sooner or later moves to another area and repeats the continuous back-and-forth movement. Thus, a mink eventually visits its entire home range, but there is considerable irregularity in the intensity of use, which may be influenced in large measure by the availability of food, as well as by suitable hunting places.

A mink that is familiar with its entire home range covers it in a more leisurely way than an individual that is not so familiar with its home range. When moving from one restricted area to another during winter, mink normally follow the banks of their home bodies of water regardless of the direction of the current. Throughout the rest of the year, however, they generally follow

Photo 141. Fresh tracks of a mink in wet sand along the edge of a small river; there are three old mink tracks along the left margin of the photograph from an earlier visit. (Photograph by Chris Maser)

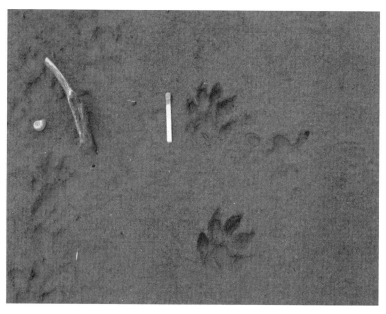

Mammals of the Pacific Northwest

the bank only when moving upstream, which means that they are hunting into the wind because the cool nighttime breezes flow downstream with the water.

The size of the home ranges of adult mink depend on the sex of an individual and vary according to season. Home ranges of males average about 2 miles in length, those of females only about 350 yards. Males cover the greatest distances during the spring, which coincides with the breeding season and their search for mates. The home ranges of juveniles are considerably smaller than those of adults, but they also vary according to the sex of the individual and the season. Home ranges of juvenile males average about one mile in length, those of females about half a mile. Males move around moderately during their first summer. As autumn approaches, their movement increases, probably as a result of dispersal. They are most sedentary during winter and most widely traveled during spring—their first breeding season. Juvenile females, on the other hand, appear to do much of their wandering during their first summer and are most stationary the following spring—their first breeding season also.

Mink examine every nook and cranny in their path, as well as some to which they have to make special trips. Their usual mode of travel is to follow the water's edge until something of interest draws them to swim and investigate. Completing an investigation, they again follow the edge of the water. Though they seldom swim great distances, they are excellent swimmers. And on occasion, a mink may float down a stream while curled up asleep.

Although mink are not particularly friendly with one another, they are not as solitary as most other members of the genus *Mustela*. When mink do meet, however, there may be a fight that consists of much shrieking, snarling, tumbling, and sometimes the discharge of the fetid odor from their anal glands. Although mink probably inflict few injuries on one another as a rule, there are fights to the death.

The reflexes of mink are very quick, yet the animals are relatively slow runners, and normally seek shelter rather than trying to outrun a pursuer.

Mink normally live in a burrow or den, which is near water, and may be in a hole under the roots of a live tree, a stump, or a fallen tree, and when possible in a hollow tree. They usually take over the abandoned or pilfered burrow of a muskrat or other mammal, but sometimes they dig their own. A mink's burrow usually is 8 to 12 or more feet long, 4 or more inches in diameter, and 2 to 3 feet under the surface of the ground, always with one or more entrances just above the level of the water.

Male mink have their own dens and do not build as complete a nest as the females, who build an enlarged nursery chamber in the burrow a few feet back from the waterside entrance. The nursery is 10 to 12 inches in diameter, and is lined with grasses, plant fibers, feathers, and fur. Adults keep their abodes clean by defecating and urinating outside of their burrows.

When snug in their dens, mink may sleep so soundly that they are difficult to awaken, but once aroused, they instantly are in full command of their senses.

Mink are primarily carnivorous, eating fish, frogs, snakes, mammals, birds and their eggs, crayfish, freshwater clams and mussels, as well as some insects. Along the coast, however, mink hunt in tide pools and also eat marine clams.

Although a mink's sense of smell may be as acute as its sense of hearing, its sight is only moderate. Mink appear to do much of their hunting by scent. When a mink tracks a rabbit or hare, the outcome seems inevitable. The outcome is not inevitable, however, when mink hunt the mountain beaver in their own burrows. It may be that a large, male mink can subdue a young mountain beaver, but an adult mountain beaver may well dictate a different ending.

The breeding season begins in late January or early February and lasts through March or early April. A male mink may visit two or more females and a female may receive two or more males; she also may share her den with a male. Because of delayed implantation, the gestation period varies from 40 to 75 days but averages 51 days. The embryo becomes implanted and grows actively for only 30 to 32 days before birth. A single litter, ranging from two to ten but usually four young, is born during April or May. The young, called "kits," are born naked, pale pinkish, with closed eyes; they weigh about a quarter of an ounce. They soon are covered with fine, short, silvery-white hair that is replaced by a dull, fluffy, reddish brown coat when the young are about two weeks old. The young grow rapidly, and by the time they are 25 days old, their eyes are open and they have a sleek coat or short hair. Weaning is begun in five or six weeks, and when about eight weeks old, the young attempt to capture their own prey. A family normally remains together until autumn, when each individual goes its own way.

THE GENERIC NAME *Spilogale* is derived from the Greek word *spilos* ("a spot") and the Latin word *gale* ("a helmet"). The name probably alludes to the white spot on the skunk's black forehead.

There are only two species of spotted skunks. The length of the head and body ranges from about 7 to almost 14 inches and their tails from 3 to 8³/₄ inches. Adults usually weigh from 8 ounces to just over 2 pounds.

Of the three genera of skunks, spotted skunks have the finest pelages. The basic color pattern consists of a black background with 6 white stripes extending along the back and sides; these stripes break into smaller stripes and spots on the rump. A triangular white spot is in the middle of the forehead, and the tail is usually tipped with white. There is infinite variation in this basic pattern, so no two individuals have exactly the same markings. Spotted skunks can be distinguished from the other two genera of skunks by their small size, the spot on the forehead, broken white pattern, and white-tipped, black tail.

Spotted skunks occur from southern British Columbia, south throughout most of the United States and Mexico into Central America. There is a single species in western Oregon.

THE SPECIFIC NAME *putorius* is the Latin word for "a foul odor" or "a stench." The first specimen came from South Carolina and was described scientifically in 1758.

THE SPOTTED SKUNK occurs in zones 1 through 4 and gets into zone 5 in the vicinity of the Oregon-California border. These skunks range in total length from 14 to 18 inches and their tails range from 4³/₈ to 7³/₄ inches. They weigh from 14 ounces to 2 pounds.

The spotted skunk is a slender, graceful, bright-eyed little creature with a small, triangular head and low, rounded ears. It has short legs, and its feet have long claws and naked soles; this skunk walks flat footedly. Its pelage is soft and full when prime, and its intricate black and white pattern is beautiful. The pelage is clear black or slightly grayish black, except for the four clear white shoulder stripes and two side stripes; there also is a white spot on the forehead, one on each cheek, and eight white spots on the rump. The basal one-half of the plumelike tail is black and the terminal one-half is white. An individual's color pattern remains the same throughout the seasons and its life.

ALTHOUGH THE SPOTTED skunk is primarily a denizen of forests and woodlands in western Oregon, it occasionally overlaps with the striped skunk along the edge of such agricultural valleys as

Spotted Skunks *Spilogale*

Spotted skunk *Spilogale putorius*

DISTRIBUTION AND DESCRIPTION

total length 351-448 mm
tail 110-192 mm
weight 400-965 g

HABITAT AND BEHAVIOR

Along the southern Oregon coast, I found one spotted skunk using the hollow base of a large Pacific madrone tree as a den. Another was using a hollow in the base of a dusky-footed woodrat's lodge; this was a large lodge constructed on the ground. Along the northern Oregon coast, these little skunks were frequently found using the large burrows of mountain beaver.

the Willamette Valley. Along the coast, they even live in the sand dunes bordering the shore and scavenge for food on the beaches.

Generally speaking, the proximity of a skunk—any skunk— invokes uneasiness in people because they know only of the unpleasant weapons of these white and black mammals. As a youth, I was fortunate enough to learn that skunks are basically peaceful and use their weapons only when under imminent duress.

While hunting, a spotted skunk may walk so that the hind feet are placed exactly in the prints of the front feet. When a skunk is hunting in the open, however, a bounding gait in which the front and hind feet work in pairs, is the most common mode of travel.

Spotted skunks are strictly nocturnal.

The living space of an individual skunk depends on the existence of several dens, available food, and safe passage between the dens and the food. The skunks are sociable with one another to the point that even dens are not the property of one particular individual. Rather, they are the communal property of the entire population, and a den may be occupied either by a single skunk or by as many as six or seven individuals. The only time a den is not shared is when a mother is using it as a nursery in which to raise her young.

Dens may be located in crevices in rocks, under piles of rocks, under or within buildings, or in hollow standing or fallen trees. Under some conditions, a skunk may dig its own den, but more often the deserted burrow of some other animal is appropriated and renovated. As variable as den sites may be, they are situated in such a way that they are totally dark within and afford protection against adverse weather. To a lesser extent, they offer protection against enemies.

A den may be a simple or complex burrow, depending on its former owner. Within a burrow, there are one or two sleeping chambers lined with grasses or other vegetation. Dens are either permanent or used only at intervals. Since a skunk moves from den to den when it is so inclined, the permanency of a den depends in part on its location.

A skunk's mobility in an area depends somewhat on its physical abilities as well as its rate of travel and endurance. Although a spotted skunk moves quickly, its speed is easily exaggerated. Even when thoroughly motivated, these skunks can be followed at a brisk walk. When traveling normally, such as while foraging, they scarcely move if food is abundant and easily obtained; otherwise they may travel all night.

I met my skunk-teacher when I was seventeen. It was late July, and I had been fishing for salmon 3 or 4 miles out in the Pacific Ocean. I had been anticipating sleeping on the beach, but a sudden storm made that out of the question. So took refuge under a huge pile of driftwood high on the beach. I made a shelter out of flat pieces of driftwood and placed my sleeping bag—the mummy type—next to a small, sandy rise with a neat, round hole in it. Other than noticing the hole and idly wondering who lived there, I ignored it.

Twilight waned, and the rainy wind blew. I added wood to my small fire, which burned cheerily as I got into bed. I had just gotten comfortable when a small head with bright, black eyes separated by a white spot appeared in the entrance of the hole, no more than a foot and a half from my head. The skunk and I regarded each other. I froze!

For the next hour, as the little spotted skunk alternately walked over, sat on, and sniffed me, I was confined to my zipped-up mummy bag. The skunk rummaged in my food, helping itself to whatever struck its fancy. Since I made no offensive move, neither did the skunk. Having established an "understanding," I was free to unzip my sleeping bag and watch the skunk, provided I did not interfere with the its comings and goings or with its meal. Toward the end of an hour, the skunk wandered off into the night, whereupon I moved the sleeping bag to allow the skunk unimpeded access to its burrow, which it entered shortly before dawn.

A spotted skunk does not have a territory that it defends; any particular part of the land where a skunk lives serves the little animal only as long as it suits him or her. When an area no longer provides the necessities of life, the skunk moves on. It may move into another part of its previous area or it may move to new and unexplored terrain. If other spotted skunks move in on it, or if it moves in on other skunks, probably makes little difference. A warm place to sleep, reasonably secure from dogs, people, and daylight; availability of a wide variety of foods, plant or animal; and chances of a reasonably safe passageway between its dens and food seem to fulfill a skunk's basic requirements.

The spotted skunk is well equipped for self-defense; its liquid musk is a potent weapon. The liquid, which contains a sulphide called butylmercaptan, looks much like skim milk with some curds of cream mixed with it. It is secreted by a gland located on either side of the anus; it is stored in vesicles, each with a capacity of about 1 teaspoon. The fluid is forcibly ejected through nipples that are hidden within the anus when the tail is down but may be everted when the tail is elevated over the back.

Since spotted skunks normally use their musk reluctantly, they give definite warning signals. One common signal is the rapid stamping or patting of the front feet on the ground that can be plainly heard 6 feet away. Such a warning may mean one of two things. It may indicate anger, especially when directed toward another skunk; when such is the case, the threat is carried out by snarling and biting. It may indicate fright, and then the stamping or patting is always accompanied by a raising of the tail; if the bluff is effective and the skunk begins to feel secure, all is well—if not, a stink ensues.

On a still day, the fluid can be discharged with amazing accuracy for a distance of almost 6 feet. On a windy day, however, a person standing downwind may receive the spray on the face from almost 15 feet away. Spotted skunks normally spray only in an emergency; whenever possible, they will climb a fencepost or a tree to avoid dogs and people rather than discharge their musk. Though not as adept in trees as are squirrels, they are rapid climbers and can remain secure in a high crotch all day if danger below persists. It is generally believed that a skunk is unable to squirt when its tail is down, but this is only partially true. If the tail cannot be raised, it may be swung to one side allowing a discharge. Although the tail is normally lifted as a warning as much as a method of preparation, a thoroughly frightened skunk will discharge its odor over its own tail if it is impossible to get it out of the way.

Another common warning of anger or fear is the handstand—a skunk balances on its front feet, tail high in the air. In this position, the animal's back is toward the disturbance and its head arched upward so that it can watch its adversary. Although they frequently do not discharge fluid in this position, they will if too closely pressed. I have seen these skunks perform the handstand on a number of occasions and have been amazed at the length of time they can retain their balance. They can even walk several yards while maintaining this position. The handstand is also used as a playful gesture, which young skunks perform while interacting with one another.

Spotted skunks are omnivorous and eat whatever is available. Mammals comprise the bulk of a skunk's diet during winter. In spring, mammals are still important, but insects are included. Insects are a prominent summer food, but fruits, mammals, birds and their eggs are also eaten. In autumn, insects again form a major portion of the diet, along with fruits, mammals, and some birds. In addition to prey they kill, spotted skunks rely on available carrion, such as brush rabbits and snowshoe hares, to supplement their diet.

The testes of male spotted skunks begin to enlarge in March and have attained maximum size by late September. Females on the Olympic Peninsula of Washington begin to breed during late September and continue into the first week of October. In more southern latitudes, however, breeding appears to occur earlier. Spotted skunks in the eastern United States breed primarily in April. Although females in the western United States have delayed implantation, those in the eastern U.S. do not. Eastern skunks have a gestation period of approximately 55 to 65 days, whereas gestation in western skunks ranges from 230 to 253 days, but the embryo becomes implanted and grows actively for only the last 28 to 31 days before birth. West of the Rocky Mountains, babies are born from late April throughout May, but east of the Continental Divide, they are born during June. Litters range from one to five young.

Newly born young weigh about half of an ounce; their eyes and ears are closed, and they are thinly covered with fine hair. The black and white markings are distinct on their smooth skin. Their teeth are not visible, but their claws are well developed. Babies can crawl feebly about the nest and can voice a distinct squeal. Although the hair is longer at seven days, it is still too thin to offer much protection.

At 24 days, they elevate their tails in a warning fashion when frightened, but no musk is emitted. The eyes open between 30 and 32 days, and when they are 35 days old, their canine teeth begin to erupt. At 42 days, they begin eating solid food, and by 46 days, they have full command of their weapon systems. Young are weaned in about 54 days and appear to be nearly full grown by the time they are three months old. Both males and females become sexually mature during their first summer of life and take part in the autumn breeding season. Spotted skunks have lived in captivity for at least 4 years.

Spotted skunks probably have few enemies other than humans and domestic dogs. Great Horned Owls occasionally kill and eat one, which I know from experience gives them really bad breath. Although these little skunks probably seldom come into conflict with porcupines, a male trapped in 1972 in the western Cascade Mountains, Lane County, Oregon, had the tip of a quill embedded in its left lung near the top of the heart. The quill did not seem to have caused permanent damage and the area surrounding the quill was not inflamed. Some skunks are killed each year by automobiles.

**Striped and
Hooded
Skunks**
Mephitis

THE GENERIC NAME *Mephitis* is a Latin word meaning "bad odor."

There are two species of skunks within the genus *Mephitis*. The length of the head and body ranges from 11 to 15 inches, and the length of the tail from 7 to 17 inches. Adults normally weigh from $1^1/2$ to just over 5 pounds.

Their contrasting black and white patterns have considerable variation. The striped skunk usually has white on the top of the head and on the nape of the neck, extending down the back as two narrowly separate stripes. In some individuals, the top and sides of the tail are white, whereas in others the white is limited to a spot on the forehead. The white areas are entirely of white hair.

Hooded skunks, on the other hand, may have a white-backed or a black-backed phase. In the former, some black hairs are intermixed with the white ones; in the latter, the two white stripes are widely separated and are situated along the sides of the body, rather than being narrowly separated and situated along the back as on the striped skunk.

Males are solitary during the summer, but in winter they may share a den with several females. In the northern parts of their geographical distribution, skunks sleep throughout the winter, with occasional forays outside the den. Resting almost anywhere they can find a dry place during the day, skunks are normally active from dusk throughout the night. They are omnivorous.

Skunks of the genus *Mephitis* breed in late winter or early spring. After a gestation period of about 63 days, four to ten (usually four or five) young are born in a den lined with vegetation. At birth, they weigh about an ounce. The young are weaned after six or seven weeks, but stay with their mother until August or September, when members of the family go their separate ways. Breeding may occur at one year. Individuals have lived ten years in captivity.

Skunks of this genus occur from southern Canada south to Central America. There is a single species in western Oregon.

Striped skunk
Mephitis mephitis

THE 1776 SCIENTIFIC DESCRIPTION of the striped skunk is based on a specimen thought to have come from Quebec, Quebec, Canada.

*DISTRIBUTION AND
DESCRIPTION*

total length 564-800 mm
tail 180-393 mm
weight 1.1-4.1 kg

YOU CAN EXPECT TO encounter striped skunks in most of zones 1, 2, and 3 (photo 142). Striped skunks range from $22^1/2$ to 32 inches in length, and weight from $2^1/2$ to 9 pounds. They are heavy-bodied animals with short legs and large, bushy tails. They have small eyes, pointed noses, and short, round ears. These skunks walk flat footedly; their feet have naked soles. The long claws on their front feet aid in digging.

Mammals of the Pacific Northwest

Although striped skunks have a characteristic black and white pelage, there is considerable variation in the proportions of white. The pelage of the body is all black except for a narrow, white stripe on the forehead between the eyes and a broad, white stripe from the top of the head to the shoulders, which then divides along the sides of the back, meeting again across the top of the tail near the middle. The tip of the tail is black. The long hairs of the tail may be white at the bases but usually have black tips. Many individuals have a light yellowish or light orangish wash to their white patterns.

STRIPED SKUNKS MAY be found anywhere around meadows, farmland, and even the suburbs of towns. They seem quite well adapted to living within the proximity of human habitation, provided the human density does not preclude quality habitat for skunks.

Skunks normally begin their activities about an hour after sunset and remain active throughout the night, but can sometimes be seen in early morning. They are placid animals, plodding along in a slow, indifferent, ambling walk, stopping here and there to inspect the ground and to dig out insect larvae.

Females may appropriate and renovate the burrow of another animal. They situate their dens, which may have multiple entrances, on a slope with good drainage. The nest is lined with grasses and dry leaves and placed in a chamber that is merely a widening of the burrow.

Males exhibit some territorial behavior by mutually avoiding one another. Females, on the other hand, move about freely within the home ranges of one or more males, and transient males move through the home ranges of other males during spring.

Photo 142. Adult striped skunk. (Oregon Department of Fish and Wildlife photograph.)

Martens, Weasels, Skunks • Mustelidae

I have found striped skunks sleeping all day in the bottoms of lodges of dusky-footed woodrats built on the ground. A striped skunk sleeping in one of these lodges taught me and a friend an important lesson. We were hunting for woodrats with bows and arrows. One of us would jump up and down on a lodge, and when the rat ran out the other shot it.

Everything went as planned for perhaps a dozen lodges; then after several jumps on one particular lodge, no rat sallied forth. But we were sure there was a rat within the lodge. When several more jumps failed to produce the rat, I tore the lodge apart while my friend waited, poised to shoot at the slightest movement. Then it happened—movement!

The arrow sped to its mark. The remainder of the lodge exploded from within, and in the same instant, as though fired from a shotgun, an accurate barrage of amber fluid struck both of us simultaneously. That the skunk died almost instantly was little consolation because we had to walked the 6 miles or so home with all our clothes, except undershorts, hung as far behind as the length of our bows would allow.

My friend lived in the woodshed for a week and I was banished to the basement. Never again have I shot anything without knowing exactly what it was!

Comparatively silent animals, skunks occasionally utter a variety of low grunts, growls, snarls, squeals, twitters, and chatters, but a series of low grunts may often accompany an individual's enjoyment of its meal.

Stiped skunks, like spotted skunks, use their contrasting black and white coloration as a warning and the odoriferous musk stored in two glands, one on either side of the anus, as a weapon. It is stored in vesicles each with a capacity of about one teaspoon. The fluid is forcibly ejected through nipples hidden within the anus when the tail is down but protruding when the tail is elevated.

With its tail erect and the black and white plume strikingly spread, the skunk makes little runs at the enemy, which, if it is wise, keeps beyond the enchanted circle. But if this warning is ignored, jets of amber liquid are jettisoned with surprising accuracy to a distance of 10 or 15, or possibly even 20 feet, filling the air with the most stifling, pungent, often gagging, odor.

A skunk's weapons are the simplest of squirt guns that can be aimed and shot from the rear of the animal with amazing precision. With a quick twist of its body, a skunk fires over one shoulder or the other at an enemy directly in front of it or aims at an object to this side or that, or in the rear, or even directly above, and generally with astonishing accuracy. The only safe place is beyond the skunk's range of fire.

Although classified as a carnivore, skunks are really more omnivorous and will eat anything that strikes their fancy, from fruit to crayfish to carrion. Much of their prey is procured by digging. They also hunt by lying in ambush or by slowly stalking an intended meal, then suddenly pouncing on it. They are simply too slow to pursue and capture prey. Prey, such as a hairy caterpillar, may be rolled on the ground with one or both forepaws in an effort to remove its hair before it was eaten.

Breeding begins during February and lasts through March, during which time the normally placid male may become nervous and aggressive. A male manifests his aggression by alternately stamping his front feet—a sign of irritation or fright— and then charging, which at times is preceded by a violent shaking of his entire body. He stops his charge, however, within a few feet of his perceived antagonist and stamps both front feet while twisting his body as if to spray. If the antagonist holds its ground, he approaches sideways, with his mouth held open. If he gets close enough, he will bite viciously.

After a gestation period of 63 days, the female gives birth to four to ten (usually four or five) skunklets in an underground nest lined with clean vegetation. The babies weigh about one ounce at birth. Their heads look much too large for their bodies, and their pinkish skin is thinly covered by very short, fine hairs. The color pattern is clearly discernible, because the white is much lighter than the rest of the animal.

The skunklets begin to fill out in about twelve days and appear to be more sturdy. Their eyes begin to open at 21 days, and they seem to be more sensitive to their surroundings. By the time they are 26 days old, their tails are well haired, and when the hairs are erected, they look quite businesslike. Both eyes and ears are open when the skunklets reach the age of one month. At this time, they can emit small quantities of musk, but their characteristic odor does not become really noticeable until they are about 50 days old.

When the skunklets are two months old, their mother leads them out at dusk to hunt, and they follow her in single file. Although the young become self-sufficient by about the middle of August, the family stays together throughout the winter.

Although the Great Horned Owl and Barn Owl prey on striped skunks, especially young of the year, domestic dogs are also among the skunks' enemies. People, their main enemy, kill many of them annually for their pelts or merely to eliminate them. In western Oregon, many striped skunks are killed by automobiles, particularly juveniles during the later summer and throughout the autumn.

Striped skunks can carry rabies, leptospirosis, and tularemia, all of which are highly infectious to humans. Any skunk that is abroad during daylight and/or appears to be sick should be left strictly alone.

River Otters
Lutra

THE GENERIC NAME *Lutra* is the Latin word for otter.

The length of the head and body of otters of the genus *Lutra* ranges from 22 to 32 inches and the tail from 12 to 20 inches. Adults weigh from 10 to 30 pounds. Otters have flat, round heads and short necks that are about as wide as the head. Their bodies are long and cylindrical and their thick, muscular, flexible tails taper to a point. They have short powerful legs and webs between their toes. Otters have small ears and nostrils that can be closed when the animals are in water. Their short, thick pelages are dark brownish above and lighter below; the lower jaw and throat may be whitish.

Superb swimmers and divers, river otters inhabit all types of inland waterways as well as estuaries. Although most active at night, they are often abroad during the day. They hunt singly or occasionally in pairs. Their diets consist of crayfish, frogs, turtles, fish, and aquatic invertebrates, in addition to any birds and mammals they can catch.

These beautiful, graceful mammals are some of the most playful mammals in the world. Otters are normally gentle, sociable animals, but males may fight one another during the breeding season. The New World otter (*Lutra canadensis*) has delayed implantation; its gestation period ranges from $9^{1}/_{2}$ to $12^{1}/_{2}$ months; the Old World otter (*Lutra lutra*) has a gestation

Photo 143. Adult river otter. (Oregon Department of Fish and Wildlife photograph.)

Mammals of the Pacific Northwest

period of only about two months. Litters, ranging from one to five (usually two or three), are born in dens lined with dry vegetation. The young are weaned in about four months; the family remains together about eight months. River otters have lived nineteen years in captivity.

River otters occur throughout most of North America and South America, most of Africa, Europe, Asia, and the large Malayan islands. There is a single species in western Oregon.

THIS SPECIES OF OTTER was named after Canada, combined with the Latin suffix, *ensis*, denoting possession. The locality given for the first specimen of the otter is "Eastern Canada." It was described scientifically in 1776.

RIVER OTTERS OCCUR IN zones 1 through 5 (photo 143). They are cylindrical carnivores with low bodies, which range from 33 to 52 inches long and weigh from almost $1^1/_2$ up to almost 6 pounds. Their legs are short, and their long, tapering tails are somewhat flattened horizontally. Their heads are small, as are their eyes and ears. The feet are webbed between the toes, and the soles are naked and may have what appear to be warts. The thick, silky underfur is wholly concealed by short, glossy guard hairs that vary from dark brown to dark reddish brown on the back and slightly paler below. The throat and cheeks are grayish brown. The pelage fades to a lighter, more reddish brown in summer.

ALTHOUGH USUALLY ASSOCIATED with streams, rivers, and lakes, river otters also frequent bays, estuaries, and even the open ocean along the coastline. The quality of much of their habitat is decreasing as a result of clearcut logging, ecologically unsound livestock grazing practices, urban sprawl, and pollution of the waterways. Though they are primarily active at dusk and throughout the night, I have on more than one occasion had the honor of seeing otters abroad during daylight hours. To be fortunate enough to observe the otters as they frolic in the bounding, rushing water, or in a placid, mountain lake is to behold the epitome of aquatic grace and beauty among mammals. Their frequent chirping gives the distinct impression that they are thoroughly enjoying a good time.

Otters are excellent and versatile swimmers. They can swim either by paddling with their limbs in any combination or by holding their limbs outstretched toward the rear, close to their sides, while they undulate their powerful bodies. This strong undulating motion is used when swimming in earnest, and

River otter
Lutra canadensis

DISTRIBUTION AND DESCRIPTION

total length 835-1300 mm
tail 300-507 mm
weight 3-13.6 kg

HABITAT AND BEHAVIOR

during such times they swim alternately on the surface of the water and beneath it.

When moving on land, they have three gaits—walking, running, and bounding. When walking, the whole body is held rigidly with the head and neck outstretched, and since the hind limbs are a little longer than the front ones, the body is inclined slightly downward. The end one-third of the tail may drag on the ground. A running otter holds its tail above the ground by slightly arching the end one-fourth of it. Bounding is an otter's fastest gait; when bounding, the front and hind feet are brought toward each other causing the back to arch and the tail to be lifted off the ground.

The antics of otters have all the appearances of play. In winter, they travel overland on snow and ice by combining running and sliding. They run, then tuck their front feet under their bodies and slide, run again and slide again. When they find a hill with good snow cover near the water that is clear of debris, they climb up the slope and coast to the bottom, do an Immelmann turn, and immediately run to the top to repeat the performance. (An Immelmann turn is a maneuver in which an airplane first completes half a loop then half a roll in order to simultaneously gain altitude and change direction in flight.) Otters toboggan again and again even though the slide may be slightly rough, so that they get repeated bumps on their stomachs. An occasional otter may not like the excessive speed on a particularly steep and slippery slide, so it will thrust its forepaws forward to slow the pace or even stop itself. But it usually does not take long before the timid one joins the others with reckless abandon. Otters seem to love nothing more than playing in the water and romping and sliding across the countryside.

Otters are great travelers. Some individuals may travel 130 to 150 miles of a particular river and its tributaries in a year, and a family may range from 8 to 25 miles in a season. A lone male or a pair may travel 20 to 30 miles from their den and return after an absence of several days. The distances that otters travel varies greatly and is related to the available supply of food, the general suitability of the habitat, and perhaps to an innate wanderlust.

Regardless of how far the otters travel, they have particular areas where they emerge to roll and rub themselves. Spending so much time in water makes it imperative that they keep their fur clean to maintain its insulating properties. The frequent rolling and rubbing appears to serve this purpose. They nearly always choose sandy areas where these are available, but they

will use grass, leaves, and even snow. All the rolling areas I have seen are on the bank farthest from a nearby well-used trail or road; human activity makes them nervous, particularly when it is some distance above the water and, therefore, out of their view. The otters invariably leave their sign at these rolling areas—both their urine and black, tarlike feces. When several otters travel together, each tries to be the last to leave its mark, even if it is only a dribble.

Otters have more than their share of curiosity. If something unusual catches their attention in the water, they will stop and watch it, or if on the bank, they may sit up full length, brace themselves with their tails, and watch. They often accompany their observations with a running commentary of chirps.

There is ample food for the otters in the rivers and tributaries of western Oregon. During winter and spring, they subsist primarily on a variety of fish (photo 144); during summer and autumn crayfish are their staple, augmented with frogs, salamanders, large aquatic beetles, a few mammals and birds, some carrion, and occasionally blueberries or huckleberries.

The breeding season occurs from November through early April. Though sociable much of the year, males fight among themselves during the breeding season, particularly when a sexually active female is in the vicinity. Some males become very rough when attempting to mate and will not tolerate the interference of another male. Others, however, remain unaggressive throughout the breeding season. E.E. Liers wrote: "One wild-trapped male, about 12 years old, was a perfect gentleman of the old school. He never handled the females roughly. If I had a female in heat in my arms, this male, Blackhawk, would come over to me, take the cuff of my trousers in his mouth, and shake it, coaxing me to put the female down on the ground."

Females come into estrus immediately after giving birth, and unless bred, they remain in estrus for about 45 days. Because they have delayed implantation, the gestation period ranges from 288 to 375 days. Implantation of the embryos occurs about the first of February. Since the embryos grow actively for only about the last two months before birth, the young are born during March and April.

The young are born in dens, cavities among the roots of trees, or in thickets of vegetation. Dens may be as much as 500 feet above the high-water mark and up to one-half mile away from water, or they may be at the water's edge. Dens normally are burrows that have been appropriated from some other animal and renovated by the otter. Otters' dens are well hidden; those located at the water's edge have the main entrance far enough below the surface to prevent their being frozen shut.

Otters normally breed every other year and produce a single litter that ranges from one to five young, but usually two or three. The process of birth lasts from three to eight hours, depending in part on the size of the litter. As soon as the newborn are cleaned, their mother curls tightly around them so that they are almost completely protected from cold air.

Babies weigh about 4 ounces at birth. They are fully covered with hair that is about a quarter of an inch long. The hair is uniformly blackish brown on the back and is lighter and more grayish on the underside. The lips, cheeks, chin, and throat are paler than they are in the adult. The babies' eyes are closed, and they are toothless. The young are helpless for five or six weeks. Although adult otters defecate and urinate outside their dens, the babies have to be cleaned by their mother until they are about seven weeks old. Their eyes open in about 35 days, and when they are five or six weeks old, they begin to play with one another and with their mother. The mother allows her youngsters to go outside of the den for the first time when they are ten to twelve weeks old.

The young have to be coaxed into the water the first time; in fact, a mother sometimes has to drag one of her offspring into the water by the scruff of its neck. They seem to have trouble keeping their heads above water when first learning to swim and so struggle and sputter.

In the beginning, the mother catches food and calls her young to come and get it. At their arrival, she releases the prey, which usually escapes into the water, followed by the floundering youngsters. Through repetition, they learn to catch and hold onto their food.

The mother generally leads the way in traveling, calling her young. Should they become too adventuresome and try to rush ahead of her, she nips their noses. When so punished, a youngster drops to the ground and lies very quietly. When punished severely, a young otter remains lying on its back and will not move forward until the mother returns and caresses it.

Otters become sexually mature in two years, but males, for some unknown reason, usually do not breed successfully until they are between five and seven years old. Otters may live to become at least eleven years old.

Otters have no natural predators. They are, however, occasionally caught and drowned in a fish net or on a set line. During the winter of 1971-72, an otter was killed by an automobile on the Oregon coast, but this is undoubtedly a rare occurrence. Humans are the only persistent enemy of the otter, trapping them for their valuable pelts or shooting them in the mistaken belief that they decimate populations of game fishes—particularly trout and salmon—and waterfowl, such as ducks.

Cats, Lynxes, and Allies
Felidae

THE FAMILIAL NAME *Felidae* is derived from the Latin word *felis* ("a cat," "the prolific one"), combined with the Latin suffix *idae* (which designates it as a family).

Although cats come in a great variety of external forms, they are all recognizable as cats. The general cat type, once established, probably has not undergone much structural modification. The differences in size and color pattern seem to have arisen mainly from the influence of the size of the prey and the local habitat.

Cats are lithe, muscular, compact animals with deep chests and short, round heads. The pupils of their eyes contract vertically; their ears vary from round to pointed, and they have well-developed whiskers. The length of the head and body of cats ranges from 1 foot 7 inches to 12 feet; the tail ranges from 4 inches to 3 feet 10 inches in length and is well haired but not bushy. The limbs of cats range from short to long and sinewy; the forefeet have five digits, the hind feet only four. Except for the cheetah, cats have retractile claws, which can be withdrawn into a sheath, thus preventing their becoming dull. The claws are large, compressed, strongly curved, and sharp. Except for the naked pads, the feet are well haired, which aids in the silent stalking of prey.

The pelage is soft and woolly. Its glossy appearance is maintained by frequent grooming with the tongue and paws. Color varies from gray to reddish brown to yellowish brown, generally with stripes, spots, or rosettes.

Most cats place the hind feet into the track left by the front feet. The possession of perfect register, as such foot placement is called, aids cats in stalking their prey because, having placed their forefeet with care, they do not have to be concerned with the placement of their hind feet, allowing them to concentrate solely on catching their next meal. Except for the cheetah, which outruns its prey, cats normally stalk and ambush their quarry.

Cats travel singly, in pairs, in family groups, or (the lion) in "prides" of about 23 individuals. Some cats are nocturnal, whereas others are primarily diurnal. Felids take shelter in trees, hollow fallen trees, caves, abandoned burrows of other animals, and amid ground vegetation. They either flee from danger or literally defend themselves with tooth and claw.

Although some cats may have a poor sense of smell, their senses of sight and hearing are acute. They prey on almost any animal that they can overpower. The "big cats," the four species in the genus *Panthera* (leopard, lion, tiger, and jaguar), are especially powerful. The leopard, for example, frequently stores its prey in a tree. One individual is known to have pulled part of a young buffalo, weighing about 100 pounds, nearly 15 feet up into a tree. A mature tiger's almost unbelievable strength is well known; a typical example is one that moved a water buffalo almost 30 feet though thirteen men could not move it.

Most cats produce one or two litters a year, but the large species may breed only every two or three years. Gestation periods range from 55 days to nine months, and normal litters range from one to six young. Babies are blind and helpless at birth, but they are well haired and their coats are often spotted. They remain with their mother until they can fend for themselves. Cats may live thirty years.

The geographical distribution of indigenous cats is almost worldwide, except for Antarctica, the Australian region, Madagascar, the West Indies, and some oceanic islands. People have introduced the domestic cat into many areas, such as Australia, where cats had not existed. There are two genera of indigenous cats in western Oregon.

Classification of cats, all kinds of cats, has been seemingly forever in disarray, which prompted Ernest P. Walker to pen the following in 1968:

Much of the classification of mammals is based on the shapes and relative proportions of the different parts of the skulls of different mammals, but the cats have a remarkable uniformity of proportions between the various parts of the skulls in spite of the great variation in sizes of the animals and the proportions of the limbs and tail, and the side variation in markings. The latter range from almost the same color over the entire animal through spots, stripes, and blotches of various colors, sizes, shapes, and arrangements. This peculiar combination of characters has led to a great diversity of opinion as to how the cats should be classified, with resultant multiplicity of names and groupings. One author used as many as twenty-three generic names for the cats, and other authors have taken the other extreme and recognized only one or three genera. Fortunately the cats are not as badly confused as the mammalogists.

True Cats
Felis

THE GENERIC NAME *Felis* has the same meaning as the familial name.

Members of the genus *Felis*, as herein treated, are medium to large carnivores, ranging in size from the puma to wild cats that are smaller than the average domestic cat. They usually have long tails and large feet. Like other members of the family, they walk on their toes. They have flat, short faces with large eyes and well-developed ears.

With the exception of the domestic cat, most members of this genus are some shade of brown, gray, or black. Members of most species have stripes, spots, or mottled patterns or dark colors on a lighter background, but there is much variation, and members of some species lack markings.

These predators are nocturnal. They seek shelter in rock crevices, in hollow trees (either standing or fallen), in holes in banks or the ground, or in tall grasses or underbrush. Cats are much alike in their habits; they prey on almost any mammal or bird that they can capture and overpower, as well as occasional reptiles. Although most cats appear to be fond of fish, only a few catch them regularly.

Females of most species become sexually mature at twelve to fifteen months of age and thereafter may come into estrus several times a year. The small species may have more than one litter per year, but the larger species may breed only every two or three years. Litters range from one to six young, usually two to four, though the domestic cat may have more than six offspring per litter. Gestation periods range from 55 days to nine months. Young are cared for by their mothers until they are old enough to fend for themselves. Members of this genus have lived to $22^1/_2$ years.

Cats of the genus *Felis* occur as indigenous mammals throughout most of the world, except the Australian region, Madagascar, the West Indies, and some other oceanic islands. People have introduced the domestic cat into many areas, such as Australia, making the genus almost worldwide in distribution. There is one species in western Oregon.

Puma, cougar, or mountain lion
Felis concolor

DISTRIBUTION AND DESCRIPTION

THE SPECIFIC NAME *concolor* is a Latin word meaning "the same color."

PUMAS OCCUR IN ALL five zones, but are least likely on the floor of the Willamette Valley proper. In the early years of European settlement in the Pacific Northwest, three wildcats lived here—the bobcat, the lynx, and largest of all, the puma, also called mountain lion or cougar. Pumas range in size from 5 to 9 feet in length and weigh from 80 to about 210 pounds.

Its coloration is such that it readily blends into the landscape through which it travels and in which it hunts. The puma's head, back, sides, and the outsides of the legs are dark tannish to reddish brown. The top of the tail is darker brown than the back, darkening to a relatively long, blackish tip. There is also black on the backs of the ears and on each side of the nose. The throat, chest, belly, and insides of the legs are whitish.

total length 1500-2734 mm

tail 534-900 mm

weight 36-95 kg

PUMAS ARE GENERALISTS and highly adaptable. Thus they can use an amazingly wide variety of habitats from the coastal forests to and above timberline in the high mountains; from the tropics to the desert; from the remotest wild places into the very edge of suburbia in towns, such as Corvallis, Oregon, where they hunt suburban deer.

HABITAT AND BEHAVIOR

A puma's normal gait is a rather long-strided walk, but it can cover 25 feet or more in long, graceful leaps. Regardless of what a puma is doing, it is always the epitome of grace and strength welded into fluid beauty.

Most adult pumas are residents, confining their movements to specific areas year after year, but there is also a contingent of younger, transient adults. Resident pumas occupy fairly distinct, contiguous winter-spring and summer-autumn home ranges; use of these areas, however, varies not only with season but also with time and individuals. Generally speaking, they use larger areas in summer than in winter, and males tend to travel more widely than females.

During winter, a male puma ranges over a minimum area of about 64 square miles and a female over a minimum area of about 13 and a maximum area of about 50 square miles (photo 145). Although resident males occupy areas that are distinct from one another, a male's area overlaps with those of females. Females, on the other hand, share common areas, and transients of both sexes move freely through occupied areas.

Pumas have a high degree of tolerance for other pumas in their areas, but are decidedly unsocial in that they avoid contact with another individual. Although pumas seldom defend an area, the behavioral mechanism of mutual avoidance keeps them distributed without injury. Pumas use all their senses in maintaining adequate distances from one another. Urine, feces, and scent from anal or other glands advertise a puma's presence, either bringing pumas together or maintaining the distance between them (photo 146), as do puma scrapes, areas where the cats scrape soil, or litter, or both into a pile in one to six places usually less than 3 feet apart. The cats may deposit feces, or urine, or both in or on the pile.

Photo 145 (left). Tracks of a puma in snow. (Oregon Department of Fish and Wildlife photograph.)

Photo 146 (right). Droppings of a puma; note segmentation, which is characteristic of the cats. (Photograph by Chris Maser)

Among pumas, land tenure is based on prior rights, and home ranges are well covered. Home ranges change after deaths or movements of the residents. Young adults establish home ranges only as vacancies occur. The system acts to maintain the density of breeding adults below the carrying capacity of the available supply of food, which is the maximum number of pumas that their prey base can sustainably support without altering the integrity of the ecosystem.

Reproduction is confined to resident pumas. Since transient pumas are nomads, the reproductive phase of their lives is restricted until they find vacancies and establish their own home ranges. Because the land tenure system is dynamic and flexible, a home range is not inherited intact, but involves a resorting of living space that takes place first among the older cats, so the younger ones must accept what is left.

The tendency of pumas to increase their movements during late winter is a result of the scarcity of food. Pumas hunt almost continuously, rarely spending more than one day in the same location. Except for the longer periods of heavy rain in spring and autumn, the activity of pumas seems largely independent of the weather.

In the short term, a puma's home range is in a state of constant change created by the availability of prey. Over the long term, however, the conditions in certain parts of the home range are such that a cat tends to be more successful there in making kills, and as a result, they spend more time there.

A puma's staple diet is deer (photos 147 and 148), elk, and porcupines, although they also eat snowshoe hares, flying squirrels, mountain beavers, and other animals. A hunting puma normally zigzags back and forth through thickets and around meadows, going up and down small draws and back and forth across creeks. This method of travel better enables it

Photo 147. A partially eaten mule deer killed by a puma and covered up to protect it for another meal. (Oregon Department of Fish and Wildlife photograph.)

Photo 148. The lower jaw and foot of a mule deer, partial remains of a puma's kill, attested by the big cat's departing tracks in the snow. (Oregon Department of Fish and Wildlife photograph.)

to detect prey and to stage a successful attack. A puma does not indiscriminately try to capture prey wherever it finds it, but being a stalker, must find prey in a location where it can stealthily approach close enough for a successful attack, which may require more than an hour.

A large male puma, which can weigh 165 pounds or more, can kill adult elk, but resident females and transient adult males and females are not large or heavy enough to do so, though a few of them are large enough, if conditions are just right, to kill large elk calves. Deer are considerably smaller than elk, and the cats are more versatile in hunting them.

Each large animal that is killed has some physical or behavioral defect with respect to its long-term survival and is the most vulnerable to predation by the cats. In addition, predation by the pumas keep the deer and elk moving, especially on the winter range. The mere presence of a puma does not usually alarm deer or elk, but when a kill is made, the reaction is striking. The deer and elk immediately leave the area, which forces a redistribution of the deer and elk and helps to prevent them from overpopulating an area and causing severe damage to their habitat.

A puma drags its kill to a protected place and a large male may eat as much as 9 pounds of meat in a meal; juveniles and females eat from 4 to 6 pounds. The puma will remain in the immediate vicinity of its kill, guarding it against scavengers, eating it over a period of one to nineteen days, depending on its size and age. It eats almost entirely any young prey, including the spinal column, skull, and feet, but about 70 percent of older prey. The 30 percent that is left includes the stomach, some viscera, bones, feet, and some of the hide.

The cats of the Cascade Mountains each kill a deer about every ten to fourteen days in winter, and a big male, if he is eating elk, kills one every twelve days or so. The winter diet is augmented with an occasional snowshoe hare, beaver, mountain beaver, northern flying squirrel, or woodrat. In summer, however, the interval between large kills is greater since other prey, such as porcupines, is more abundant.

Pumas are variously adept at killing and eating porcupines. They often manage to avoid most of the quills, but sometimes get them in their paws, shoulders, face, and mouth. They eat everything except the head; once eaten the quills begin to soften in the stomach within one hour.

Although the pumas breed at any time of the year, individual females breed only every other year. Just before and just after the young become independent, a female associates with, perhaps only tolerating, adult males and even adult females more frequently and for longer periods than at any other time. Such tolerance reaches its peak during estrus when a pair remain in one another's company, traveling together for eight to sixteen days.

After a gestation period of 91 to 97 days, one to five, but usually three, kittens are born, each weighing just about a pound. They are covered with short, soft hair that is dull tannish with darker blotches on the body and bands on the tail. Their eyes are closed and will not open for nine to fourteen days. The kittens grow rapidly and weigh between 8 and 10 pounds in two months; they are weaned shortly after, but the mother is restricted in the use of her home range and her pattern of movement will be more complex until the kittens are one year old. She leaves them in a protected place for a day or two at a time while she hunts in a loop away from them and back again. When she moves with her family, she seldom moves more than a mile at a time. As the kittens mature, the family wanders farther, and during the kittens' second year of life, they will begin to cover their mother's home range. They will become independent during their second winter and will finally break the family ties the following spring. The mother will leave them at a kill and simply not return. The siblings will remain together for a while and then meet for short periods before going their separate ways, becoming transients and wandering about until they find a vacancy, take up residence, and thus achieve breeding status.

Humans probably are the pumas' main enemy, shooting them for sport as well as attempting to control or eliminate them. Although not predation in the strict sense, adult male pumas occasionally kill young kittens, even their own; thus females with small kittens are sensitive about the presence of another puma.

Lynxes, Bobcats, and Caracals
Lynx

THE GENERIC NAME *Lynx* is the Greek word for the lynx.

The head and body of members of this genus range from $30^1/_2$ to 40 inches and the tail from 4 to 12 inches. Lynxes weigh from 12 to about 24 pounds, but occasionally almost 40 pounds. On average, they are larger than bobcats. Lynxes have relatively short bodies and long, heavy legs. Their well-haired feet are of particular benefit to individuals living in cold, snowy, northern climates, such as Canada. Except for the caracal of Africa and southern Asia, which has a short, smooth pelage, members of this genus have a long, soft pelage. Pelages vary from tannish gray to reddish brown, and except for the caracal, all have dark markings on light backgrounds.

Members of this genus occur from the Arctic to the deserts. They are expert tree climbers and swimmers. Although not large, they are powerful fighters, using both teeth and claws as weapons. They are most active after dark, hunting by sight and sound.

A population of lynxes, whose principal prey is the snowshoe hare, fluctuates according to their supply of food. Bobcats, on the other hand, replace the lynx in the more southern climates, such as Oregon; they have an extremely variable diet and do not exhibit such fluctuations. Lynxes and bobcats eat almost any bird or mammal they can kill, and their habits are much like those of other cats.

Lynxes and bobcats breed in late winter. After a gestation period of about sixty days, one to four young are born. Young weigh about $13^1/_2$ ounces at birth; their eyes open in about ten days, and they are weaned in about two months. Their lifespan is ten to twenty years.

There is one species in western Oregon.

Bobcat
Lynx rufus

THE SPECIFIC NAME *rufus* is the Latin word for "reddish." The 1777 scientific description of the bobcat is based on a specimen from New York

DISTRIBUTION AND DESCRIPTION

total length 710-1252 mm
tail 95-195 mm
weight 5.4-31.0 kg

THE BOBCAT OCCURS throughout all five zones of western Oregon (photo 149). They range in total length from 2 feet 4 inches to 4 feet, and their tails range from 3³/4 to 7 inches in length. They weigh from 12 to 68 pounds.

Bobcats are short-tailed, long-legged animals with relatively small feet. The ears are slightly tufted and blackish with a white spot near the tip. The eyelids are white. The hairs along the sides of the face, from the ears to the throat, are long and like sideburns. The upper parts of the body are grayish, tannish, or reddish, usually with irregular black spots; the color is most intense along the middle of the back, becoming lighter on the sides. The rump and hind legs are tannish. The underparts, including the inner sides of the legs, are whitish with black spots. There are indistinct black rings on the tail, but the tip is black above and whitish below.

HABITAT AND BEHAVIOR

BOBCATS USE ALMOST all habitats in western Oregon, at least to some extent, from sea level up to and beyond timberline. They are active mainly at night but on occasion may be abroad during the day. In western Oregon, they occur most commonly in the brushy areas, where they often travel along old logging roads.

Photo 149. An adult bobcat. (Oregon Department of Fish and Wildlife photograph by Mark Henjum.)

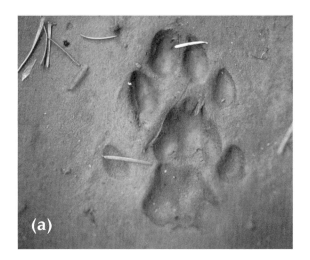

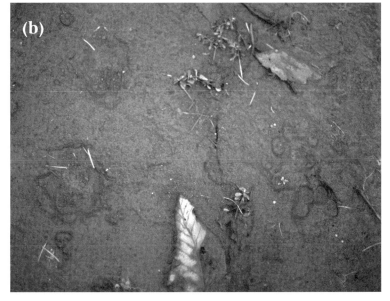

Photo 150. (a) Fresh tracks of a bobcat in soft, but firm mud; note that the hind foot has been placed partially in the track of the front foot; note also the irregular margins of the hind pads on both feet, which is characteristic of the cats. (b) Old tracks of a coyote traveling south along the left of the photograph; note the mud was soft enough to squish outward under the weight of the coyote's front foot, lower left; note also the squirish margin of the hind pad on the front foot, lower left, which is characteristic of the dogs. In contrast, are the old tracks of a bobcat traveling north along the right of the photograph; note the irregular margin of the hind pad. (Photographs by Chris Maser)

They are seldom seen, and the main evidence of a bobcat's being in the area is tracks (photo 150), droppings (photo 151 a & b), scrapes (photo 152), and marks on the trunks of trees, where they sharpen their claws (photo 153).

Their sight and hearing are particularly well developed. They possess exceptionally sensitive whiskers that help them gain a "feel" of their surroundings. Like other cats, they have excellent balance. Bobcats are quick, active, and lithe. Although their endurance does not seem great, physically they are exceedingly well coordinated and strong. The usual gait of these medium-sized predators is a stiff-legged walk, but they can cover the ground in bounds of 6 to 8 feet when they are in a hurry.

Under normal circumstances, bobcats are silent, retiring animals, but when cornered, there is no mistaking their

Photo 151. (a) Fresh droppings of a bobcat; note the segmentation, which is characteristic of the cats. (b) Old droppings of a bobcat; again note the segmentation. (Photographs by Chris Maser)

intention to fight for survival. With ears laid back flat against their heads and eyes flashing, they snarl and spit their defiance.

The den of a bobcat may be in a crevice among rocks or in a hollow standing or fallen tree or stump. The den is lined with a shallow layer of leaves, mosses, and other vegetation scratched into shape by the cat. A den is occupied as a shelter, but during fair weather, a bobcat more often lies quietly concealed amid protective vegetation.

Bobcats are stealthy and patient hunters and masters of ambush. They appear to hunt primarily by sight and sound, which means they spend much of their time sitting, watching, and listening.

Bobcats in western Oregon follow more or less regular circuits when they hunt. Their tracks and feces show that they travel the same general routes over and over again, but not necessarily in a particular number of days.

Their main prey along the southern Oregon coast is brush rabbits, which inhabit the dense thickets along the margins of roads and trails, and are easily ambushed. These cats are generalists and therefore seldom go hungry. In addition to brush rabbits, bobcats in western Oregon also eat snowshoe hares, black-tailed jackrabbits, eastern cottontail rabbits (although there may be few left compared to earlier years), deer mice, creeping voles, Pacific jumping mice, Townsend voles, dusky-footed woodrats, mountain beaver, chipmunks, tree squirrels, ground squirrels, quail, ruffed grouse, deer, and even domestic sheep on occasion. They also kill house cats; whether they eat them, I do not know. In northeastern Oregon, the red squirrel is main dietary staple of the bobcat in winter, along with deer mice.

The bobcat's breeding season commences in January and extends until at least July. The gestation period is about 63 days. A single litter is usually born from March through July but most arrive in April and May. Litters range from one to six young, but three or four is the usual number.

Young are born in caves, hollow fallen trees, or some other type of den. They are well clothed in short hair when they are

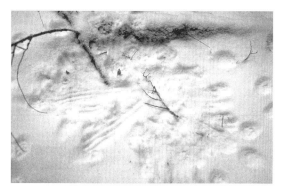

Photo 152. A territorial "scrape" made by a bobcat. (Photograph by Chris Maser)

Photo 153. Red alder trees on which a bobcat has often honed its claws and marked its territory. (Photograph by Chris Maser)

born. Newly born young are about 7 inches long and weigh about 3 ounces. Their eyes open in nine or ten days. Males seem not to help care for the young, but females appear to be model mothers. There are several instances of females, caught in steel traps, giving birth and giving the young the best care possible under the circumstances. Nevertheless, their fear of humans and dogs may cause them to abandon their young without apparent willingness to fight. But when the kittens are large enough to follow their mothers, at least some mothers display hostility if their kittens are threatened by people.

Young are weaned in sixty to seventy days, but they often stay with their mothers until they are two-thirds to three-fourths grown. They become mature and essentially full grown when about one year old. One captive bobcat lived for fifteen years and ten months, another for twenty-five years.

There are at least three records of male bobcats mating with female domestic cats; some of the resultant hybrid offspring showed definite characteristics of the bobcats.

Although the Great Horned Owl occasionally preys on young bobcats, the main enemies of this species are humans and their domesticated dogs. Bobcats are hunted for sport, for their pelts, or for elimination.

Even-toed Hoofed Mammals
Artiodactyla

THE ORDINAL NAME Artiodactyla is derived from the Greek words *artios* ("even numbered") and *dactylos* ("finger" or "toe"). The even-toed hoofed mammals are characterized by two or, more often, four toes on each foot, each toe terminating in a hoof. Artiodactyla encompasses such diverse members as deer, ox, bison, goat, sheep, antelope, camel, pig, giraffe, and hippopotamus.

In artiodactyls, the upper incisors are reduced or absent. The canines are also usually reduced or absent, but in some species, such as some wild pigs, they are large and tusklike. The space between the upper incisors and the cheek teeth (premolars and molars) is greatest in the "cud-chewers" or ruminants. Those that ruminate, such as cows and deer, have stomachs with as many as four chambers. Those with one, two, or three chambers, such as pigs, are nonruminating. The molars are more complex than the premolars. Some species have horns, others have antlers. Antlers are usually present only on males and are normally shed in winter. Horns, however, are frequently found on both sexes and are retained throughout the animal's life.

Gestation periods range from 112 days in the pig family to 450 days in the giraffe family. Numbers of young range from one or two in most artiodactyls to as many as fourteen in the pig family. Young artiodactyls are precocial and capable of running shortly after birth. Many have protective coloration in the form of spots or stripes that disrupt their outlines and make them difficult to see and less vulnerable to predation.

As a whole, artiodactyls are gregarious herd animals that occupy a wide variety of habitats from arctic regions, to deserts, to jungles. Some are diurnal, others nocturnal, and some are more or less active at any time.

Wild artiodactyls are indigenous worldwide, except in Antarctica, Australia, and most isolated islands, such as New Zealand. Humans have introduced them as domestic animals almost everywhere in the world, except Antarctica. Many have escaped from captivity and are now feral.

There is one family of artiodactyls indigenous in western Oregon.

THE FAMILIAL NAME Cervidae is derived from the Latin word *cervus* ("deer"), combined with the Latin suffix *idae* (which designates it as a family).

Members of the deer family are proportionately slim, long-legged animals, best characterized by the presence of antlers in males only, except for caribou and reindeer, in which both sexes bear antlers. Antlers differ in size and shape from one species to another.

Antlers are not the same as horns. Antlers are shed annually; horns remain permanently affixed to the skull and are composed of bony cores that grow directly out of the skull near the top of the head. They are covered with a hard sheath of horny material. Antlers are supported on permanent skin-covered pedicels. In most species, they are shed each year during the winter and spring and regenerated during the summer. Deer indigenous to the temperate zones grow new antlers each year in synchrony with the annual light-dark cycle, thus ensuring the maturation of antlers in time for the breeding season. But males of tropical species can grow antlers at any time of the year, though they shed and replace their antlers every twelve months. While the antlers are growing, they are soft, tender, and well supplied with blood and encased in a thin layer of skin covered with short, fine hairs called "velvet." Antlers attain maximum size in temperate regions by late summer, when the supply of blood gradually decreases, then terminates. The velvet dries, loosens, and drops off; by this time the antlers are hard and dead. Once the velvet is off, the antlers serve as sexual characteristics as well as weapons.

Antlers begin to develop when an individual is one or two years old; the first-year antlers are generally short, almost straight spikes. They become larger and acquire more points until the animal attains maturity, by which time the antlers have assumed the shape typical for the species. Normal antler growth depends mainly on adequate diet, and if certain minerals or vitamins are lacking, the antlers may be stunted or dwarfed.

The length of the head and body of cervids ranges from 29 inches to 9 feet 5 inches. Females generally are somewhat smaller and more delicately built than males; this is particularly noticeable in the neck. Necks of females are not as heavily haired as those of males nor do they become as large and heavy during the breeding season. The coloration varies from brownish to reddish; some species are spotted as adults. The young are usually spotted, but most lose their spots as they mature.

Nearly all cervids have facial glands situated in a pit just in front of the eyes and lined with the skin of the face. Glands

North American Elk, Deer, and Allies Cervidae

also occur on the limbs. The upper canine teeth are larger and saber shaped in members of some species.

Cervids usually associate in groups, but a few appear to be solitary, at least during the nonbreeding season. Within a few species, males fight for possession of a harem. During such combat, their antlers sometimes become locked and both opponents may die.

Cervids are herbivores, feeding on grasses, lichens, bark, twigs, and other plants and plant parts.

Cervids mate during the late autumn and early winter in temperate climates. Gestation periods range from 160 days to about ten months. One species is known to have delayed implantation. The usual number of young is one or two, but three or four are sometimes born. Cervids in temperate regions give birth only once a year.

Cervids occur from the arctic to the tropics, throughout North America, south to 40° latitude in South America, in northwestern Africa, Eurasia, Japan, the Philippines, and most of Indonesia. They have been introduced into New Guinea, Australia, New Zealand, New Caledonia, Mariana Islands, Mauritius, Cuba, and the Hawaiian Islands; and have been reintroduced into Great Britain from where they had been exterminated. There are two genera of cervids in western Oregon.

North American Elk and Allies
Cervus

THE GENERIC NAME *Cervus* is the Latin word for deer.

Of the approximately fifteen species of *Cervus*, *Cervus elaphus* (North American elk and the red deer of Europe) is the best known species. The height at the shoulder ranges from 1 foot 6 inches to almost 5 feet. Their color generally varies from brown to reddish brown. Different species occur from mountainous terrain to lowland swamps.

Members of the genus are indigenous to North America, Great Britain, Europe, Asia, the East Indies, and the Philippine Islands. Some species have been introduced into other areas, such as New Zealand. There is one species of *Cervus* in Oregon

North American elk
Cervus elaphus

THE SPECIFIC NAME *elaphus* is the Greek word for deer or stag. The 1777 scientific description of the North American elk is from a specimen thought to be from Quebec, Quebec, Canada.

DISTRIBUTION AND DESCRIPTION

ELK (PHOTO 154) OCCUR IN zones 1 through 5, except in the floor of the Willamette Valley proper, although in the past they lived there also. Elk are slender-legged animals with necks that are thick in proportion to their heads. The hair along the sides of the neck is long and dark, forming a dark brown mane on the throat. The hair on the back and sides is shorter and varies from light grayish, yellowish gray, yellowish brown, to brown. The head, neck, mane, and legs are dark brown to almost blackish; the underparts are darker than the back. The rump patch and tail vary from light yellowish to dark yellowish or tannish yellow. Cows and calves are darker than the bulls. Young calves are brownish with large yellowish white blotches. There are whitish metatarsal glands below the hocks on the outsides of the hind legs. Males develop large, widely branched antlers. The main backward sweeping beam may reach 3 feet 6 inches in length. Adults have a well-developed brow tine or "dog killer" and normally five other tines.

total length 2032-2972 mm

tail 80-213 mm

weight 159-454.5 kg

Habitat and Behavior

Elk have a matriarchal society in which the adult bulls live separately during the nonbreeding season. A matriarchal herd is composed of cows, which weigh an average of about 400 pounds, their calves, and subadults (adolescents) of both sexes. The degree to which a member of any particular sex and age class associates with the central cow-calf unit is determined by the behavioral interrelationships of that individual with the other members of the herd. Yearling bulls, for example, show a strong cow-herd attachment at times of the year when they are not driven out by an adult bull during the breeding season or when the cows are not giving birth to their calves. The composition of a cow-herd dominated by an old lead cow is most stable from November to May, and the association of subadult bulls with the cow-herd reaches a peak during the winter. The duration of their visits decreases annually, however, as the bulls approach maturity.

Adult bulls, weighing an average of about 500 pounds and as much as 1,000 pounds, join a cow-herd only temporarily during the breeding season. At other times, bulls gather into groups.

A cohesive herd has a central area that it uses to the exclusion of other individuals or herds even though the area is not actively defended. As the distance from the central area increases, use by the resident herd decreases and competition with other groups intensifies. A herd's strong orientation toward its central area is probably based on the availability of preferred food, water, the herd's knowledge of escape routes, and two kinds of cover. Hiding cover is capable of concealing 90 percent of a standing elk at a distance of 200 feet or less. Thermal cover, on the other hand, can act as hiding cover but also is large enough in size and dense enough to produce its own internal climate—cool in summer and warm in winter—that allows an elk to maintain a nearly constant core body temperature, which conserves its energy.

Although precise boundaries do not exist, there is seldom any overlap between closely adjoining herds and little trespassing, perhaps because the resident herd marked trees most frequently within its central area, communicating its monopoly of the area. A cow marks by carefully drawing her nose several times up and down a small tree as though sniffing it. She then scrapes the tree with her lower front teeth, drawing them in deliberate vertical strokes from the bottom to the top of the marked area, letting the shavings fall to the ground. She then deliberately rubs the sides of her muzzle and chin against her flanks.

Bulls five years of age and older also mark, but they use the bases of their antlers to scrape the trees. I had long heard as a boy that bulls rub their antlers to rid them of the velvet because it itched, but that is not true. Bulls may go for many days with bothersome masses of stripped velvet hanging over their eyes without making any attempt to remove it even though it causes them a great deal of annoyance. It does not itch because the antlers are dead and hard when the velvet begins to come off. It comes off by accident or when the bulls start to mark early enough to rub it off before it simply falls by itself.

Such early rubbing is always rather gentle and hesitant as though the antlers are not yet completely hardened and capable of withstanding heavy pressure and rough use. They are a dingy brown when the velvet has just dropped off, but as breeding intensifies and bulls begin to thrash vegetation, their antlers become polished and gleaming. But not all bulls have antlers of the same color because they become stained by whatever vegetation they most frequently attack.

Photo 154. Mature elk. (Oregon Department of Fish and Wildlife photograph.)

If a big bull thrusts his head against a young tree around 2 inches in diameter and shakes his head vigorously, he can break the tree. When attacking a shrub, a bull may unsheath his penis and eliminate copious amounts of urine in spurts that can carry over 3 feet. He directs the urine along his belly, thoroughly soaking the dark brown hair, and then lowers his head and saturates the long, dark mane on his throat and the sides of his face. He may back out of the shrub after a few minutes and dig his antlers into the urine-soaked grass and herbs, flinging them over his back. Finally, he may lay down in the urine-soaked area and rolled several times before leaving. A bull may also wade into the stagnant water and foul-smelling mud of a wallow and submerge his head and neck in the water, then kneel and rub his chest, neck, and face in the slimy mud on the bottom.

Bulls sometimes get hurt, even killed, when they fight, usually when a young bull approaches a herd of cows that belongs to a large bull. Each of the bulls will bugle several times and circle one another about 50 feet apart, stopping suddenly and charging, their antlers coming together with a terrific crash. They push and twist, each trying to take advantage of an opening and knock the other bull off his feet.

By the end of October, most of the cows are bred and on their way to the winter range. Some of the bulls follow them to lower elevations, but the biggest ones stay in the high country until the snow is so deep they are forced to leave or die.

The cows carry their calves for 255 to 275 days and give birth during the last week in May and the first week in June. Just before giving birth, a cow leaves the herd and selects a place where she gives birth to a single young, seldom two. The mother and her new offspring rejoin the herd in a week or so, at which time the calf is well coordinated and able to keep up with its mother.

When alarmed or frightened, cows emit a call similar to the bark of a dog; this call warns the calves, especially before a mother and her new offspring return to the safety of the herd. A newborn calf reacts immediately to the bark of a cow. It conceals itself in any available vegetation by dropping to its belly, stretching its head out flat on the ground, and remaining motionless (photo 155). The light blotches on its overall dark coat help to disrupt its outline, making it difficult to see. Once a mother and her calf have returned to the herd, however, the youngster does not hide in response to a bark, but it does direct its attention to whatever alarmed the cow.

Calves depend mainly on their mothers' milk for the first four to six weeks and may nurse five or six times a day. They are tended by a babysitter cow who keeps them together while

Photo 155. Young calf elk assuming protective posture. (Oregon Department of Fish and Wildlife photograph.)

their mothers feed. When it is time for a youngster to nurse, its mother calls her baby with a high-pitched neigh. When the nursing period is over, the mother simply walks away and starts to eat. Further attempts by the calf to nurse may bring a resounding whack across its back from a front hoof or a butt from the side of her head, but her youngster normally returns to the babysitter without hesitation. As the calves become less dependent on milk, the babysitter becomes more lax in her efforts to keep them together. By autumn they are feeding with the herd, but still tend to remain together.

Mule Deer and White-tailed Deer *Odocoileus*

THE GENERIC NAME *Odocoileus* is derived from the Greek words *odontos* ("tooth") and *koilos* ("hollow"), combined with the Greek suffix *eus* ("pertaining to"). The name alludes to the deep pits in the surface of the cheek (grinding) teeth.

There are two species of deer within this genus. The length of the head and body ranges from 5 feet 2 inches to 6 feet 10 inches; and the length of the tail from 4 to 11 1/4 inches. The height at the shoulder ranges from 2 feet 7 inches to 3 feet 7 inches. Adults weigh from 100 to 455 pounds.

Autumn and winter coats are usually brownish gray with whitish underparts. Summer coats are brownish to reddish above and whitish below. The hairs, particularly those of the winter coat, are tubular and somewhat stiff and brittle. Because of the tubular nature of the hair of the thick coats of deer killed in winter, their skins float on water and have been used as life preservers.

Photo 156. Branching antlers of a mule deer. (Oregon Department of Fish and Wildlife photograph.)

In North America, these deer shed their antlers from January to March; new antlers begin to grow about April or May and lose their velvet in August or September. Bucks attain full-sized antlers when they are four or five years old. The antlers of mule deer branch into two nearly equal parts (photo 156), whereas those of white-tailed deer have a main beam with minor branches.

Strictly speaking, these deer are not herd animals, but they do congregate on favorable winter ranges. Their diet includes a wide variety of plants, such as grasses, forbs, twigs of shrubs, fungi, nuts, acorns, and lichens.

The breeding season, or rut as it is called, usually occurs during November and December. The gestation period ranges from 196 to 210 days. Does give birth once a year; they usually have one fawn during their first pregnancy and twins thereafter, occasionally three or four young. A fawn weighs from about 3 to almost 5 pounds at birth and is able to walk in a short time. These deer have lived twenty years in captivity.

Deer of the genus *Odocoileus* occur from as far north as the southern Yukon and Mackenzie districts of northwestern Canada, south across southern Canada, throughout the lower 48 United States and Mexico into Patagonia, South America.

Both the mule deer and the white-tailed deer occur in western Oregon.

Mule deer
Odocoileus hemionus

THE SPECIFIC NAME *hemionus* is derived from the Greek prefix *hemi* ("half") and the Greek work *onos* ("an ass"); hence the common name "mule deer." The first specimen came from the mouth of the Big Sioux River in South Dakota and was described scientifically in 1817.

DISTRIBUTION AND
DESCRIPTION

total length 1345-
1800 mm
tail 106-230 mm
weight 45.4-207.3
kg

MULE DEER ARE REFERRED to as "black-tailed deer" in western Oregon because deer west of the crest of the High Cascades have black on the tops of their tails, while the tails of those east of the crest have only a black tip. They occur in all five zones.

These deer are well-muscled animals; the females may weigh around 100 pounds and the males around 150 to 200 pounds. They have relatively large ears, long, slender legs, and slender, pointed hoofs. Both sexes have metatarsal glands, which are longitudinal glands located on the hind legs in the area of the hock or heel and are visible as tufts of long, stiff hairs. Their coats, which are composed of hollow hairs, vary from dull yellowish brown to reddish brown in summer and dark grayish to rich brownish gray in winter. The upper throat, insides of the ears, and insides of the legs are whitish; the belly varies from white to tannish. The nose, forehead, and chest are dark brown to blackish. The rump patch is white, and the rather bushy tail is brown on top with a black tip. Youngsters are thickly spotted with white over tannish to reddish brown upper parts. Males grow branching antlers each summer and shed them in winter.

**HABITAT AND
BEHAVIOR**

THE SUBSPECIES OF MULE deer in western Oregon, called black-tailed deer, are versatile when it comes to habitat. Although primarily animals of the edge or ecotone between grassy areas and forest, some live in deep, old-growth coniferous forest. But they are also at home in small islands of trees or shrubs, including those in suburbia. Deer occur wherever there is enough cover to protect them from the heat of summer and from predators. Because they are not herd animals, they require only scattered cover of sufficient size to protect at most a small family group. A family group includes at least a doe and a fawn; at most a doe, two fawns, and two yearlings.

Newborn fawns lie flat against the ground with their necks outstretched and their ears laid back against their heads. This flat, immobile posture, their white spots that disrupt the outlines

Photo 157. Recently born fawn mule deer in protective posture. (Oregon Department of Fish and Wildlife photograph.)

of their body contours (photo 157), and the fact that they have little odor all serve as protection from predators.

Mother and young learn to recognize each other by sniffing one another's metatarsal glands. The fawns bleat for their mother during the day, but at night they "rub-urinate" to communicate as well as to give a distress signal, rhythmically rubbing their hocks together while slowly releasing urine over them. As the youngsters mature, they will use rub-urinating to threaten other deer. Smelling the metatarsal glands is normally the first stage of aggressive behavior. As the young male matures, he will erect and rhythmically move the long hairs around his metatarsal glands, exposing the scent, and he will often combine this with rub-urination.

Mutual grooming begins between the mother and her youngest fawns and helps to establish strong bonds among members of a family and reestablishes them if they are weakened by frequent contact with a relatively large number of deer during the gregarious wintering period. In addition to grooming, members of a family group sniff the metatarsal glands

---

Like almost any baby animals, fawns are playful. One day in early June 1956, I headed for Marys Peak in Oregon's Coast Range to think about my decision to enter the military. It was a clear, cool night with a full moon. I lay against a large log under a small, bark lean-to at the edge of a little, grassy clearing in a virgin forest. Since I had not built a fire and I was downwind, the doe did not detect my presence as she quietly crossed the meadowy area and stopped within 20 to 30 feet of my head. Between her and me, a bright shaft of moonlight filtered through the tall firs and lighted a small place in the meadow.

As the doe started feeding, her fawn discovered the moonbeam. With head outstretched and nostrils flaring, it approached the light, its legs stiff and tucked well beneath its body for instant flight. The youngster suddenly snorted and fled to its mother, whose head snapped erect at the sound. The fawn nursed, and its mother returned to eating.

During the next half hour or so the fawn alternately struck the light with its forefeet, kicked it with its hind feet while twisting sideways in the air, and raced through it. These feats of daring were often interrupted by sudden returns to its mother. At such times, the fawn usually went to the far side of the doe and, with its head under her belly, observed the moonbeam from safety. After a few seconds, however, it again engaged the light. As the doe gradually drifted away from the light, the fawn ceased its play.

---

of one another's legs once or twice an hour during the day and as often as six times an hour during the night.

Adult females are mutually antagonistic much of the year, and conflicts may arise when they come together. Such antagonism results in a fairly regular spacing of the females' centers of activity. A female occasionally remembers a conflict and avoids the area of another female even if she is dead. Although family ties weaken during late winter and spring when fawns have been weaned and when females congregate in a choice feeding area, the birth of new fawns renews the mutual antagonism and thereby spaces the females throughout the habitat.

Most males are solitary, but some have a strong tendency to congregate throughout much of the year. They usually disperse with the onset of the breeding season but may gather into groups again during winter and spring. The males' centers of activity are often clustered in close proximity to one another, and other than maintenance of social rank and sexual aggressiveness, no antagonistic behavior occurs. Several families and groups of males may come together in spring and form large feeding bands. Although these bands resemble a social herd, each small group retains its integrity. Conflict often arises when these small groups approach one another too closely. No permanent social herd forms, however, because each group goes its own way as the feeding period ends.

The deer sniff and rub their foreheads against trunks of trees, branches, twigs, and occasionally other things throughout the year, forming an intricate system of communication. Males apply scent on sign posts, and both males and females sniff these sites. The material from the scent glands on their foreheads

---

I often hunted in an orchard—long abandoned by early settlers and of late well hidden in the forest. It was raining lightly as I slowly worked my way through wet vegetation that was soaking me to the skin. Cresting the knoll, I saw two deer looking at me. I slowly raised my rifle. I was just taking up the slack in the trigger when there issued from the orchard a most fantastic belch—then another and another. Lowering my rifle, I studied the deer.

It slowly dawned on me that they were not standing under the apple trees—but leaning against them! After a few moments observation, I walked up to them, slowly at first and ready to shoot, then with casual abandon. Both deer gazed stupidly at me with glassy, sightless eyes, and slobbering muzzles. They were totally drunk from eating fermenting apples. So I left them to their hangovers, if deer get such things.

---

is apparently washed off these marked areas by rain and melting snow because there is an increase in rubbing activity following precipitation.

Males share rubbing sites as well as using some that are strictly their own. Rubbing sites are established at strategic places, such as resting areas and along commonly used trails. The dominant male frequently marks rubbing sites but seldom sniffs, whereas lower ranking males sniff more and mark less. From spring to autumn, there is an increase in the frequency and intensity with which males mark shared rubbing sites and a decrease in the number of exclusive sign posts that an individual maintains. Rubbing, which reaches its peak during the breeding season, seems to advertise the presence and physiological state of a particular male because both females and males sniff sign posts more frequently during the breeding season than at other times (photo 158). Forehead rubbing and thrashing vegetation with the antlers may help breeding males establish dominance, express a threat, and simultaneously avoid unnecessary conflict.

Yearling and mature males begin to grow antlers in spring in response to daylength; this timing insures the maturation of antlers in time for the breeding season and therefore that the young will be born within a certain period the following spring.

Antlers contain a good deal of calcium and are shed in winter, during a hard time of the year for most active rodents. So it is a fortunate deer mouse, chickaree, or flying squirrel that finds a discarded antler on which to sharpen its evergrowing front teeth and obtain a welcome supply of calcium and other minerals (photo 159).

Although pumas, bobcats, and coyotes are the main predators of mule deer throughout the year, humans take a heavy toll during hunting seasons.

Photo 158. Sign post of a buck mule deer; note where the buck's antlers have shredded the bark of this red alder. (Photograph by Chris Maser)

Photo 159. Shed antler of a mule deer chewed by deer mice. (Photograph by Chris Maser)

Photo 160. (a) Front foot, larger on left, and hind foot, smaller on right, of a mule deer. The front feet are larger than the hind feet because a deer carries most of its weight from the mid- section forward; the tracks of the front feet are therefore correspondingly larger than those of the hind feet. (b) Fresh tracks of a bounding mule deer. (Photographs by Chris Maser)

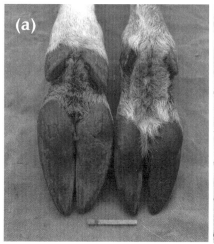
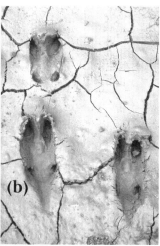

I have always found it ironic that most hunters cannot read deer tracks. I have more than once heard one hunter tell another that one track he was studying was that of a buck, while another was that of a doe, which was determined by their respective sizes, when both belonged to the same animal. One must consider that most of a deer's weight is carried forward, toward the front end of the animal, and so the front feet are both larger and heavier than the hind feet because they must support the additional weight (photos 160). Also the dew claws are both larger and closer to the hoof on the front feet than the hind feet. In addition, young animals tend to almost match or even over-step the print of the front foot with the hind foot, but as an individual ages, the hind foot lags further back until, in old animals, it does not even reach the track of the front foot—again, the same animal.

White-tailed deer
Odocoileus virginianus

DISTRIBUTION AND DESCRIPTION

total length 1,575-1,935 mm
tail 178-190 mm
weight 72.7 kg

THIS DEER WAS NAMED after the state of Virginia, where the first specimen was collected; it was described scientifically in 1780.

THE WHITE-TAILED DEER is found in zones 1 and 2 on some of the islands in the Columbia River west of the Cascade Mountains, and in the vicinity of Roseburg in the northern-most part of zone 3 (photo 161). The length of an adult male is from about 5 feet 2 inches to 6 feet 5 inches and that of their tails is about 7 to 7 5/8 inches. The weight of an adult male is around 160 pounds.

The white-tailed deer differs from other members of the genus in that the main beam of its antlers is directed forward and bears the several tines coming off the main beam. In addition, the long, rather bushy tail is brown above and clear white below and is held erect and wags from side to side when a deer is running. The large, white underside of the tail is one of its most remarkable features and helps to distinguish it from the mule deer.

The color and character of the pelage has marked differences by season. In summer, the coat is bright tan to slightly reddish tan, and is darker along the mid-part of the back. The color is paler on the face, throat, and chest. There is a prominent white band just above the nose, white rings around the eyes, and pure white inside the ears, on the insides of the legs, and on the belly. In addition, there is a black spot on either side of the chin. The young are more reddish yellow, spotted with white, but the spots disappear by late summer.

In autumn, the reddish pelage of summer is replaced by a dense, somewhat more brittle, and dark brownish gray, slightly darker along the back of the neck. The forehead is dark brown, whereas the top of the tail and the outer parts of the legs remain reddish tan, and the areas of white remain the same.

WHITE-TAILED DEER were once abundant along the river bottoms, islands in the Columbia River, thickets of willows, and open prairies of the valley bottoms of western Oregon, occasionally going into the foothills. In 1806, for example, Lewis and Clark reported seeing upward of a hundred on Deer Island in the Columbia River, below the present site of Portland. Once found throughout the interior valleys of western Oregon and still common between 1860 and 1875 in the foothills around Beaverton, in Washington County, white-tailed deer began to disappear rapidly after 1875. The late D.D. Hickleman of Albany told Vernon Bailey that the last white-tailed deer he knew to have been killed in the Willamette Valley was in about 1898 near Sweet Home, in Linn County.

HABITAT AND BEHAVIOR

White-tailed deer are active at all hours, although much less so during daylight hours. They normally commence feeding with the approach of dusk and continue to move around throughout the night, normally bedding down in some concealed place with the approach of dawn.

Although primarily browsers, these deer also graze; in fact much of the food of the few remaining deer in western Oregon is probably succulent herbs that grow in the interface between the river-bottom forest and the agricultural lands. With the approach of autumn, they undoubtedly fatten themselves on the acorns of the white oak trees. In winter, they browse on buds and twigs of both deciduous and coniferous trees.

As the daylight hours of autumn lessen, reproductive activities in white tailed deer begin. At this time, bucks have hardened, velvet-stripped antlers, and those 28 months old or

Photo 161. Female white-tailed deer. (Oregon Department of Fish and Wildlife photograph by the late John Ely.)

older develop sexual readiness. By November, bucks with polished antlers search for does, which become most attractive to the persistent, belligerent, swollen-necked bucks for at least three days prior to and perhaps two days after receptiveness. During this time, a buck becomes so single-minded he frequently follows a doe, disregarding both nourishment and potential danger, until she, at her time of maximum receptiveness, also becomes highly active sexually. If she does not conceive in the 24 to 36 hours during which she is in estrus, she will again become receptive in about 28 days. Does can and may conceive while they are still nursing their young, or at least while there is still some milk in their udders.

With the rut nearly complete by late December, does generally separate from the bucks, except for intermingled groups during periods of feeding. Pregnant does follow their normal activities until late May or early June, when, about 200 days after conception, they disassociate themselves from other does and give birth to their young in some secluded place. First-year does, if well nourished, give birth to a single fawn, whereas healthy older does generally produce two, occasionally three, young. The production of healthy babies is closely related to the physical condition the doe has been able to maintain throughout the winter.

At birth, a fawn is well marked with white spots on its back and flanks, which serve to disrupt the reddish background of its dorsal coloration; the spots are significantly larger on a male than a female. It has fine, silky hair, good vision, and up-standing ears. It can, within an hour or so after birth, stand on its long, spindly legs. Receiving a long and continuous licking by its mother, it can bleat and seems to have little odor. A female weighs as little as 5 pounds and a male as much as $7^1/2$ pounds. Within its first twelve hours, it will start to suckle once or twice daily in a standing position. When not nursing, it remains hidden in a relatively motionless position lying flat on the ground, its white spots blending well in the dapple of light and shadows. In three or four days, however, the youngster is able to travel considerable distances.

At 21 days of age, the fawn nibbles the first tips of grass and follows its mother as she joins the rest of her family group. The fawn may suckle for up to 120 days, although it is eating solid food long before that. The fawn grows its winter coat, losing its spots between 90 and 120 days of age, usually by October.

It the fawn is a male, small buttons of antlers begin to show beneath the skin of its forehead around 180 days of age. This growth is preparation for the young buck's second summer of life, when it develops "spike" antlers that may reach 6 inches in length. If, however, the buck has a good and nutritious supply of food at its disposal, it may develop forked antlers or even antlers with several points. The size, symmetry, and number of normal points on each of a buck's antlers are, consequently, an expression of his physical wellbeing and not age alone. Antlers may be shed as early as late December and as late as early April. Older bucks may finish shedding their antlers before the younger ones do.

Pumas, bobcats, and coyotes prey on white-tailed deer. But it is humans, through over-hunting in the early years and through the continued destruction of their potential habitat, who have brought the white-tailed deer in western Oregon to the point that it is today listed among the threatened and endangered species of mammals within western Oregon.

Epilogue

WHY DOES A SPECIES of indigenous wild animal, such as the white-tailed deer in western Oregon, become threatened or endangered? The primary reason is human-introduced fragmentation of habitat, which imposes stress on the indigenous animals with which they are ill adapted to cope. Biogeographical studies show, for example, that *connectivity* of habitats within the context of the landscape is of prime importance to the persistence of indigenous plants and animals in biologically viable numbers within their respective habitats. The landscape must be considered a mosaic of interconnected patches of habitats—like vegetated fencerows—that act as corridors or routes of travel between other patches of habitat, such as small acreages of forest associated with farms.

I grew up in Corvallis, Oregon, which was still surrounded in those days by fields and forest, connected by swift forest streams that fed meandering valley rivers. And I was free to wander over hill and dale without running into a "No Trespassing" sign on every gate and seemingly every other fence post. The unspoken code was to leave open any gate that was open and to close after one's passage any gate that was closed. It was also understood that one was free to cross a farmer's property as long as one respected it by walking around planted fields rather than through them. And, if I asked permission, I could wander, hunt, fish, and trap almost anywhere I wished.

Much of the Coast Range and most of the Cascade Mountains of Oregon that I knew as a youth were covered with unbroken ancient forest and clear, cold streams from which it was safe to drink. Although the streams were filled with trout and salmon, the forests and mountain meadows were already devoid of wolf and grizzly bear.

In the valley that embraced my home town, the farmers' fields were small and friendly, surrounded by fencerows sporting shrubs and trees, including apples and pears that proffered delicious fruit, each in its season. In spring, summer, and autumn, the fencerows were alive with the colors of flowers and butterflies and the songs of birds. They harbored woodrats and rabbits, pheasants and deer, squirrels and red valley foxes. The air was clean, the sunshine bright and safe, and the drinking water amongst the sweetest and purest in the world.

Over the years, however, I have watched helplessly as the small protected fields of the personable family farms around my home town and throughout the Willamette Valley increasingly gave way to the larger and larger homogeneous fields of corporate-style farms as fencerows were cleared to maximize the amount of tillable soil. With the loss of habitat along each fencerow, the bird song of the valley was diminished in like measure, as were all the other creatures that lived there.

Gone are the fencerows with their rich, fallow strips of grasses and herbs, of shrubs and trees, which interlaced the valley in such beautiful patterns of flower and leaf with the changing seasons. Gone are the burrowing owls from the quiet secluded fields that I once knew. Gone is the liquid melody of the meadowlark that I so often heard as a boy. Gone is the fencerow trill of the towhee. Gone are the song sparrows, Bewick's wrens, yellow warblers, and MacGillivray's warblers. Gone are the woodrat nests, the squirrels, and the rabbits. Although these species may still occur along the edge of the valley and in isolated patches of habitat, they are gone with the fencerows from the agricultural fields of the valley floor.

As I now approach sixty years of age, the valley's floor, compared with the time of my youth, offers little in the way of habitat, other than a great depersonalized open expanse of silent, naked fields in winter and a monotonous sameness under the sun of summer (photo 162).

Photo 162. The barn, some miles east of Corvallis, says it all: "Think habitat; pheasants forever," all the while the barn is bordered by enormous monocultural fields that years ago were stripped of every vestige of fencerow habitat.

It is important to understand that whether populations of indigenous wild plants and animals survive in a particular landscape depends on the rate of local extinctions from this patch of habitat or that and on the rate at which an organism can move among existing patches. Those species living in habitats isolated as a result of fragmentation are less likely to persist. Fragmentation of habitat is not only the primary cause of the often-discussed *global* crisis in the rate of biological extinctions but also the less-discussed crisis in the rate of *local* extinctions.

Modification of the connectivity among patches of habitat strongly influences the abundance of species and their patterns of movement. The size, shape, and diversity of patches also influence the patterns of species' abundance, and the shape of a patch may determine the species that can use it as habitat. The interaction between the processes of a species' dispersal and the pattern of a landscape determines the temporal dynamics of its populations. Local populations of organisms that can disperse great distances may not be as strongly affected by the spatial arrangement of patches of habitat as are more sedentary species.

Our responsibility now is to make decisions about patterns across the landscape of western Oregon while considering the consequences of our decisions on the potential quality of life for the generations of the future. Although the decisions are up to us, one thing is clear: the current trend toward homogenizing the landscape, which may help maximize short-term monetary profits, progressively degrades the long-term biological adaptability of the land and thus the long-term sustainability of society as we know it.

The stability of an ecosystem flows from the patterns of relationship that have evolved among the various species living in it. A stable, culturally oriented system, such as the Willamette Valley, even a very diverse one, that fails to support these coevolved relationships has little chance of being sustainable. Ecological sustainability and adaptability depend on the connectivity of the landscape; we must ground our culturally designed landscapes within Nature's evolved patterns and take advantage of them if we are to have a chance of creating a quality environment that is both pleasing to our cultural senses and ecologically adaptable—and thus sustainable.

To do this, we must move purposefully, consciously toward connectivity by shifting our focus to two primary things. We must consciously account for the sustainable connectivity and biological richness between such areas as forest clearcuts, agricultural fields, and urban development within the context of the landscape as a whole. And we must protect existing biodiversity—including habitats—as the price for the long-term sustainability of the ecological wholeness of the patterns we create across the landscape of western Oregon. The decision about the quality of habitat in western Oregon belongs to us, of course, the adults of the world. But the consequences of our decisions we irrevocably bequeath to those generations we call "the future."

Appendix 1: Common and Scientific Names

Common Name	Scientific Name
Plants	
Lichens	
Lichen	*Bryoria fremonti sp.*
Lichen	*Usnea* sp.
Trees and Shrubs	
Alder	*Alnus* sp.
Apple	*Malus* sp.
Bigleaf maple	*Acer macrophyllum*
Bitter cherry	*Prunus emarginata*
Black cottonwood	*Populus trichocarpa*
Black walnut	*Juglans nigra*
Blueberries	*Vaccinium* sp.
Blue elderberry	*Sambucus cerulea*
Buckbrush	*Ceanothus cuneatus*
California black oak	*Quercus kelloggii*
California-laurel	*Umbellularia californica*
Canyon live oak	*Quercus chrysolepis*
Cascara	*Rhamnus purshiana*
Cottonwood	*Populus* sp.
Currant	*Ribes* sp.
Douglas fir	*Pseudotsuga menziesii*
Engelmann spruce	*Picea engelmannii*
Evergreen or "shot" huckleberry	*Vaccinium ovatum*
Golden chinkapin	*Castanopsis chrysophylla*
Grand fir	*Abies grandis*
Himalayan blackberry	*Rubus discolor*
Huckleberries	*Vaccinium* sp.
Jeffery pine	*Pinus jefferyi*
Lodgepole pine	*Pinus contorta*
Long-leaved Oregon grape	*Mahonia nervosa*

Common Name	Scientific Name
Manzanita	*Arctostaphylos* sp.
Mountain hemlock	*Tsuga mertensiana*
Mountain huckleberry	*Vaccinium membranaceum*
Noble fir	*Abies procera*
Ocean-spray	*Holodiscus discolor*
Oregon ash	*Fraxinus latifolia*
Oregon grape	*Mahonia* sp.
Oregon white oak	*Quercus garryana*
Pacific madrone	*Arbutus menziesii*
Pacific silver fir	*Abies amabilis*
Pear	*Pyrus communis*
Poison oak	*Rhus diversiloba*
Ponderosa pine	*Pinus ponderosa*
Port-Orford-cedar	*Chamaecyparis lawsoniana*
Quaking aspen	*Populus tremuloides*
Queen Anne's lace	*Daucus carota*
Rabbitbrush	*Chrysothamnus* sp.
Red alder	*Alnus rubra*
Red elderberry	*Sambucus racemosa*
Red huckleberry	*Vaccinium parvifolium*
Red willow	*Salix lasiandra*
Salal	*Gaultheria shallon*
Salmonberry	*Rubus spectabilis*
Scotch broom	*Cytisus scoparius*
Serviceberry	*Amelanchier alnifolia.*
Sitka spruce	*Picea sitchensis*
Snowberry	*Symphoricarpos albus*
Snowbush	*Ceanothus velutinus*
Stink currant	*Ribes bracteosum*
Subalpine fir	*Abies lasiocarpa*
Sugar pine	*Pinus lambertiana*
Tanoak	*Lithocarpus densiflorus*

Common Name	Scientific Name
Thimbleberry	*Rubus parviflorus*
Trailing blackberry	*Rubus parviflorus*
Tulip tree	*Liriodendron tulipifera*
Vine maple	*Acer circinatum*
Wax myrtle	*Myrica californica*
Western hemlock	*Tusga heterophylla*
Western juniper	*Juniperus occidentalis*
Western redcedar	*Thuja plicata*
White-barked pine	*Pinus albicaulis*
Willow	*Salix* sp.

Ferns

Common Name	Scientific Name
Bracken fern	*Pteridium aquilinum*
Deer fern	*Blechnum spicant*
Parsley fern	*Cryptogramma crispa*
Sword fern	*Polystichum munitum*

Grasses and Grasslike Plants

Common Name	Scientific Name
European beachgrass	*Ammophila arenaria*
Horsetail	*Equisetum* sp.
Rush	*Juncus* sp.
Sedge	*Carex* sp.
Velvet-grass	*Holcus lanatus*

Herbs

Common Name	Scientific Name
Alfalfa	*Medicago sativa*
American twinflower	*Linnea borealis*
Angled bittercress	*Cardamine angulata*
Bear grass	*Xerophyllum tenax*
Cattail	*Typha latifolia*
Clover	*Trifolium* sp.
Douglas iris	*Iris douglasiana*
False dandelion	*Hypochoeris radicata*
Lupine	*Lupinus* sp.
Mint	Lamiaceae
Northwest nettle	*Urtica dioica*
Oregon oxalis	*Oxalis oregana*

Common Name	Scientific Name
Pacific bleeding heart	*Dicentra formosa*
Pacific waterleaf	*Hydrophyllum tenuipes*
Pearly everlasting	*Anaphalis margaritacea*
Plantain	*Plantago lanceolata*
Poison oak	*Rhus diversiloba*
Skunk cabbage	*Lysichitum americanum*
Stonecrop	*Sedum* sp.
Strawberries	*Fragaria* sp.
Western golden saxifrage	*Chrysosplenium gelchomaefolium*
Western spring beauty	*Montia sibirica*
Wild onion	*Allium amplectens*
Wild garlic	*Allium* sp.
Wild rose	*Rosa* sp.
Wood groundsel	*Senecio sylvaticus*

Invertebrates

Insects

Common Name	Scientific Name
Bald-faced black hornet	*Vespula maculata*
Blister beetle	*Epicauta puncticollis*
Bumble bee	*Bombus* sp.
Carpenter ant	*Camponotus* sp.
Caddisflies	Trichoptera
Cascade tiger beetle	*Cicindela longilabris ostenta*
Craneflies	Tipulidae
Damp-wood termite	*Zootermopsis angusticollis*
Darkling beetles	Tenebrionidae
Flightless tiger beetle	*Omus dejeani*
Honey bee	*Apis* sp.
Jellowjacket	*Vespula maculifrons*
Jerusalem cricket	*Stenopelmatus* sp.
Long-horned grasshopper and katydids	Tettigoniidae

Common Name	Scientific Name	Common Name	Scientific Name

Common Name	Scientific Name
Midges	Chironomidae
Mosquitos	Culicidae
Mutillid wasps	*Dasymutilla* sp.
Red forest ants	*Formica* sp.
Scarab beetles	Scarabaeidae

Millipedes and Centipedes

Large millipede	*Californibolus oregonus*
Large centipede	*Scolopocryptops sexspinosa*

Snails

Thin-shelled, green land snail	*Haplotrema vancouverense*

Spiders and Allies

Harvestmen or daddy-longlegs	Phalangida
Scorpion	*Uroctonus mordax*

Isopods

Pillbugs	*Armadillidium vulgare*
Sowbugs	*Porcellio scaber*

Vertebrates

Fish

Bluegill	*Lepomis macrochirus*
Catfish	Siluriformes
Cutthroat trout	*Salmo clarki*
Large-mouthed bass	*Micropterus salmoides*
Salmon	*Onorhynchus* sp.
Sculpin	*Cottus* sp.
Squaw fish	*Ptychocheilus* sp.
Trout	*Salmo* sp.

Amphibians

Bullfrog	*Rana catesbeiana*
Northern leopard frog	*Rana pipiens*
Pacific giant salamander	*Dicamptodon tenebrosus*
Pacific treefrog	*Hyla regilla*

Reptiles

Alligator lizard	*Gerrhonotus* sp.
Gopher snake	*Pituophis melanoleucus*
Northwestern garter snake	*Thamnophis ordinoides*
Pacific rattlesnake	*Crotalus viridis*
Pond turtle	*Clemmeys marmorata*
Ring-necked snake	*Diadophis punctatus*
Rubber boa	*Charina bottae*
Sharp-tailed snake	*Contia tenuis*

Birds

Band-tailed Pigeon	*Columba fasciata*
Barbets	Capitonidae
Barn Owl	*Tyto alba*
Barred Owl	*Strix varia*
Belted Kingfisher	*Ceryle alcyon*
Bewick's Wren	*Thryomanes bewickii*
Blue Grouse	*Dendragapus obscurus*
Burrowing Owl	*Athene cunicularia*
Common Flicker	*Colaptes auratus*
Cooper Hawk	*Accipiter cooperii*
Golden Eagle	*Aquila chrysaetos*
Goshawk	*Accipiter gentilis*
Great Blue Heron	*Ardea herodias*
Great Horned Owl	*Bubo virginianus*
Long-eared Owl	*Asio otis*
MacGillivray's Warbler	*Oporornis tolmiei*
Marbled Murrelet	*Brachyramphus marmoratus*
Meadow Lark	*Sturnella neglecta*
Northern Spotted Owl	*Strix occidentalis*
Ring-necked Pheasant	*Phasianus colchicus*
Pileated Woodpecker	*Dryocopus pileatus*
Raven	*Corvus corax*

Common Name	Scientific Name	Common Name	Scientific Name
Red-tailed Hawk	*Buteo jamaciensis*	Sea otter	*Enhydra lutris*
Rough-legged Hawk	*Buteo lagopus*	Spiny anteater	*Tachyglossus aculeatus*
Ruffed Grouse	*Bonasa umbellus*		
Rufous-sided Towhee	*Pipilo erythrophthalmus*	Tiger	*Panthera tigris*
		Townsend pocket gopher	*Thomomys townsendi*
Saw-whet Owl	*Aegolius acadicus*		
Screech Owl	*Otus kennicottii*	Ungava vole	*Phenacomys ungava*
Snowy Owl	*Nyctea scandiaca*	Water buffalo	*Bubalus bubalis*
Song Sparrow	*Melospiza melodia*	Wolf	*Canis lupus*
Swainson's Thrush	*Catharus ustulatus*	Woolly mammoth (extinct)	*Elephas? columbi*
Wood Duck	*Aix sponsa*		
Yellow Warbler	*Dendroica petechia*		

Mammals

Caracal	*Lynx caracal*
Cave bear (extinct)	*Arctotherium* sp.
Cheetah	*Acinonyx jubatus*
Chinchilla	*Chinchilla laniger*
Dire-wolf (extinct)	*Canis* cf. *dirus*
Domestic cat	*Felis catus*
Duck-billed platypus	*Ornithorhynchus anatinus*
European badger	*Meles meles*
European beaver	*Castor fiber*
Himalayan weasel	*Mustela siberica*
Grizzly bear	*Ursus horribilis*
Jaguar	*Panthera onca*
Lemming	*Lemmus*
Leopard	*Panthera pardus*
Lion	*Panthera leo*
Lynx	*Lynx canadensis*
North American badger	*Taxidea taxus*
Northern pocket gopher	*Thomomys talpoides*
Polar bear	*Thalarctos maritimus*
Pronghorned antelope	*Antilocapra americana*
Pygmy shrew	*Micosorex hoyi*
Raccoon-dog	*Nyctereutes* sp.
Saber-toothed tiger (extinct)	*Smilodon* sp.

Appendix 2:
Table of Mammals' Location in the Northwest

Of the 89 species of mammals in western Oregon, 67 occur in British Columbia, 70 in Washington, and 83 in northwestern California: • means the mammal occurs within B.C., western Washington, western Oregon, or northwestern California; + means the mammals occurs within the state but outside the designated area.

Species of Mammal	B.C.	Washington	Oregon	California
Opossum	•	•	•	•
Wandering shrew	•	•	•	•
Dusky shrew	•	•	•	•
Yaquina shrew			•	
Pacific shrew			•	•
Marsh shrew	•	•	•	•
Northern water shrew	•	•	•	•
Trowbridge shrew	•	•	•	•
American shrew-mole	•	•	•	•
Townsend mole	•	•	•	•
Coast mole	•	•	•	•
Broad-handed mole			•	•
Little brown bat	•	•	•	•
Yuma bat	•	•	•	•
Long-eared bat	•	•	•	•
Fringed bat	•	+	•	•
Long-legged bat	•	•	•	•
California bat	•	•	•	•
Silver-haired bat	•	•	•	•
Big brown bat	•	•	•	•
Hoary bat	•	•	•	•

Mammals of the Pacific Northwest

Species of Mammal	B.C.	Washington	Oregon	California
Western big-eared bat	•	•	•	•
Pallid bat	•	+	•	•
Brazilian free-tailed bat			•	•
Pika	•	•	•	•
Snowshoe hare	•	•	•	•
Black-tailed jackrabbit		+	•	•
Brush rabbit			•	•
Eastern cottontail	•	•	•	
Mountain beaver	•	•	•	•
Townsend chipmunk	•	•	•	•
Yellow-pine chipmunk	•	•	•	•
Beechey ground squirrel		+	•	•
Belding ground squirrel			•	+
Mantled ground squirrel	•		•	•
Yellow-bellied marmot	•	•	•	•
Western gray squirrel		•	•	•
Eastern gray squirrel	•	•	•	?
Eastern fox squirrel	•	•	•	?
Chickaree	•	•	•	•
Northern flying squirrel	•	•	•	•
California kangaroo rat			•	•
Camas pocket gopher			•	
Mazama pocket gopher			•	•
Botta pocket gopher			•	•
North American beaver	•	•	•	•
Western harvest mouse	•	+	•	•
Deer mouse	•	•	•	•
Piñon mouse			•	•
Dusky-footed woodrat			•	•
Bushy-tailed woodrat	•	•	•	•

Species of Mammal	B.C.	Washington	Oregon	California
California red-backed vole			•	•
Heather vole	•	•	•	•
White-footed vole			•	•
Oregon red tree vole			•	•
Townsend vole	•	•	•	•
Long-tailed vole	•	•	•	•
Creeping vole	•	•	•	•
Gray-tailed vole		•	•	
Montane vole	•	+	•	•
California vole			•	•
Water vole	•	•`	•	
Muskrat	•	•	•	•
Black rat	•	•	•	•
Norway rat	•	•	•	•
House mouse	•	•	•	•
Pacific Jumping mouse	•	•	•	•
North American porcupine	•	•	•	•
Coypu	•	•	•	•
Coyote	•	•	•	•
Red fox	•	•	•	•
Gray fox			•	•
North American black bear	•	•	•	•
Ringtail			•	•
Raccoon	•	•	•	•
North American marten	•	•	•	•
Fisher	•	•	•	•
Wolverine	•	•	•	•
Short-tailed weasel	•	•	•	•
Long-tailed weasel	•	•	•	•

Species of Mammal	B.C.	Washington	Oregon	California
Mink	•	•	•	•
Spotted skunk	•	•	•	•
Striped skunk	•	•	•	•
River otter	•	•	•	•
Puma	•	•	•	•
Bobcat	•	•	•	•
North American elk	•	•	•	•
Mule deer	•	•	•	•
White-tailed deer	•	•	•	

Glossary

Abdominal pouch—fur-lined pouch on the belly of a female opossum, which is designed for carrying newborn young.

Allies—as used here, it means other, unspecified members of the Order, Family, of genus.

Antler—bony growth (erroneously called "horn") on the heads of members of the deer family (usually males only); they are shed annually.

Arboreal—tree-dwelling.

Arid—describes a habitat with a scanty supply of water.

Bacteria (plural; singular is bacterium)—any of numerous one-celled microorganisms of the class Schizomycetes, occurring in a wide variety of forms, existing as free-living organisms or as parasites.

Biodiversity—the diversity of living organisms.

Browser—a herbivorous animal that specializes in eating twigs, leaves, bark, and buds of woody plants.

Calcar—in bats, a process connected with the heel, which helps support the membrane between the leg and the tail.

Canine—of, or pertaining to, or designating the tooth next to the incisors in mammals; designed for grabbing and holding on.

Carnassial—pertaining to or designating certain teeth in mammals of the order Carnivora, which are usually adapted for cutting.

Carnivorous—eating flesh; preying or feeding on animals, as opposed to eating plants.

Carrying capacity—the maximum number of animals that a habitat can sustainably support without altering the integrity of the ecosystem.

Castorum—the waxy secretion of the anal glands of the beaver; once used extensively in the fixing of perfumes.

Cecal—of or pertaining to the cecum.

Cecal pellet—a pellet produced in the cecum that is expelled and reingested, as opposed to a fecal pellet that is defecated as waste material.

Cecum—a blind pouch or saclike extension of the digestive tract.

Chitin—a semitransparent horny substance forming the principal component of an insect's external skeleton.

Chitinous—of or pertaining to chitin.

Claw—a pointed, curved, horny structure at the end of a toe.

Color phase—the genetic circumstance in which members of the a given species can consistently exhibit more than one color, such as black bears, which have both totally black and totally brown individuals, often in the same families.

Competition—a relationship among members of the same or different species such that use of a particular available resource by one party reduces the amount of the resource that is available to the other party.

Conifer—a cone-bearing tree, such as fir, hemlock, spruce, and pine.

Coprophagy—the eating of feces.

Crepuscular—active during twilight.

Deciduous—falling off, shedding, or falling out at maturity, at certain seasons, or at certain ages; said of the leaves of certain trees, the antlers of deer, and the first set of teeth of most kinds of mammals.

Deciduous teeth—the first set of teeth (milk teeth), subsequently replaced by the permanent set of teeth.

Delayed implantation—a condition in some mammals, such as members of the weasel family, where a fertilized egg will develop to a certain point, then lie quiescent, perhaps for months, before it becomes implanted in the wall of the uterus and begins actively to grow.

Diurnal—active during the day.

Domestic mammal—a mammal that, through direct selection by humans, has certain inherent biological characteristics by which it differs from its wild ancestors.

Dorsal—pertaining to or situated on the back of a mammal, as opposed to the ventral.

Droppings—feces.

Echolocation—a guidance system in which an animal, such as a bat, emits sound waves, sometimes ultrasonic, which are then reflected back after hitting objects.

Ecology—the study of the relationships between living things and their environments.

Ecosystem—the complex of a community of organism and its environment functioning as an ecological unit in nature.

Ectomycorrhiza (singular; plural ecotmycorrhizae)—fungi that form a symbiotic relationship with the roots of plants where the fungus mantles the surface of the host rootlet with fungal tissue and grows between the cells of the rootlet.

Emigration—the act of leaving an area.

Estivation—a form of torpor, usually a response to high temperature and/or or scarcity of water.

Estrus—a period in the reproductive cycle of the female when the animal is most physiologically and psychologically receptive to mating and subsequent conception.

Evert—to turn inside out.

Exoskeleton—the external skeleton of chitin in such organisms as insects.

Feces—intestinal excrement.

Feral—a domestic animal that has reverted to the wild.

Fossorial—fitted for digging.

Fungus (singular; plural fungi)—mushrooms, truffles, molds, yeasts, etc.; simply organized plants, unicellular or made of cellular filaments or strands called hyphae; lacking chlorophyll; reproduce asexually and sexually with the formation of spores.

Gestation period—the time between fertilization of an egg and birth of the young.

Gland—an organ of secretion or excretion.

Grazer—an animal that feeds on grasses.

Guard hairs—the stiffer, longer hairs that grow up through the limber, shorter hairs (fur) of a mammal's pelage.

Habitat—a particular area in which one actually finds a given species, which includes food, cover, water, and space.

Herbivorous—feeding chiefly on grasses or other plant material.

Hibernacula (plural; singular hibernaculum)—a place where animals hibernate.

Hibernation—the torpidity of an animal, especially in winter when the temperature of the body approximates that of the surroundings and rate of both the respiration and heartbeat are much slower than in an active animal.

Hock—the joint on the hind leg of mammals, such as deer, elk, and horses that corresponds to the human ankle.

Home range—the entire area in which an individual moves about.

Hoof—the horny covering that protects the end of the digit and supports the weight of an animal.

Hyphae (plural; singular hypha)—the mold part of the fungus.

Hypogeous—hypo (below) + geous (ground) = belowground.

Immigration—the act of moving into a new area.

Implantation—the act of a fertilized egg attaching itself to the wall of the uterus.

Incisor—pertaining to or designating the tooth in front of and between the canine teeth or fangs.

Inner bark—part of the living tissue of a tree situated just under the nonliving outer bark.

Insectivorous—eating, preying, or feeding on insects.

Keel—In bats, refers to the free flap of skin attached posteriorly along the calcar (*see* calcar).

Lateral—of or pertaining to the side.

Lax—fur that is soft and loose, bordering on fluffy.

Lichen—actually two plants in one. A fungus forms the outside body and an alga that can make its own food grows inside the body. Thus while the fungus protects the alga, the alga feeds the fungus.

Litter—two or more babies brought forth at one birth by a female mammal.

Mamma (singular, plural is mammae)—the glandular organ for secreting milk, characteristic of all mammals.

Mammary gland—the milk-producing organ of mammals, inactive in males.

Mane—the long hair along the top and sides of the neck of such mammals as a horse or male lion; in this book pertaining to the long, coarse, black hair on the top of the tail of the gray fox or the long hair atop the neck of the weasel and elk.

Midden—a refuse pile.

Migration—a periodic movement away from and back to a given area.

Milk teeth—the baby or deciduous teeth

Molar—one of the back teeth, not preceded by deciduous teeth.

Molt—in a mammal, the process of shedding hair.

Mutation—a hereditary change, passed from parent to offspring.

Mycophagous—one who eats fungi.

Mycorrhiza—literally means "fungus-root" and denotes the symbiotic relationship between certain fungi and plant roots.

Nail—the horny plate on the upper surface of the end of fingers and toes of the human and some other animals; differs from claw and hoof only in shape and form.

Nape—the back part of the neck.

Nictitating membrane—refers to an inner eyelid that helps to keep an animal"s eyes clean.

Nitrogen fixation—the transformation of gaseous, atmospheric nitrogen into a form available and useable by plants.

Nitrogen-fixing bacteria—bacteria that can fix nitrogen (*see* nitrogen fixation).

Nocturnal—active at night.

Nose pad—the bare, thickened skin surrounding the nostrils.

Omnivorous—eating plant and animal material.

Ovulation—shedding of a ripe egg (or eggs) from ovary to fallopian tubes.

Pedicel—a small stalk, part, or organ, especially one serving as a support.

Pelage—the covering or coat of a mammal, as of hair, fur, or wool.

Placenta—organ of the developing embryo that provides for its nourishment and the elimination of waste products by interchange through the membranes of the uterus of the mother.

Plantigrade—walking on the sole of the foot with the heel touching the ground, as in humans, skunks, raccoon, and bears.

Polyestrus—having more than one heat or reproductively receptive period in females per year.

Polygamous—a male that breeds with more than one female.

Population—a group of individuals that are interfertile and that regularly contribute germ cells to the formation of fertile offspring.

Posterior—at or toward the rear of the body.

Precocial—in mammals, capable of moving about at birth.

Predator—an animal that kills and consumes animal prey.

Prehensile—adapted for seizing or grasping.

Premolar—designating or pertaining to one of the teeth in front of the true molars. There are a maximum of four on each side of the upper and lower jaws of placental mammals, 16 in all. When canine teeth are present, premolars are behind the canines but in front of the molars. They are preceded by deciduous teeth.

Rabies—a usually fatal viral disease affecting the nervous system.

Retractile—capable of being drawn back or in, as of the claws of cats.

Riparian—an area identified by the presence of vegetation that requires free or unbound water or conditions more moist than normally found in the area.

Ruminant—any of various hoofed mammals, such as cattle and deer, that characteristically have a four-chambered stomach and chew a cud.

Rut—the breeding season, usually in reference to hoofed mammals.

Saprophyte—a living plant that derives its nutrients from dead or decaying organic material.

Scat—a fecal dropping.

Sporocarp—the belowground fruiting body of mycorrhizal fungi.

Tactile—of, pertaining to, or relating to the sense of touch.

Talus—the accumulation of broken rock that occurs at the base of cliffs or other steep slopes.

Terrestrial—inhabiting the land, rather than water, trees, or air.

Territory—an area, generally surrounding the home, that is defended against other individuals of the same species.

Thallus—the nonreproductive part of a fungus.

Thorax—the chest in mammals; the second or middle region of the body of insects that bears the true legs and wings.

Tine—any of the branches that come off the main beam of an antler.

Torpid—devoid of the power of movement; sluggish in function or action.

Truffle—the belowground fruiting body of mycorrhizal fungi.

Underfur—the thick, soft fur lying beneath the longer, coarser hair of a mammal.

Vascular plant—any plant that contains vessels.

Venter—the underside, abdomen, or belly.

Ventral—of or pertaining to the underside, abdomen, or belly.

Vesicles (plural; singular vesicle)—a small bladder of sac, especially one containing fluid.

Vessel—one of the tubular conductive structures of woody plants that circulate fluids, consisting of cylindrical, often dead, cells that are attached end to end.

Vibrissa (plural, vibrissae)—one of the stiff tactile hairs on a the face of a mammals, whiskers.

Windfall—trees blown over by wind, often into piles of crisscrossed trunks.

Yeast—any of various one-celled fungi that reproduce by budding—dividing their bodies into new, one-celled individuals—and that are capable of fermenting carbohydrates.

Selected References

Albright, R. 1959. Bat banding at Oregon Caves. *Murrelet* 40:26-27.

Aldous, S.E., and J. Manweiler. 1942. The winter food habits of the short-tailed weasel in northern Minnesota. *Journal of Mammalogy* 23:250-55.

Aleksiuk, M. 1968. Scent-mound communication, territoriality, and population regulation in beaver (*Castor canadensis* Kuhl). *Journal of Mammalogy* 49:759-62.

Aleksiuk, M. 1968. The metabolic adaptation of the beaver (*Castor canadensis* Kuhl) to the Arctic energy regime. Ph. D. thesis. University of British Columbia, Vancouver, Canada. 123 pp.

Allen, D.L. 1938. Notes on the killing technique of the New York weasel. *Journal of Mammalogy* 19:225-29.

Allen, G.M. 1939. *Bats.* Harvard University Press, Cambridge, MA. 368 pp.

Altmann, M. 1952. Social behavior of elk, *Cervus canadensis nelsoni*, in the Jackson Hole area of Wyoming. *Behavior* 4:116-43.

Anderson, S., and J. N. Jones, Jr. 1967. *Recent Mammals of the World, a Synopsis of Families.* Ronald Press Co., New York 453 pp.

Anthony, H.E. 1916. Habits of Aplodontia. American Museum of Natural History *Bulletin* 35:53-63.

Arthur, S.M., and W.B. Krohn. 1991. Activity patterns, movements, and reproductive ecology of fishers in southcentral Maine. *Journal of Mammalogy* 72:379-85.

Asdell, S.A. 1964. *Patterns of mammalian reproduction.* Cornell University Press, Ithaca, NY. 670 pp.

Atwood, E.L. 1950. Life history studies of nutria or coypu, in coastal Louisiana. *Journal of Wildlife Management* 14:249-65.

Bailey, V. 1923. The combing claws of the beaver. *Journal of Mammalogy* 4:77-79.

Bailey, V. 1936. The Mammals and Life Zones of Oregon. *North American Fauna* 55:1-416.

Barbour, R.W., and W.H. Davis. 1969. *Bats of America.* University of Kentucky Press, Lexington, KY. 286 pp.

Basey, J.M., and S.H. Jenkings. 1995. Influences of predation risk and energy maximization of food selection by beavers (*Castor canadensis*). *Canadian Journal of Zoology* 73:2197-208.

Ben-David, M. 1997. Timing of reproduction in wild mink: the influence of spawning Pacific salmon. *Canadian Journal of Zoology* 75:376-82.

Bird, E.A.R. 1987. The social construction of nature: Theoretical approaches to the history of environmental problems. *Environmental Review* 11:255-64.

Bogan, M.A. 1972. Observations on parturition and development in the hoary bat, *Lasiurus cinereus. Journal of Mammalogy* 53:611-14.

Brassard, J.A., and R. Bernard. 1939. Observations on breeding and development of marten, *Martes a. americana* (Kerr). *Canadian Field-Naturalist* 53:15-21.

Brigham, R.M., H.D.J.N. Aldridge, and R.L. Mackey. 1992. Variation in habitat use and prey selection by Yuma bats, *Myotis yumanensis. Journal of Mammalogy* 73:640-45.

Chapman, J.A. 1971. Orientation and homing of the brush rabbit (*Sylvilagus bachmani*). *Journal of Mammalogy* 52:686-99.

Clark, F.W. 1972. Influence of jackrabbit density on coyote population change. *Journal of Wildlife Management* 36:343-56.

Clark, W.R. 1994. Habitat selection by muskrats in experimental marshes undergoing succession. *Canadian Journal of Zoology* 72:675-80.

Clothier, R.R. 1955. Contribution to the life history of *Sorex vagrans* in Montana. *Journal of Mammalogy* 36:214-21.

Constantine, D.G. 1958. Ecological observations on lasiurine bats in Georgia. *Journal of Mammalogy.*39:64-70.

Constantine, D.G. 1959. Ecological observations on lasiurine bats in the North Bay area of California. *Journal of Mammalogy* 40:13-15.

Constantine, D.G. 1966. Ecological observations on lasiurine bats in Iowa. *Journal of Mammalogy* 47:34-41.

Coulter, M.W. 1966. Ecology and management of fishers in Maine. Ph. D. thesis. Syracuse University, Syracuse, NY. 183 pp.

Cowan, I. McT., and C.J. Guiguet. 1965. The mammals of British Columbia. *British Columbia Provincial Museum* 11:1-141.

Crabb, W.D. 1948. The ecology and management of the prairie spotted skunk in Iowa. *Ecological Monographs* 18:203-32.

Dalquest, W.W. 1942. Geographic variation in northwestern snowshoe hares. *Journal of Mammalogy* 23:166-83.

Dalquest, W.W. 1943. Seasonal distribution of the hoary bat along the Pacific coast. *Murrelet* 24:20-24.

Dalquest, W.W. 1947. Notes on the natural history of the bat, *Myotis yumanensis*, in California, with a description of a new race. *American Midland Naturalist* 38:224-47.

Dalquest, W.W. 1947. Notes on the natural history of the bat *Corynorhinus rafinesquii* in California. *Journal of Mammalogy* 28:17-30.

Dalquest, W.W. 1948. *Mammals of Washington*. University of Kansas Museum of Natural History Publication 2:1-4444.

Dalquest, W.W., and D.R. Orcutt. 1942. The biology of the least shrew-mole, *Neurotrichus gibbsii minor*. *American Midland Naturalist* 27:387-401.

Dasmann, R.F., and R.D. Taber. 1956. Behavior of Columbian black-tailed deer with reference to population ecology. *Journal of Mammalogy* 37:143-64.

Davis, W.B. 1966. The mammals of Texas. Texas Parks and Wildlife Department *Bulletin* 41:5-267.

Davis, W.B., C.F. Herreid II, and H.L. Short. 1962. Mexican free-tailed bats in Texas. *Ecological Monographs* 32:311-46.

Davis, W.H., R.W. Barbour, and M.D. Hassell. 1968. Colonial behavior of *Eptesicus fuscus. Journal of Mammalogy* 49:44-50.

deVos, A. 1952. Ecology and management of fisher and marten in Ontario. Ontario Department of Lands and Forest *Technical Bulletin* 9. 90 pp.

Dice, L.R. 1932. The songs of mice. *Journal of Mammalogy* 13:187-96.

Dice, L.R. 1949. Variations of *Peromyscus maniculatus* in parts of western Washington and adjacent Oregon. *Contribution* of the Laboratory of Vertebrate Biology University of Michigan 44:1-33.

Dixon, J. 1919. Notes on the natural history of the bushy-tailed woodrats of California. University of California *Publication in Zoology* 21:49-74.

Dixon, J. 1934. A study of the life history and food habits of mule deer in California. *California Fish and Game* 20:6-144.

Dixon, J.S. 1933. Red fox attacked by a golden eagle. *Journal of Mammalogy* 14:257.

Dorney, R.S. 1954. Ecology of marsh raccoons. *Journal of Wildlife Management* 18:217-25.

Eadie, W.R. 1954. Skin gland activity and pelage descriptions in moles. *Journal of Mammalogy* 35:186-96.

Eadie, W.R., and W.J. Hamilton, Jr. 1958. Reproduction in the fisher in New York. *New York Fish and Game Journal* 5:77-83.

Edson, J.M., 1933. A visitation of weasels. *Murrelet* 14:76-77.

Eisenberg, J.F. 1964. Studies on the behavior of *Sorex vagrans. American Midland Naturalist* 72:417-25.

Elton, C. 1942. *Voles, Mice, and Lemmings: Problems in Population Dynamics.* Oxford University Press, London. 496 pp.

Emmons, L H. 1989. Tropical rain forests: why they have so many species, and how we may lose this biodiversity without cutting a single tree. *Orion* 8:8-14.

Ender, R.K. 1952. Reproduction in the mink (*Mustela vison*). *Proceedings* of the American Philosphical Society 96:691-755.

English, P.F. 1923. The dusky-footed woodrat (*Neotoma fuscipes*). *Journal of Mammalogy* 4:1-9.

Erickson, A.W., and J.E. Nellor. 1964. Breeding biology of the black bear. Michigan State University *Research Bulletin* 4:1-45.

Erickson, A.W., J.E. Nellor, and G.A. Petrides. 1964. The black bear in Michigan. Michigan State University *Research Bulletin* 4:1-102.

Errington, P.L. 1943. An analysis of mink predation upon muskrats in north-central United States. Iowa Agricultural Experiment Station *Research Bulletin* 320:798-924.

Evans, J. 1970. About nutria and their control. U.S. Department of the Interior, Bureau of Sport Fisheries and Wildlife *Resource Publication* 86. 65 pp.

Ferrel, C.M., H.R. Leach, and D.F. Tillotson. 1953. Food habits of the coyote in California. *California Fish and Game* 39:301-41.

Field, R.J. 1970. Winter habits of the river otter (*Lutra canadensis*) in Michigan. *Michigan Academy* 3:49-58.

Findley, J.S., and C. Jones. 1964. Seasonal distribution of the hoary bat. *Journal of Mammalogy* 45:461-70.

Findley, J.S., E.H. Studier, and D.E. Wilson. 1972. Morphologic properties of bat wings. *Journal of Mammalogy* 53:429-44.

Fitch, H.S. 1948. Ecology of the California ground squirrel on grazing lands. *American Midland Naturalist* 39:513-96.

Forbes, R.B., and L.W. Turner. 1972. Notes on two litters of Townsend's chipmunks. *Journal of Mammalogy* 53:355-59.

Franklin, W.L. 1968. Herd organization, territoriality, movements and home ranges in Roosevelt elk. M.S. Thesis. Humboldt State College, Arcata, CA. 89 pp.

Franklin, W.L., A.S. Mossman, and M. Dole. 1975. Social organization and home range of Roosevelt elk. *Journal of Mammalogy* 56:102-18.

Gabrielson, I.N. 1923. Notes on Thomomys in Oregon. *Journal of Mammalogy* 4:189-90.

Gabrielson, I.N. 1928. Notes on the habits and behavior of the porcupine in Oregon. *Journal of Mammalogy* 9:33-38.

Gashwiler, J.S. 1971. Deer mouse movement in forest habitat. *Northwest Science* 45:163-70.

Gashwiler, J.S. 1972. Life history notes on the Oregon vole, Micro*tus oregoni*. *Journal of Mammalogy* 53:558-69.

Gashwiler, J.S., W.L. Robinette, and O.W. Morris. 1961. Breeding habits of bobcats in Utah. *Journal of Mammalogy* 42:76-84.

Giger, R.D. 1973. Movements and homing in Townsend's mole near Tillamook, Oregon. *Journal of Mammalogy* 54:648-59.

Giles, L.W. 1939. Fall food habits of the raccoon in central Iowa. *Journal of Mammalogy* 20:68-70.

Gill, D. 1972. The evolution of a discrete beaver habitat in the Mackenzie River Delta, Northwest Territories. *Canadian Field-Naturalist* 86:233-39.

Gill, D., and L.D. Cordes. 1972. Winter habitat preference of porcupines in the southern Alberta foothills. *Canadian Field-Naturalist* 86:349-55.

Gillesberg, A.M., and A.B. Carey. 1991. Arboreal nests of *Phenacomys longicaudus* in Oregon. *Journal of Mammalogy* 72:784-87.

Goehring, H.H. 1972. Twenty-year study of Epte*sicus fuscus* in Minnesota. *Journal of Mammalogy* 53:201-7.

Goertz, J.W. 1964. Habitats of three Oregon voles. *Ecology* 45:846-48.

Gordon, K. 1943. The natural history and behavior of the western chipmunk and mantled ground squirrel. Oregon State Monographs, *Studies in Zoology*, No. 5, Oregon State College Press.

Goss, R.J. 1963. The deciduous nature of deer antlers. pp. 339-369. In: *Mechanisms of Hard Tissue Destruction*. R. Sognnaes, ed. American Academy for the Advancement of Science Publication 75, Washington, D.C.

Goss, R.J. 1968. Inhibition of growth and shedding of antlers by sex hormones. *Nature* 220:83-95.

Goss, R.J. 1969. Photoperiodic control of antler cycles in deer. I. Phase shift and frequency changes. *Journal of Experimental Zoology* 170:311-24.

Goss, R.J. 1970. Photoperiodic control of antler cycles in deer. II. Alterations in amplitude. *Journal of Experimental Zoology* 171:223-34.

Gould, E. 1955. The feeding efficiency of insectivorous bats. *Journal of Mammalogy* 36:399-407.

Graf, W. 1955. The Roosevelt elk. Port Angeles *Evening News*, Port Angeles, Washington. 105 pp.

Graf, W. 1956. Territorialism in deer. *Journal of Mammalogy* 37:165-70.

Graham, R.E. 1966. Observations on the roosting habits of the big- eared bat, Pleco*tus townsendii*, in California limestone caves. *Cave Notes* 8:17-22.

Grange, W.B. 1932. Observations on the snowshoe hare, *Lepus americanus phaeonotus* Allen. *Journal of Mammalogy* 13:1-19.

Grinnell, J., J.S. Dixon, and J.M. Linsdale. 1937. *Fur-bearing Mammals of California: Their Natural History, Systematic Status, and Relations to Man.* University of California Press, Berkeley. 375 pp.

Hafner, D.J., and R.M. Sullivan. 1995. Historical and ecological biogeography of Nearctic pikas. *Journal of Mammalogy* 76:302-15.

Hagmeier, E.M. 1956. Distribution of marten and fisher in North America. *Canadian Field-Naturalist* 70:149-68.

Hall, D.S. 1991. Diet of the northern flying squirrel at Sagehen Creek, California. *Journal of Mammalogy* 72:615-17.

Hall, E.R. 1951. American weasels. University of Kansas Museum of Natural History *Publication* 4:1-446.

Hamilton, W.J., Jr. 1931. Skunks as grasshopper destroyers. *Journal of Economic Entomology* 24:918.

Hamilton, W.J., Jr. 1933. The weasels of New York. *American Midland Naturalist* 14:289-344.

Hamilton, W.J., Jr. 1936. The food and breeding habits of the raccoon. *Ohio Journal of Science* 36:131-40.

Hamilton, W.J., Jr. 1937. Winter activity of the skunk. *Ecology* 18:326-27.

Hamilton, W.J., Jr. 1940. The summer food of mink and raccoons on the Montezuma Marsh, New York. *Journal of Wildlife Management* 4:80-84.

Hamilton, W.J., Jr. 1961. Late fall, winter, and early spring foods of 141 otters from New York. *New York Fish and Game Journal* 8:106-9.

Hamilton, W.J., Jr., and W.R. Eadie 1964. Reproduction in the otter, Lutra can*adensis*. *Journal of Mammalogy* 45:242-52.

Hamilton, W.J., III. 1962. Reproductive adaptations of the red tree mouse. *Journal of Mammalogy* 43:486-504.

Hammer, E.W., and C. Maser 1973. Distribution of the dusky-footed woodrat, *Neotoma fuscipes* Baird, in Klamath and Lake Counties, Oregon. *Northwest Science* 47:123-27.

Hansen, E.L. 1965. Muskrat distribution in southcentral Oregon. *Journal of Mammalogy* 46:669-71.

Harper, J.A. 1964. Movement and associated behavior of Roosevelt elk in southwestern Oregon. *Proceedings of the Western Association of State Game and Fish Commissions* 44:139-41.

Harper, J.A. 1971. *Ecology of Roosevelt Elk.* Oregon State Game Commission, Portland, OR. 44 pp.

Harper, J.A., J.H. Harn, W.W. Bentley, and C.F. Yocom. 1967. The status and ecology of the Roosevelt elk in California. *Wildlife Monographs* 16:1-49.

Hawley, V.D. 1955. The ecology of the marten in Glacier National Park. M.S. Thesis. Montana State University, Missoula. 131 pp.

Hawley, V.D. and F.E. Newby. 1957. Marten home ranges and population fluctuations. *Journal of Mammalogy* 38:174-84.

Hayes, J.P., E.G. Horvath, and P. Hounihan. 1995. Townsend's chipmunk populations in Douglas-fir plantations and mature forests in the Oregon Coast Range. *Canadian Journal of Zoology* 73:67-73.

Hayward, G.D., and R. Rosentreter. 1994. Lichens as nesting material for northern flying squirrels in the northern Rocky Mountains. *Journal of Mammalogy* 75:663-73.

Hibben, R.C. 1937. A preliminary study of the mountain lion, Fel*is oregonensis.* University of New Mexico *Biological Series* 5:1-59.

Hodgson, R.G. 1937. *Fisher Farming.* Fur Trade Journal of Canada, Toronto. 104 pp.

Hollister, N. 1911. A systematic synopsis of the muskrats. *North American Fauna* 32:1-47.

Hollister, N. 1913. A synopsis of the American minks. *Proceedings* of the United States National Museum 44:471-80.

Hooven, E.F., R.F. Hoyer, and R.M. Storm. 1975. Notes on the vagrant shrew, S*orex vagrans*, in the Willamette Valley of western Oregon. *Northwest Science* 49:163-73.

Horner, M.A., and R.A. Powell. 1990. Internal structure of home ranges of black bears and analyses of home-range overlap. *Journal of Mammalogy* 71:402-10.

Hornocker, M.G. 1969. Winter territoriality in mountain lions. *Journal of Wildlife Management* 33:457-64.

Hornocker, M.G. 1970. An analysis of mountain lion predation upon mule deer and elk in the Idaho Primitive Area. *Wildlife Monograph* 21, 39 pp.

Horvath, O. 1966. Observation of parturition and maternal care of the bushy-tailed woodrat (N*eotoma cinerea occidentalis* Baird). *Murrelet* 47:6-8.

Howell, A.B. 1919. *Microtus townsendi* in the Cascade Mountains of Oregon. *Journal of Mammalogy* 1:141-42.

Howell, A.B. 1926. Voles of the genus *Phenacomys. North American Fauna* 48:1-66.

Howell, A.B. 1928. The food and habitat preferences of *Phenacomys albipes. Journal of Mammalogy* 9:153-54.

Ingles, L.G. 1961. Home range and habitats of the wandering shrew. *Journal of Mammalogy* 42:455-62.

Ingles, L.G. 1965. *Mammals of the Pacific States.* Stanford University Press, Stanford, CA. 506 pp.

Jackson, H.H.T. 1961. *Mammals of Wisconsin.* University Wisconsin Press, Madison. 504 pp.

Jiang, Z., and R.J. Hudson. 1992. Estimating forage intake and energy requirements of free-ranging wapiti (*Cervus elaphus*). *Canadian Journal of Zoology* 70:675-79.

Johnson, C.E. 1921. The "hand-stand" habit of the spotted skunk. *Journal of Mammalogy* 2:87-89.

Johnson, M.L. 1973. Characters of the heather vole, Phena*comys*, and the red tree vole, *Arborimus. Journal of Mammalogy* 54:239-44.

Johnson, M.L., and C. Maser. 1982. Generic Relationships of *Phenacomys albipes. Northwest Science* 56:17-19.

Jones, G. J.O. Whitaker, Jr., and Chris Maser. 1978. Food Habits of Jumping Mice (*Zapus trinotatus* and *Z. Princeps*) in Western North America. *Northwest Science* 52:57-60.

Jonkel, C.J. 1959. Ecological and physiological study of the pine marten. M.S Thesis. Montana State University, Bozeman. 81 pp.

Jonkel, C.J., and I. McT. Cowan. 1971. The black bear in the spruce- fir forest. *Wildlife Monographs* 27:1-57.

Jonkel, C.J., and R.P. Weckwerth. 1963. Sexual maturity and implantation of blastocysts in the wild pine marten. *Journal of Wildlife Management* 27:93-98.

Kalcounis, M.C., and R.M. Brigham. 1995. Intraspecific variation in wing loading affects habitat use by little brown bats (*Myotis lucifugus*). *Canadian Journal of Zoology* 73:89-95.

Kebbe, C.E. 1959. The nutria in Oregon. Oregon State Game Commission *Bulletin* 14:8.

Kebbe, C.E. 1961. Return of the fisher. Oregon State Game Commission *Bulletin* 16:2, 6-7.

Kilpatrick, H.J., and P.W. Rego. 1994. Influence of season, sex, and site availability on fisher (*Martes pennanti*) rest-site selection in the central hardwoods forest. *Canadian Journal of Zoology* 72:1416-19.

King, J., Ed. 1968. Biology of *Peromyscus* (Rodentia). The American Society of Mammalogists *Special Publication* 2. 593 pp.

Koehler, G.M., and M.G. Hornocker. 1991. Seasonal resource use among mountain lions, bobcats, and coyotes. *Journal of Mammalogy* 72:391-96.

Krutzsch, P.H. 1954. Notes on the habits of the bat *Myotis californicus. Journal of Mammalogy* 35:539-45.

Krutzsch, P.H. 1954. North American jumping mice (genus *Zapus*). University of Kansas Museum of Natural History *Publication* 7:349-72.

Krutzsch, P.H. 1955. Observations on the Mexican free-tailed bat, T*adarida mexicana*. *Journal of Mammalogy* 36:236-42.

Kuhn, L.W., W.Q. Wick, and R.J. Pedersen. 1966. Breeding nests of Townsend's mole in Oregon. *Journal of Mammalogy* 47:239-49.

Larrison, E.J. 1943. Feral coypus in the Pacific Northwest. *Murrelet* 24:3-9.

Layne, J.N., and W.H. McKeon. 1956. Some aspects of red fox and gray fox reproduction in New York. *New York Fish and Game Journal* 3:44-74.

Leege, T.A. 1968. Natural movements of beavers in southeastern Idaho. *Journal of Wildlife Management* 32:973-76.

Li, C.Y., C. Maser, Z. Maser, and B.A. Caldwell. 1986. Role of three rodents in forest nitrogen fixation in western Oregon: Another aspect of mammal-mycorrhizal fungus-tree mutualism. *Great Basin Naturalist* 46:411-14.

Liers, E.E. 1951. Notes on the river otter (Lutra *canadensis*). *Journal of Mammalogy* 32:1-9.

Linsdale, J.M. 1946. *The California Ground Squirrel*. University of California Press, Berkeley and Los Angeles, CA. 475 pp.

Linsdale, J.M., and L.P. Tevis. 1951. *The Dusky-footed Woodrat*. University of California Press, Berkeley, CA. 664 pp.

Linsdale, J.M., and Q.P. Tomich. 1953. *A Herd of Mule Deer*. University of California Press, Berkeley, CA. 567 pp.

Livezey, R., and F. Evenden, Jr. 1943. Notes on the western red fox. *Journal of Mammalogy* 24:500-501.

McGrew, P.O. 1941. The Aplodontoidea. Field Museum of Natural History *Geological Series* 9:1-30.

Macnab, J.A., and J.C. Dirks. 1941. The California red-backed mouse in the Oregon Coast Range. *Journal of Mammalogy* 22:174-80.

Marinelli, J., and F. Messier. 1995. Parental-care strategies among muskrats in a female-biased population. *Canadian Journal of Zoology* 73:1503-10.

Marshall, A.D., and J.H Jenkins. 1966. Movements and home ranges of bobcats as determined by radio-tracking in the upper coastal plain of west-central South Carolina. *Proceedings* of the Conference of the Southeastern Association of Game and Fish Commissions 20:206-14.

Marshall, W.H. 1936. A study of the winter activities of the mink. *Journal of Mammalogy* 17:382-92.

Marshall, W.H. 1946. Winter food habits of pine marten in Montana. *Journal of Mammalogy* 27:83-84.

Martin, P. 1971. Movements and activities of the mountain beaver (Ap*lodontia rufa*). *Journal of Mammalogy* 52:717-23.

Maser, C. 1966. Life histories and ecology of *Phenacomys albipes*, *Phenacomys longicaudus*, *Phenacomys silvicola*. M.S. Thesis. Oregon State University, Corvallis, OR. 221 pp.

Maser, C. 1966. Notes on a captive *Sorex vagrans*. *Murrelet* 47:51-53.

Maser, C. 1967. Black bear damage to Douglas-fir in Oregon. *Murrelet* 48:34-38.

Maser, C., and E.F. Hooven. 1974. Notes on the Behavior and Food Habits of Captive Pacific Shrews, *Sorex pacificus pacificus*. *Northwest Science* 48:81-95.

Maser, C., and J.F. Franklin. 1974. Checklist of vertebrate animals of the Cascade Head Experimental Forest. USDA Forest Service *Resource Bulletin* PNW-51, Pacific Northwest Forest and Range Experiment Station, Portland, OR. 32 pp.

Maser, C., and M.L. Johnson. 1967. Notes on the white-footed vole (*Phenacomys albipes*). *Murrelet* 48:24-27.

Maser, C., and R.M. Storm. 1970. *A Key to Microtinae of the Pacific Northwest* (Oregon, Washington, Idaho). Oregon State University Book Stores, Inc., Corvallis, OR. 162 pp.

Maser, C., J.M. Trappe, and R.A. Nussbaum. 1978. Fungal-small mammal interrelationships with emphasis on Oregon coniferous forests. *Ecology* 59:799-809.

Maser, C., B.R. Mate, J.F. Franklin, and C.T. Dyrness. 1981. Natural History of Oregon Coast Mammals. USDA Forest Service *General Technical Report* PNW-133. Pacific Northwest Forest and Range Experiment Station, Portland, OR. 496 pp.

Maser, C., and R.S. Rohweder. 1983. Winter food habits of cougars from northeastern Oregon. *Great Basin Naturalist* 43:425-27.

Maser, C., and J.M. Trappe. Technical Editors. 1984. The seen and unseen world of the fallen tree. USDA Forest Service, *General Technical Report* PNW-164. Pacific Northwest Forest and Range Experiment Station, Portland, Oregon, in cooperation with the U.S. Department of the Interior, Bureau of Land Management. 56 pp.

Maser, C., and Z. Maser 1987. Notes on mycophagy in four species of mice in the genus *Peromyscus*. *Great Basin Naturalist* 47:308-12.

Maser, C., and Z. Maser. 1988. Mycophagy of red-backed voles, *Clethrionomys californicus* and *C. gapperi*. *Great Basin Naturalist* 48:269-73.

Maser, C., and Z. Maser. 1988. Interactions among squirrels, mycorrhizal fungi, and coniferous forests in Oregon. *Great Basin Naturalist* 48:358-69.

Maser, C., Z. Maser, J.W. Witt, and G. Hunt. 1986. The northern flying squirrel: a mycophagist in southwestern Oregon. *Canadian Journal of Zoology* 64:2086-89.

Maser, Z., and C. Maser. 1987. Notes on mycophagy of the yellow- pine chipmunk (Eu*tamias amoenus*) in northeastern Oregon. *Murrelet* 68:24-27.

Mead, R.A. 1968. Reproduction in western forms of the spotted skunk (genus *Spilogale*). *Journal of Mammalogy* 49:373-90.

Meredith, D.H. 1972. Subalpine cover associations of E*utamias amoenus* and *Eutamias townsendii* in the Washington Cascades. *American Midland Naturalist* 88:348-57.

Messier, F., and J.A. Virgl. 1992. Differential use of bank burrows and lodges by muskrats, *Ondatra zibethicus*, in a northern marsh environment. *Canadian Journal of Zoology* 70:1180-84.

Millar, J.S., E.M. Derrickson, and S.T.P. Sharpe. 1992. Effects of reproduction on maternal survival and subsequent reproduction in northern *Peromyscus maniculatus*. *Canadian Journal of Zoology* 70:1129-34.

Miller, F.L. 1965. Behavior associated with parturition in black-tailed deer. *Journal of Wildlife Management* 29:629-31.

Miller, F.L. 1970. Distribution patterns of black-tailed deer (Odo*coileus hemionus columbianus*) in relation to environment. *Journal of Mammalogy* 51:248-60.

Miller, F.L. 1971. Mutual grooming by black-tailed deer in northwestern Oregon. *Canadian Field-Naturalist* 85:295-301.

Miller, F.W. 1930. The spring moult of *Mustela longicauda*. *Journal of Mammalogy* 11:471-73.

Miller, F.W. 1931. A feeding habit of the long-tailed weasel. *Journal of Mammalogy* 12:164.

Miller, F.W. 1931. The fall moult of M*ustela longicauda. Journal of Mammalogy* 12:150-52.

Miller, R.C., R.W. Ritcey, and R.Y. Edwards. 1955. Live-trapping marten in British Columbia. *Murrelet* 36:1-8.

Montgomery, G.G. 1968. Pelage development of young raccoons. *Journal of Mammalogy* 49:142-45.

Montgomery, G.G. 1969. Weaning of captive raccoons. *Journal of Wildlife Management* 33:154-59.

Montgomery, G.G. 1969. Tooth emergence and raccoon weaning. *American Midland Naturalist* 82:285-87.

Moore, J.C. 1945. Life history notes on the Florida weasel. *Proceedings* of the Florida Academy of Sciences 7:247-63.

Mossman, A.S. 1955. Reproduction of the brush rabbit in California. *Journal of Wildlife Management* 19:177-84.

Müller-Schwarze, D. 1969. Pheromone function of deer urine. *American Zoologist* 9:21.

Müller-Schwarze, D. 1972. Social significance of forehead rubbing in black-tailed deer (*Odocoileus hemionus columbianus*). *Animal Behaviour* 20:788-97.

Murie, A. 1940. Ecology of the coyote in the Yellowstone. *Fauna Series* 4:1-206.

Muul, I. 1969. Mating behavior, gestation period, and development of *Glaucomys sabrinus. Journal of Mammalogy* 50:121.

Novakowski, N.S. 1969. The influence of vocalization on the behavior of beaver, *Castor canadensis* Kuhl. *American Midland Naturalist* 81:198-204.

Nussbaum, R.A., and C. Maser. 1975. Food habits of the Bobcat, *Lynx rufus*, in the Coast and Cascade ranges of Western Oregon in relation to present management policies. *Northwest Science* 49:261-66.

Orr, R.T. 1940. The rabbits of California. California Academy of Sciences *Occasional Papers* 19:1-227.

Orr, R.T. 1942. Observations of the growth of young brush rabbits. *Journal of Mammalogy* 23:299-302.

Orr, R.T. 1954. Natural history of the pallid bat, A*ntrozous pallidus* (LeConte). *Proceedings* of the California Academy of Sciences 28:165-246.

Pattie, D.L. 1969. Behavior of captive marsh shrews (*Sorex bendirii*), *Murrelet* 50:27-32.

Pearl, D.C., and M.B. Fenton. 1996. Can echolocation calls provide information about group identity in the little brown bat (*Myotis lucifugus*)? *Canadian Journal of Zoology* 74:2248-53.

Pearson, O.P., and R.K. Enders. 1943. Ovulation, maturation and fertilization in the fox. *Anatomical Record* 85:69-83.

Pearson, O.P., M.R. Koford, and A.K. Pearson. 1952. Reproduction of the lump-nosed bat (*Corynorhinus rafinesquii*) in California. *Journal of Mammalogy* 33:273-320.

Pimentel, R.A. 1949. Black rat and roof rat taken in the central Oregon coast strip. *Murrelet* 30:52.

Po-Chedley, D.S., and A.R. Shadle. 1955. Pelage of the porcupine, *Erethizon dorsatum dorsatum. Journal of Mammalogy* 36:84-95.

Poelker, R.J., and H.D. Hartwell. 1973. Black bear of Washington. Washington State Game Department *Biological Bulletin* 14:1-180.

Polderboer, E.B., L.W. Kuhn, and G.O. Hendrickson. 1941. Winter and spring habits of weasels in central Iowa. *Journal of Wildlife Management* 5:115-19.

Pool, K.G., and R.P Graf. 1996. Winter diet of marten during a snowshoe hare decline. *Canadian Journal of Zoology* 74:456-66.

Preble, E.A. 1899. Revision of the Jumping mice of the genus Za*pus*. *North American Fauna* 15:1-43.

Quick, H.F. 1944. Habits and economics of the New York weasel in Michigan. *Journal of Wildlife Management* 8:71-78.

Quick, H.F. 1953. Wolverine, fisher, and marten studies in a wilderness region. *Transactions* of the North American Wildlife Conference 18:513-32.

Quick, H.F. 1953. Occurrence of porcupine quills in carnivorous mammals. *Journal of Mammalogy* 34:256-59.

Quick, H.F. 1955. Food habits of marten in northern British Columbia. *Canadian Field-Naturalist* 69:144-47.

Reid, E.D., and R.J. Brooks. 1994. Effect of water retention time and food consumption in deer mice. *Canadian Journal of Zoology* 72:1711-14.

Remington, J.D. 1952. Food habits, growth, and behavior of two captive pine martens. *Journal of Mammalogy* 33:66-70.

Richards, S.H., and R.L. Hine. 1953. Wisconsin fox populations. Wisconsin Conservation Department *Technical Wildlife Bulletin* 6:1-78.

Richardson, W.B. 1942. Ring-tailed cats (*Bassariscus astutus*): Their growth and development. *Journal of Mammalogy* 23:17-26.

Roest, A.I. 1951. Mammals of the Oregon Caves area, Josephine County. *Journal of Mammalogy* 32:345-51.

Rogers, L.L. 1974. Shedding of foot pads by black bears during denning. *Journal of Mammalogy* 55:672-74.

Rolseth, S.L., C.E. Koehler, and R.M.R. Barclay. 1994. Differences in the diets of juvenile and adult hoary bats, *Lasiurus cinereus*. *Journal of Mammalogy* 75:394-98.

Salsbury, C.M., and K.A. Armitage. 1995. Reproductive energetics of adult male yellow-bellied marmots (*Marmota flaviventris*). *Canadian Journal of Zoology* 73:1791-97.

Scheffer, V.B. 1938. Notes on the wolverine and fisher in the State of Washington. *Murrelet* 19:8-10.

Scheffer, V.B. 1950. Notes on the raccoon in southwestern Washington. *Journal of Mammalogy* 31:444-48.

Scheibe, J.S., and M.J. O'Farrell. 1995. Habitat dynamics in *Peromyscus truei*: eclectic females, density dependence, or reproductive constraints? *Journal of Mammalogy* 76:368-75.

Seidensticker, J.C., IV, M.G. Hornocker, W.V. Wiles, and J.P. Messick. 1973. Mountain lion social organization in the Idaho Primitive Area. *Wildlife Monograph* 35, 60 pp.

Seton, E.T. 1928. *Lives of Game Animals.* 4 vols. Doubleday, Doran and Company, Inc., Garden City, NY.

Seton, E.T. 1932. The song of the porcupine (E*rethizon epixanthum*). *Journal of Mammalogy* 13:168-69.

Shadle, A.R. 1944. The play of American porcupines (E*rethizon d. dorsatum* and *E. epixanthum*). *Journal of Comparative Psychology* 37:145-50.

Shadle, A.R. 1946. Copulation in the porcupine. *Journal of Wildlife Management* 10:159-62.

Shadle, A.R. 1950. Feeding, care, and handling of captive porcupines (E*rethizon*). *Journal of Mammalogy* 31:411-16.

Shadle, A.R. 1951. Laboratory copulations and gestations of porcupine, *Erethizon dorsatum*. *Journal of Mammalogy* 32:219-21.

Shadle, A.R., and W.R. Ploss. 1943. An unusual porcupine parturition and development of the young. *Journal of Mammalogy* 24: 492-96.

Shadle, A.R., and D. Po-Chedley. 1949. Rate of penetration of porcupine spine. *Journal of Mammalogy* 30: 172-73.

Shaw, W.T. 1921. The nest of the Washington weasel (Mu*stela washingtoni*). *Journal of Mammalogy* 2:167-68.

Skupski, M.P. 1995. Population ecology of the western harvest mouse, *Reithrodontomys megalotis*: a long-term perspective. *Journal of Mammalogy* 76:358-67.

Smith, C.C. 1970. The coevolution of pine squirrels (*Tamiasciurus*) and conifers. *Ecological Monographs* 40:349-71.

Sperry, C.C. 1941. Food habits of the coyote. U.S. Department of the Interior, *Wildlife Research Bulletin* 4:1-70.

Storer, T.I. 1930. Summer and autumn breeding of the California ground squirrel. *Journal of Mammalogy* 11:235-37.

Svihla, A., and R.D. Svihla. 1933. Notes on the jumping mouse, Za*pus trinotatus trinotatus* Rhoads. *Journal of Mammalogy* 14:131-34.

Sweitzer, R.A., and J. Berger. 1993. Seasonal dynamics of mass and body condition in Great Basin porcupines (*Erethizon dorsatum*). *Journal of Mammalogy* 74:198-203.

Tallmon, D., and L.S. Mills. 1994. Use of logs within home ranges of California red-backed voles on a remnant forest. *Journal of Mammalogy* 75:97-101.

Teferi, T., and J.S. Millar. 1993. Early maturation by northern *Peromyscus maniculatus*. *Canadian Journal of Zoology* 71:1743-47.

Toweill, D.E. 1974. Winter food habits of river otters in western Oregon. *Journal of Wildlife Management* 38:107-11.

Toweill, D.E., and C. Maser. 1985. Food of cougars in the Cascade Range of Oregon. *Great Basin Naturalist* 45:77-80.

Toweill, D.E., C. Maser, L.D. Bryant, and M.L. Johnson. 1988. Reproductive characteristics of Eastern Oregon cougars. *Northwest Science* 62:147-49.

Trainer, C.E. 1971. The relationship of physical condition and fertility of female Roosevelt elk (*Cervus canadensis roosevelti*) in Oregon. M.S. Thesis. Oregon State University, Corvallis, OR. 93 pp.

Trapp, G.R. 1972. Some anatomical and behavioral adaptations of ringtails, *Bassariscus astutus*. *Journal of Mammalogy* 53:549-57.

Tyson, E.L. 1950. Summer food habits of the raccoon in southwestern Washington. *Journal of Mammalogy* 31:448-49.

Van Zyll De Jong, C.G. 1972. A systematic review of the Nearctic and Neotropical river otters (Genus *Lutra*, Mustelidae, Carnivora). Royal Ontario Museum of Life Sciences *Contribution* 80:1-104.

Vaughan, T., and P. Krutzsch. 1954. Seasonal distribution of the hoary bat in southern California. *Journal of Mammalogy* 35:431-32.

Verts, B.J. 1967. *The Biology of the Striped Skunk*. University of Illinois Press, Urbana, IL 218 pp.

Verts, B.J., and L.N. Carraway. 1998. Land Mammals of Oregon. University of California Press, Berkeley, CA.

Vispo, C., and I.D. Hume. 1995. The digestive tract and digestive function in the North American porcupine and beaver. *Canadian Journal of Zoology* 73:967-74.

Voth, E.H., 1968. Food habits of the Pacific mountain beaver, *Aplodontia rufa pacifica* Merriam. Ph. D. Thesis. Oregon State University, Corvallis. 263 pp.

Voth, E.H., C. Maser, and M.L. Johnson. 1983. Food Habits of *Arborimus albipes*, the White-footed Vole, in Oregon. *Northwest Science* 57:1-7.

Walker, A. 1930. The "hand-stand" and some other habits of the Oregon spotted skunk. *Journal of Mammalogy* 11:227-29.

Walker, A. 1942. The fringed bat in Oregon. *Murrelet* 23:62.

Walker, E.P., F. Warnick, S. Hamlet, and others. 1975. *Mammals of the World.* 3d ed. Johns Hopkins University Press, Baltimore, MD. 1,5000 pp.

Walker, K.M. 1949. Distribution and life history of the black pocket gopher, *Thomomys niger* Merriam. M.S. Thesis. Oregon State College [University], Corvallis. 94 pp.

Walker, K.M. 1955. Distribution and taxonomy of the small pocket gophers of northwestern Oregon. Ph. D. Thesis. Oregon State College [University], Corvallis. 200 pp.

Walters, R.D. 1949. Habitat occurrence and notes on the history of the dusky-footed woodrat, Neotoma *fuscipes* Baird. M.S. Thesis. Oregon State College [University], Corvallis. 122 pp.

Wechwerth, R.P. 1957. The relationship between the marten population and the abundance of small mammals in Glacier National Park. M.S. Thesis. Montana State University, Missoula. 76 pp.

Wechwerth, R.P., and V.D. Hawley. 1962. Marten food habits and population fluctuations in Montana. *Journal of Wildlife Management* 26:55-74.

Whitaker, J.O., Jr. and C. Maser. 1976. Food Habits of Five Western Oregon Shrews. *Northwest Science* 50:102-7.

Whitaker, J.O., Jr., C. Maser, and L.E. Keller. 1977. Food Habits of Bats of Western Oregon. *Northwest Science* 51:46-55.

Whitaker, J.O., Jr., C. Maser, and R.J. Pedersen. 1979. Food and Ectoparasitic Mites of Oregon Moles. *Northwest Science* 53:268-73.

Whitaker, J.O., Jr., C. Maser, and S.P. Cross. 1981. Food Habits of Eastern Oregon Bats, Based on Stomach and Scat Analyses. *Northwest Science* 55:281-92.

Whitaker, J.O., Jr., C. Maser, and S.P. Cross. 1981. Foods of Oregon Silver-haired Bats, Lasiony*cteris noctivagans.* *Northwest Science* 55:75-77.

Whitaker, J.O., Jr., S.P. Cross, and C. Maser. 1981. Food of Vagrant Shrews (*Sorex vagrans*) from Grant County, Oregon, as Related to Livestock Grazing Pressures. *Northwest Science.* 57:107-11.

Wight, H.M. 1928. Food habits of Townsend's mole, S*capanus townsendii* (Backman). *Journal of Mammalogy* 9:19-33.

Wimsatt, W.A. 1944. Further studies on the survival of spermatozoa in the female reproductive tract of the bat. *Anatomical Record* 88:193-204.

Wimsatt, W.A. 1960. An analysis of parturition in Chiroptera, including new observations on *Myotis l. lucifugus.* *Journal of Mammalogy* 41:183-200.

Winchell, J.M., and T.H. Kunz. 1993. Sampling protocols for estimating time budgets of roosting bats. *Canadian Journal of Zoology* 71:2244-49.

Windberg, L.A. 1995. Demography of a high-density coyote population. *Canadian Journal of Zoology* 73:942-54.

Witt, J.W. 1991. Fluctuations in the weight and trap response for *Glaucomys sabrinus* in western Oregon. *Journal of Mammalogy* 72:612-15.

Witt, J.W. 1992. Home range and density estimates for the northern flying squirrel, *Glaucomys sabrinus*, in western Oregon. *Journal of Mammalogy* 73:921-29.

Wood, J.E. 1958. Age structure and productivity of a gray fox population. *Journal of Mammalogy* 39:74-86.

Wood, W. 1974. Muskrat origin, distribution, and range extension through the coastal areas of Del Norte Co., California, and Curry Co., Oregon. *Murrelet* 55:1-4.

Wright, P.L. 1942. Delayed implantation in the long-tailed weasel (*Mustela frenata*), the short-tailed weasel (*Mustela ciognani*), and the marten (*Martes americana*). *Anatomical Record* 83:341-53.

Wright, P.L. 1942. A correlation between the spring moult and spring changes in the sexual cycle in the weasel. *Journal of Experimental Zoology* 91:103-10.

Wright, P.L. 1947. The sexual cycle of the male long-tailed weasel (*Mustela frenata*). *Journal of Mammalogy* 28:343-52.

Wright, P.L. 1948. Breeding habits of captive long-tailed weasels (*Mustela frenata*). *American Midland Naturalist* 39:338-44.

Wright, P.L. 1948. Preimplantation stages in the long-tailed weasel (*Mustela frenata*). *Anatomical Record* 100:593-608.

Wright, P.L., and M.W. Coulter. 1967. Reproduction and growth in Maine fishers. *Journal of Wildlife Management* 31:70-87.

Yocom, C.F. 1974. Status of marten in northern California, Oregon, and Washington. *California Fish and Game* 60:54-57.

Yocom, C.F., and M.T. McCollum. 1973. Status of the fisher in northern California, Oregon, and Washington. *California Fish and Game* 59:305-9.

Young, S.P. 1958. *The Bobcat of North America*. Stackpole Co., Harrisburg, PA. 193 pp.

Young, S.P., and H.H.T. Jackson. 1951. *The Clever Coyote*. Stackpole Co., Harrisburg, PA., and The Wildlife Management Institute, Washington, D.C. 411 pp.

Zwickel, F.G.J., and H. Brent. 1953. Movement of Columbian black-tailed deer in the Willapa Hills area, Washington. *Murrelet* 34:41-46.

Index

Numbers in italics designate a photograph.